IEE Electromagnetic Waves Series 11

Series Editors: Professor P. J. B. Clarricoats
Dr Y. Rahmat-Samii
Professor J. R. Wait

Adaptive Array principles

Other volumes in this series

Volume 1 Geometrical theory of diffraction for electromagnetic waves
Graeme L. James
Volume 2 Electromagnetic waves and curved structures
Leonard Lewin, David C. Chang and Edward F. Kuester
Volume 3 Microwave homodyne systems
Ray J. King
Volume 4 Radio direction-finding
P. J. D. Gething
Volume 5 ELF communications antennas
Michael L. Burrows
Volume 6 Waveguide tapers, transitions and couplers
F. Sporleder and H. G. Unger
Volume 7 Reflector antenna analysis and design
P. J. Wood
Volume 8 Effects of the troposphere on radio communications
Martin P. M. Hall
Volume 9 Schumann Resonances in the earth-ionosphere cavity
P. V. Bliokh, A. P. Nikolaenko and Y. F. Filippov
Volume 10 Aperture antennas and diffraction theory
E. V. Jull
Volume 11 Adaptive array principles
J. E. Hudson
Volume 12 Microstrip antenna theory and design
J. R. James, P. S. Hall and C. Wood
Volume 13 Energy in electromagnetism
H. G. Booker
Volume 14 Leaky feeders and subsurface radio communications
P. Delogne
Volume 15 The handbook of antenna design, volume 1
Editors: A. W. Rudge, K. Milne, A. D. Olver, P. Knight
Volume 16 The handbook of antenna design, volume 2
Editors: A. W. Rudge, K. Milne, A. D. Olver, P. Knight
Volume 17 Surveillance radar performance prediction
P. Rohan
Volume 18 Corrugated horns for microwave antennas
P. J. B. Clarricoats and A. D. Olver
Volume 19 Microwave antenna theory and design
Editor: S. Silver
Volume 20 Advances in radar techniques
Editor: J. Clarke
Volume 21 Waveguide handbook
N. Marcuvitz
Volume 22 Target adaptive matched illumination radar
Dag T. Gjessing
Volume 23 Ferrites at microwave frequencies
A. J. Baden Fuller
Volume 24 Propagation of short radio waves
Editor: D. E. Kerr
Volume 25 Principles of microwave circuits
Editors: C. G. Montgomery, R. H. Dicke, E. M. Purcell
Volume 26 Spherical near-field antenna measurements
Editor: J. E. Hansen
Volume 27 Electromagnetic radiation from cylindrical structures
J. R. Wait

Adaptive Array principles

J.E. Hudson

Peter Peregrinus Ltd. on behalf of the Institution of Electrical Engineers

Published by: Peter Peregrinus Ltd., London, United Kingdom

Reprinted 1989, 1991

ISBN 0 86341 247 5

Printed in Great Britain by
Short Run Press Ltd., Exeter

Contents

Preface

This book originated, somewhat indirectly, from the deliberations of the Propagation, Antennas, and Waveguides committee of the Electronics Research Council which, during an investigation of the state of adaptive array processing for Electromagnetic and Acoustic applications in the UK, observed that an impediment to development was the absence of an explanatory text which would help introduce the concepts and mathematics to engineers. The present text is intended to rectify this anomaly and, given the background, is intended to be simultaneously introductory, comprehensive, and mathematical in so far as such a goal is attainable.

The introductory phase is directed at those who are already in the radio antenna or sonar areas and familiarity with the general engineering lore in the design of array patterns is assumed. The historical development of adaptive antennas, somewhat shrouded in a mist of rather inaccessible defence research reports, is not covered; further details may be found in References 56 and 67. Likewise the hardware aspects of design are described only briefly.

The aim of comprehensiveness is restricted to encompass the field of narrowband gradient-based adaptive arrays. Some important topics are inadequately treated in the literature; these include main-lobe constraints, finite bandwidth effects, and multipath. Not a little effort has been required to discuss these topics in a coherent and useful fashion and some generally unfamiliar mathematical tools, such as the pseudoinverse and various vector space methods, have been introduced in one or two places but it is believed that these techniques are the most economical and, in the final analysis, the simplest. The amount of material for inclusion, even with this restriction, is almost overwhelming and the discussion of broadband arrays, convergence acceleration, and spectrum analysis is correspondingly curtailed. Since these areas are, at the time of writing, passing through a fluid state of development, the reader is perhaps additionally best advised to read the appropriate journals and await further developments.

Finally, as a philosophical point, the text is founded on mathematical modelling and solutions as far as is feasible; indeed many interesting but completely intractable problems are omitted for this reason. The alternative—demonstration of performance by simulation—leads to difficulties of interpretation, does little to improve insight, and the results are not easily extrapolated to new situations except by even more simulation. Nevertheless, the analytic approach is only effective in idealised interference scenarios and the closer one gets to the real world and the more realistic the modelling so the greater is the requirement for numerical techniques.

Organisation of the book

The material is arranged to a large extent as a text book with a gradually increasing level of difficulty so that the reader who wishes to follow a continuous line of development can do so. In this case, Sections 2.5, 2.7, 3.2, 3.5, 3.7, 3.8, 4.9–4.11, 5.3, and some of the mathematics in Chapters 6 and 7 could be omitted at first reading. A standard engineering mathematics text at hand might be useful* for basic definitions excluded to save space.

Acknowledgments

Discussions with a number of people have upgraded the author's perceptions on several matters, particularly the optimisation of functions of complex variables. D. Brandwood, P. Hargrave, and Prof. L. Lewin are prominent among these. P. Barton and A. Ratner have read the manuscript and conveyed improvements while, for suggesting the creation of the book, thanks go to S. Reynolds and Prof. P. Claricoats.

Research in this area was supported initially by the Admiralty Underwater Weapons Establishment and more recently by the Royal Aircraft Establishment and the Admiralty Surface Weapons Establishment. Some of the topics are abstracted from work done under a consultancy agreement with the Standard Telecommunications Laboratories, Harlow.

* Kreysig, E.: 'Advanced Engineering Mathematics', Wiley.

Introduction

by Prof. J. W. R. Griffiths

An array comprises a set of sensors, the outputs of which are combined in some way to produce a desired effect, e.g. a set of beams 'looking' in various directions. The sensors may be of many forms, e.g. acoustic transducers for sonar, monopole aerials for h.f. reception, microwave horns in a radar system, the influence of the application on the processing required concerns the technology rather than the principles. Array processing systems which can respond to an unknown interference environment are currently of considerable interest. The fundamentals of such systems are by no means new—basically they depend on Weiner's filter theory—but application in practice has been limited both by technology and by the lack of robust algorithms rapid enough for real-time operation. Rapid strides in the past decade or so in the twin fields of electronic components and computer technology have changed the situation considerably by offering the possibility of complicated signal processing in real time at economic costs. This has led to an increased interest in adaptive processing and although it is probably still finding its main application in the defence field the civilian applications are growing. As a consequence the literature on this subject has expanded rapidly and there has been a need for sometime for an authoritive book to lay down the fundamentals.

Dr. Hudson is widely known for his substantial contribution to the subject of adaptive array processing and has taught on a number of courses on this and allied subjects. He therefore seems an ideal person to carry out the task of writing such a book and I am sure the reader will find the material well explained and logical in presentation.

List of principle symbols and uses

d	interelement spacing
$d(t)$	reference signal
k	wavenumber, $2\pi/\lambda$
$g(\theta)$	amplitude gain of an array relative to a single element
N	number of elements in array
p	an array output power
Q	output signal to interference plus noise ratio
SIR	signal-to-interference ratio
SNR	signal-to-noise ratio
SNIR	signal-to-noise plus interference ratio
JNR	jammer-to-noise ratio
LMS	least mean square
PI	power inversion
t	time
$y(t)$	output signal of array
α	source bearing (usually relative to broadside)
β	normalised feedback coefficient
λ	wavelength or, when subscripted, an eigenvalue
π	power at an array element
μ	absolute feedback coefficient
θ	sin (α), related to spatial frequency
ω	angular frequency in time
δ_{ij}	$\begin{cases} = 1, i = j \\ = 0, i \neq j \end{cases}$

σ^2	variance
ψ	covariance
ρ	normalised version of $g(\theta)$, or normalised version of ψ
$E(.)$	expectation, or ensemble average
$\mathbf{B}$	beamformer matrix
$\mathbf{C}$	constraint vector
$\mathbf{G}$	gradient vector
$\mathbf{I}$	identity or unit matrix
$\mathbf{M}$	covariance matrix of noise and interference
$\mathbf{P}$	projection matrix
$\mathbf{R}$	total covariance matrix, i.e. $\mathbf{M}$ plus signal covariance
$\mathbf{S}$	space vector
$\mathbf{U}$	eigenvector
$\mathbf{W}$	weighting vector
$\mathbf{W}_c$	conventional (cophasal) version of $\mathbf{W}$
$\mathbf{W}_{\text{opt}}$	optimised weighting vector
$\mathbf{X}(t)$	input data at array
$\mathbf{\Lambda}$	a diagonal matrix
∇	del; differential vector operator
$(.)^t$	simple transposition
$(.)^*$	simple conjugation
$(.)^H$	Hermitian transpose
$(.)^+$	pseudoinverse

Chapter 1

Introductory ideas

1.1 Antenna null rotation

Adaptive antennas adjust their directional patterns so as to maximise the signal-to-noise ratio, at the receiver output, of some desired signal which is received in the presence of interference and noise. The simplest adaptive antenna, stretching the adjective slightly, is a rotatable fixed pattern antenna. If the only noise present were due to the receiver's circuit noise, the signal-to-noise ratio would be maximised when the main lobe of the pattern was directed at the desired signal bearing since this action maximises the signal power, the noise power remaining constant. This is, for example, the natural orientation of a portable radio when one is trying to receive a weak signal.

However, it sometimes happens that co-channel interference is present at an objectionable level. If the interference arises from a source whose bearing is significantly different from that of the desired source then the signal-to-noise plus interference ratio is maximised for a rather different orientation of the antenna and will occur when a null of the antenna directional pattern is directed toward the interference even though this action decreases the absolute power of the desired signal to some extent and reduces the signal to receiver-noise ratio below that obtainable in the absence of interference. The situation is shown in Fig. 1.1.1 where a cos (α) pattern typical of a ferrite rod is assumed.

A practical demonstration of the technique is obtained by tuning to a weak medium-wave signal with a portable radio and orienting for maximum signal-to-noise ratio. Switching on a television set a few metres away provides a convenient interference source and a fairly sharply defined orientation of the radio should be found where the signal-to-interference ratio is maximised.

1.2 Electronic null steering

Physical orientation of an antenna is not particularly convenient or rapid when adaptation is to be done automatically though if mechanical steering is already provided in an antenna it might be used. More rapid and flexible results are

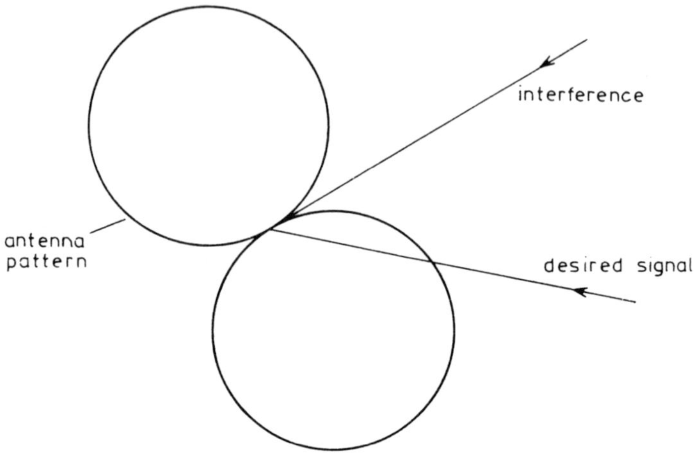

Fig. 1.1.1 *Reception of a signal in the presence of an interference*

obtained by weighting the outputs of a multielement array electrically before summation when adjustment of the weighting coefficients will change the pattern shape. The simplest case is that of a two element array, shown in Fig. 1.2.1, in which phase shifters are used in the channels to affect the necessary variation. It is assumed that the array elements are omnidirectional in the horizontal plane of interest and could consist of vertical dipoles. In the acoustic case the elements may be pressure sensing hydrophones or microphones.

For a sinusoidal plane wave arriving from a direction $\theta = \sin(\alpha)$ from broadside the output of the summer is given by

$$\begin{aligned} y(t) &= e^{j\omega t}[e^{-j(1/2kd\theta - \phi)} + e^{j(1/2kd\theta - \phi)}] \\ &= 2e^{j\omega t} \cos[\tfrac{1}{2}kd\theta - \phi] \end{aligned} \tag{1.2.1}$$

where $k = 2\pi/\lambda$ is the wave number, d is the interelement spacing, and ϕ is the phase advance introduced by the shifters. If d is less than one half-wavelength and $\phi = \pi/2$, the array has a dipole-like pattern in the horizontal plane as shown in Fig. 1.2.2(a) and for small d values the pattern can be approximated as

$$y(t) = e^{j\omega t}kd \sin(\alpha) \qquad \phi = \pi/2 \tag{1.2.2}$$

If ϕ is set to a phase other than $\pi/2$ the nulls in the pattern are steered to different directions and it is readily shown from eqn. 1.2.1 that for small d/λ the nulls, if they exist, occur at a direction α relative to broadside such that

$$\tfrac{1}{2}kd \sin(\alpha) - \phi = \pm\pi/2 \tag{1.2.3}$$

The exact patterns for $\phi = \pi/3$, $\pi/4$, and 0 with $d = \lambda/4$ are shown in Fig. 1.2.2(b), (c), and (d).

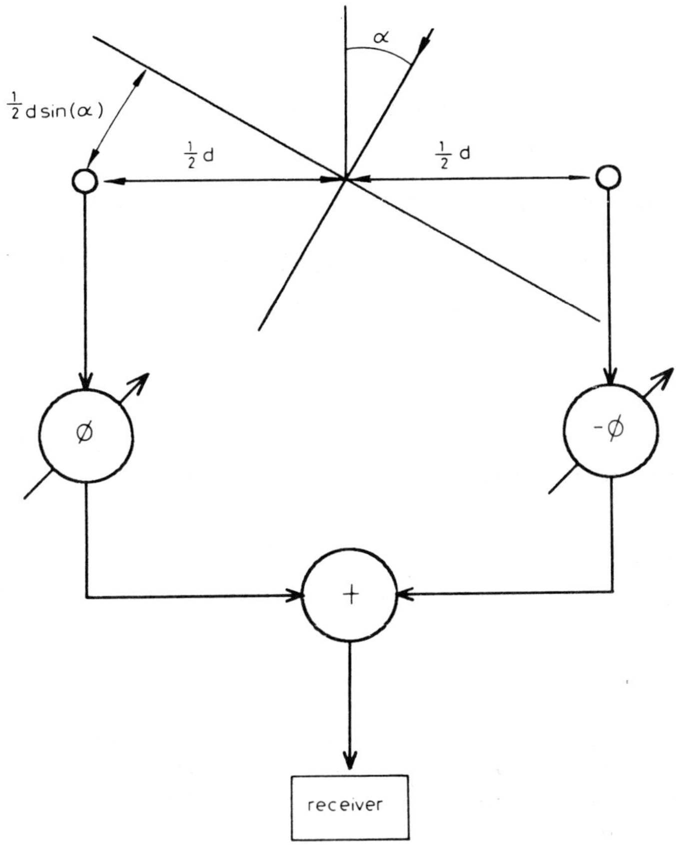

Fig. 1.2.1 *Two-element array with steerable null*

The electrically phased two-element pattern varies in a rather different way from the ferrite rod pattern in which the whole rotates together. The two-element array has mirror symmetry about the line joining the elements. It is possible to get a pattern rotation effect with purely electrical weighting but it is necessary to use two crossed dipole arrays, a total of four vertical dipoles with some fairly complicated law defining the four weighting values. The goniometer once popular in direction finding achieves this weighting for crossed loop antennas for low frequencies and arrays of four vertical monopoles in the HF band.

Although the ferrite rod loop antenna or the two-element array posssess two nulls, these cannot be independently oriented. If the position of one is fixed the other is known. Consequently, the pattern only has one degree of freedom in this respect and the two-element array is of little use for cancellation of two interference sources. It is perhaps interesting to speculate that a rotatable two

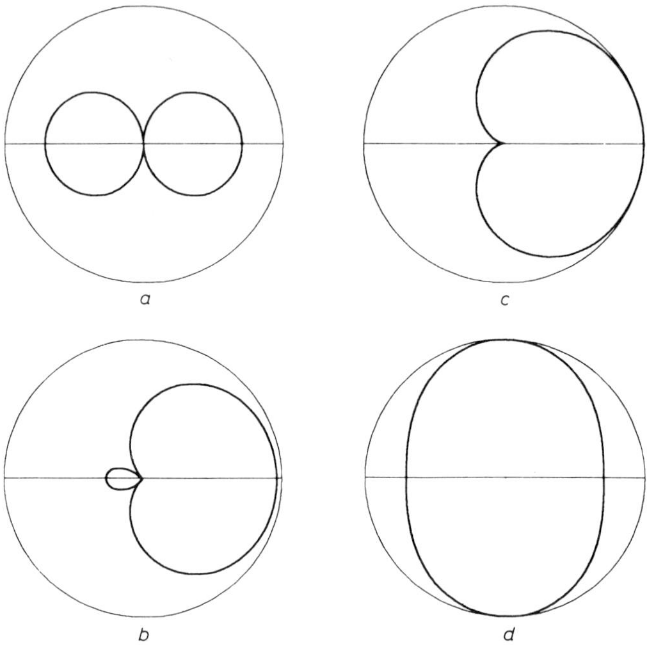

Fig. 1.2.2 *Direction patterns for a quarter wave spaced two-element array*
a $\phi = \pi/2$
b $\phi = \pi/3$
c $\phi = \pi/4$
d $\phi = 0$

element array together with electrical phase shifters would have two degrees of freedom and could steer the two nulls quite independently.

In order to compensate slight differences in the sensitivities of the two array elements in the electrically phased case, it is necessary to include an attenuator in one or both channels. Otherwise, if the element voltage outputs differ by as little as a few percent there will be an appreciable effect on the depth of the null obtainable. Such errors must be trimmed out *in situ* and, if deep nulls are required, retrimmed every time the null is moved. However, the addition of this new component does not impart two degrees of freedom to the array; in fact, two degrees of freedom are required to form just one null since the in-phase and quadrature pattern amplitudes must be independently set to zero at the chosen null bearing.

The two-element array can be rather more than just a theoretical example. Because there are few variables the system is steered manually quite feasibly and it can form a low cost adaptive interference canceller. So long as there is only one interference source, probably the most likely event, its performance is not excessively inferior to the multielement arrays described later. It is also a useful tool in the investigation of the stability and coherence of propagation paths where the installation of an adaptive antenna is proposed.

Automatic null steering: Having illustrated the possibility of null steering for a two-element array, attention is now directed to making it steer its nulls automatically so that it becomes self-adaptive. The feasibility of doing this is always the foremost problem in design and if the desired transmission is present alongside the interference, the task may be rather difficult since an automatic system may have difficulty in distinguishing between cancellation of the interference and cancellation of the desired signal. Although most adaptive antenna theory is aimed toward maximisation of the output signal-to-noise ratio, this parameter is seldom available from the receiver and one must make do with somewhat inferior data, contriving to maximise SNR indirectly. One of the easier quantities to measure is the output power of the antenna at receiver RF or IF level and within limits this can be used automatically to improve output SNR as discussed below. Another measurable quantity for binary modulation systems may be an error signal between some feature of the code and a locally generated reference.

1.3 Constrained power minimisation

The two element array shown in Fig. 1.3.1 has four variable coefficients each comprising either part of a quadrature coefficient with independent weighting of in-phase (real) and quadrature (imaginary) components of the RF or IF signal, as shown in Fig. 1.3.2(*a*) or a cascaded attenuator and phase shifter as in Fig. 1.3.2(*b*). Either type of weighting is equivalent to a complex multiplication in a bandpass system. For a unity amplitude complex input wave arriving from a bearing of $\theta = \sin(\alpha)$ the antenna output signal $y(t)$ is given by

$$\begin{aligned} y &= w_1 e^{-1/2jkd\theta} + w_2 e^{1/2jkd\theta} \\ &= (w_1 + w_2)\cos(\tfrac{1}{2}kd\theta) - j(w_1 - w_2)\sin(\tfrac{1}{2}kd\theta) \end{aligned} \tag{1.3.1}$$

where k and d are defined as in eqn. 1.2.1. The time dependence $e^{j\omega t}$ has been omitted. For the stated conditions, y can be defined as the amplitude gain of the antenna for the signal and as θ is varied defines the beam pattern $g(\theta)$. If there are two signals present at different angles θ_s (desired signal) and θ_i (interference) with uncorrelated waveforms $f_s(t)$ and $f_i(t)$ such that

$$\overline{f_s(t) f_i(t)^*} = 0 \tag{1.3.2}$$

then the total array output power is given by

$$\begin{aligned} p_{tot} &= \overline{|f_s(t)g(\theta_s) + f_i(t)g(\theta_i)|^2} \\ &= \overline{|f_s(t)|^2}\,|g(\theta_s)|^2 + \overline{|f_i(t)|^2}\,|g(\theta_i)|^2 \\ &\quad + 2\mathrm{Re}\,[\overline{f_s(t)f_i(t)^*}g(\theta_s)g(\theta_i)^*] \\ &= \pi_s|g(\theta_s)|^2 + \pi_i|g(\theta_i)|^2 \end{aligned} \tag{1.3.3}$$

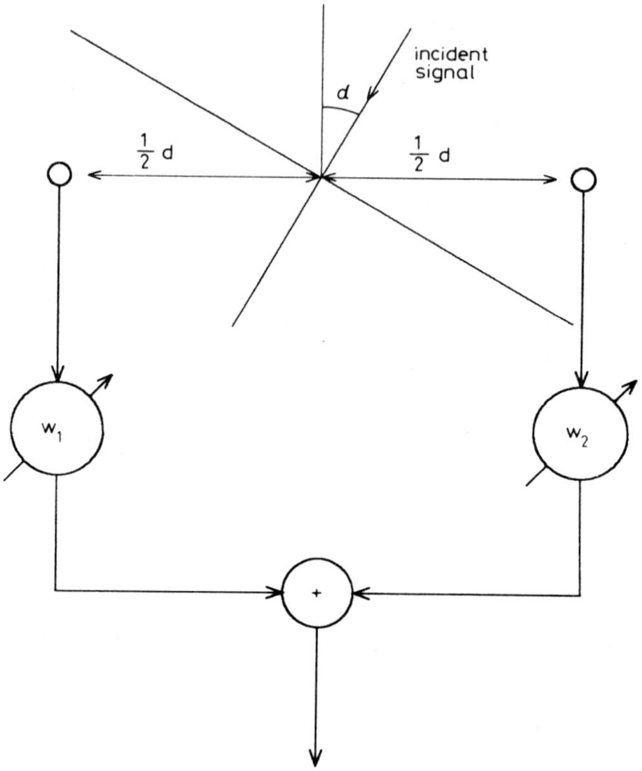

Fig. 1.3.1 *Two-element adaptive antenna*

where π_s is the signal power $\overline{|f_s(t)|^2}$ and π_i is the interference power $\overline{|f_i(t)|^2}$. Eqn. 1.3.3 indicates that the output power is a linear summation of the powers due to the signal and interference when each is presented alone to the antenna. The output signal-to-interference-power ratio is defined as

$$\mathrm{SIR} = \frac{\pi_s |g(\theta_s)|^2}{\pi_i |g(\theta_i)|^2} \tag{1.3.4}$$

and is easily seen to go to a maximum of infinity when $g(\theta_i)$ goes to zero, a condition which can be shown to correspond to

$$\frac{w_1}{w_2} = -e^{jkd\theta_i} \tag{1.3.5}$$

from eqn. 1.3.1. This corresponds to steering a null toward the interference. It is not necessary to worry about the pattern behaviour at bearings between the signal and interference since no source are present there in the model. In practice the output signal to noise ratio would not become infinitely high when a

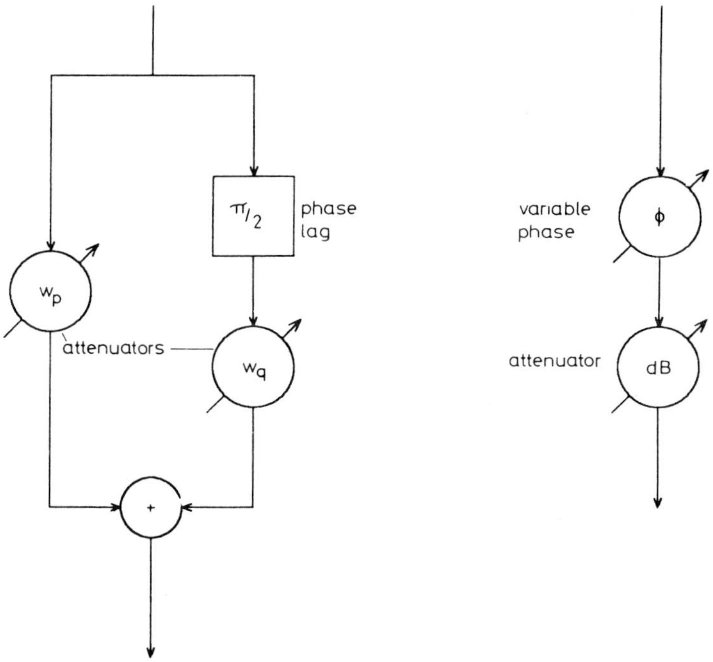

a *b*

Fig. 1.3.2 *Alternative implementations of variable coefficients*

null was formed since there is always a residue of noise from the receiver itself but this does affect the present argument.

Suppose the signal bearing θ_s is known, as would normally be the case in ground based communications systems, and for convenience is initially assumed to be broadside, $\theta_s = 0$. The amplitude gain of the antenna for the desired signal can then be held constant by suitably constraining the values which the coefficients can assume, such a constraint being $w_1 = 1 + a$, $w_2 = 1 - a$ where a is an arbitrary complex variable. The broadside pattern amplitude in the direction of the signal is now fixed since $w_1 + w_2 = 2$ and this can be proved by substituting $\theta = 0$ in eqn. 1.3.1. A solution can still be found for eqn. 1.3.5:

$$\frac{1 + a}{1 - a} = -e^{jkd\theta_i} \tag{1.3.6}$$

so that a null can nevertheless be trained on the interference. From eqn. 1.3.1 it is seen that the constrained pattern has the form, in general, of

$$g(\theta) = 2 \cos\left(\tfrac{1}{2}kd\theta\right) - 2aj \sin\left(\tfrac{1}{2}kd\theta\right) \tag{1.3.7}$$

and, since the first term in eqn. 1.3.7 is purely real, in order for $g(\theta_i)$ to be equated to zero the value of a must be purely imaginary. The solution for a obtained from eqn. 1.3.7 is

$$a = -j \cot \tfrac{1}{2}kd\theta_i \tag{1.3.8}$$

There is no solution if $\theta_i = 0$, as may be expected. Automatic searching for the value of a could proceed as follows. If the total output power of the array is determined by substituting eqn. 1.3.7 into 1.3.3 we obtain

$$p_{tot} = 4\pi_s + 4\pi_i |\cos(\tfrac{1}{2}kd\theta_i) - ja \sin(\tfrac{1}{2}kd\theta_i)|^2 \tag{1.3.9}$$

and it can be seen that the variable part of p_{tot} corresponds to the term containing $g(\theta_i)$ in eqn. 1.3.3. If p_{tot} is minimised by variation of the value of a, then $|g(\theta_i)|^2$ is minimised too and vice versa, the signal amplitude gain and output power remaining constant due to the constraint system. If a square law detector is used in the receiver to measure p_{tot}, a feedback system can be created which minimises it automatically by varying a and which at the same time converges to the coefficient solution eqn. 1.3.8.

The contours of output power in the domain of the complex variable a are circular as shown in Fig. 1.3.3 with a minimum central value of $4\pi_s$ which lies on the imaginary axis. The direction in which a should be varied to move to the optimum point can be sensed by measuring the gradient of the output power function, for example by perturbing a slightly. Such a minimisation technique forms the fundamental basis of a number of adaptive antenna algorithms and can be generalised to cover optimisation by the variation of several coefficients simultaneously.

It is not necessary to know the direction of the interference source for the minimisation of output power. This is an important feature of the technique, one which enables it to track slowly moving interferences or interference arising from an unknown bearing. However, it is essential to know the direction of the desired signal. The broadside assumption treated above is easily generalised to cover arbitrary directions of incidence, θ_s say, and the amplitude gain for the desired signal is then

$$g(\theta_s) = w_1 e^{-1/2jkd\theta_s} + w_2 e^{1/2jkd\theta_s} \tag{1.3.10}$$

The desired constraint is simply obtained by using this equation as a constraint equation, forcing the amplitude gain to 2 at θ_s:

$$w_1 e^{-1/2jkd\theta_s} + w_2 e^{1/2jkd\theta_s} = 2 \tag{1.3.11}$$

Combining this equation with eqn. 1.3.5,

$$w_1 e^{-1/2jkd\theta_i} + w_2 e^{1/2jkd\theta_i} = 0 \tag{1.3.12}$$

we obtain two simultaneous linear equations which can be solved for w_1 and w_2. However, demonstration of the existence of a solution is little help since in practice θ_i is unknown. For automatic output power minimisation we wish to express the coefficients as a function of a single parameter a as before, so that a

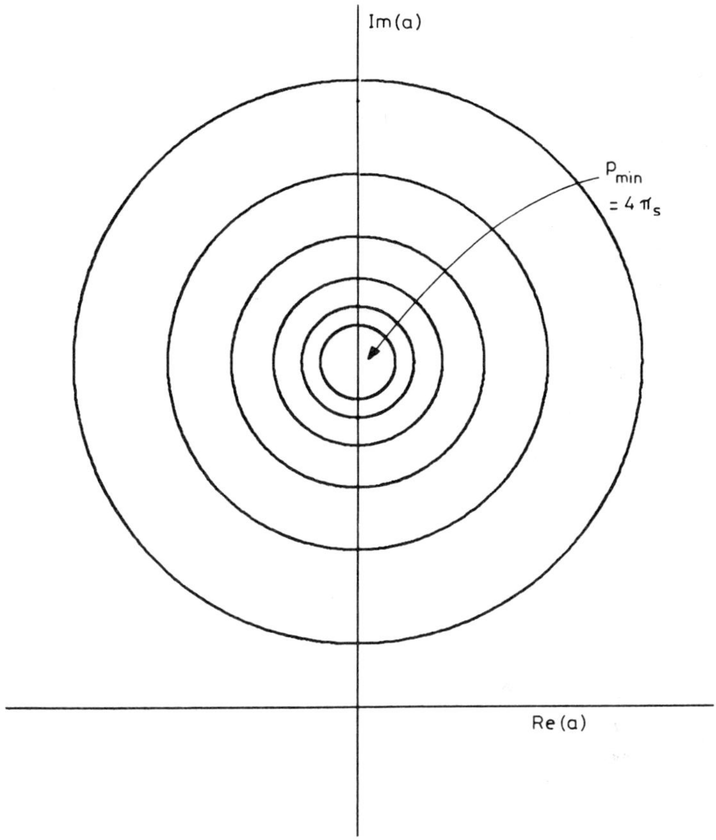

Fig. 1.3.3 *Variation of P_{tot} with variable a*

single variable feedback loop can be created. This leads to

$$w_1 = (1 + a)e^{1/2jkd\theta_s}$$
$$w_2 = (1 - a)e^{-1/2jkd\theta_s} \quad (1.3.13)$$

by analogy with eqn. 1.3.6. An alternative approach which is more hardware oriented is to keep the broadside steering constraint but phase the element outputs so that the desired signal appears to arrive from the broadside direction and Fig. 1.3.4 shows a suitable system for doing this. The apparent direction of the interference is also changed.

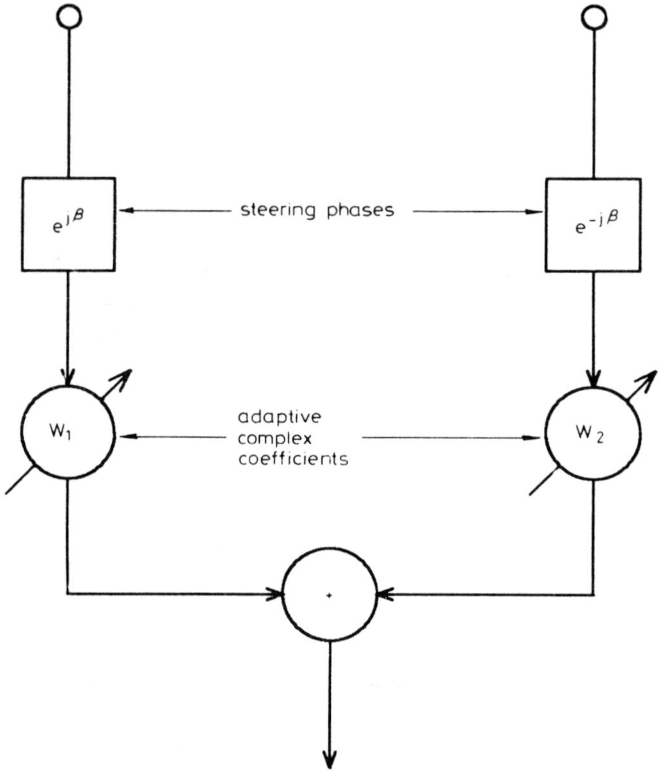

Fig. 1.3.4 *Presteering of desired signal to broadside direction*
$\beta = \frac{1}{2}kd \sin \alpha_s$

1.4 Weak signal adaptation

The constrained system can often be simplified if it is known that the desired signal is very weak, because there is then no tendency for accidental nulling of the desired signal when output power is minimised. Indeed under these circumstances the only necessary constraint is one which prevents the weighting coefficients all falling to zero, the ideal minimum output power solution but one which also guarantees loss of the desired signal. The simplest one is to clamp a single coefficient in the array to unity while allowing the remainder unrestricted variation. The choice of the end element appears to produce the narrowest null widths. For a two-element array, the antenna system is now of the form shown in Fig. 1.4.1 and is known as a power inversion array since it can be shown that the interference output power tends to be the reciprocal of interference input power (shown in Section 7.3). As long as the interference direction is a reasonable distance from the desired signal bearing, the amplitude gain for the latter input does not vary very much from unity when the interference is cancelled.

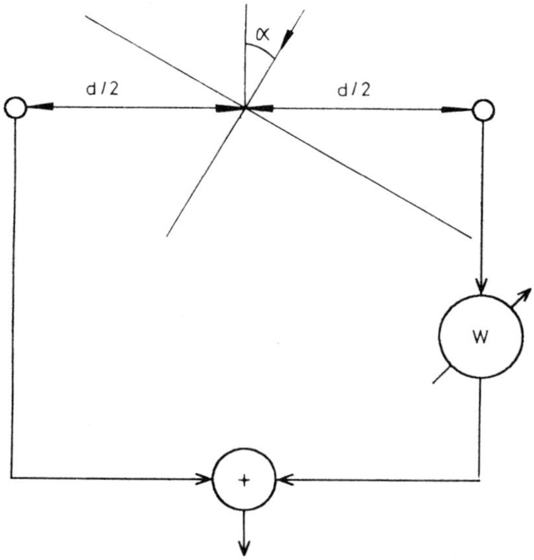

Fig. 1.4.1 *Two-element power-inversion array*

Should the interference direction be very close to that of the desired signal the latter may be suppressed along with the interference. This might seem to be a drawback when compared with the linearly constrained system above but that is also rather disrupted by the same conditions and the performances are quite comparable.

Examples where a weak signal occurs are first in radar receivers where, once clutter has decayed due to increasing range, target echoes are few and widely spaced and of insignificant power in comparison to a steady jamming source. The second type of weak signal occurence is in communication systems where spread spectrum techniques or frequency hopping may be used and spread the transmitter power over such a large bandwidth that the signal to noise ratio at the receiver front end may well be negative before signal processing. If the adaptive antenna operates at the front end of the receiver any interference which is sufficiently strong to be troublesome will exceed the receiver noise and is easily detectible.

Signal-blocking techniques

In order to gain the advantages of weak signal adaptation when the signal is intrinsically strong one may introduce circuits which block the desired signal from the power measurement. One such technique is frequency filtering. Fig. 1.4.2 shows the power density spectrum of an AM signal which it is desired to receive in the presence of wideband interference. If the filter with response shown is used prior to demodulation in the receiver, then a substantial reduction in the desired signal output power is obtained and the weak signal

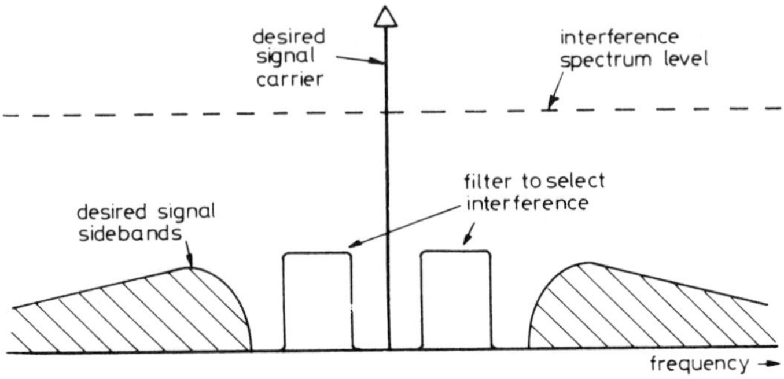

Fig. 1.4.2 *Filtering of an AM signal to block desired signal components*

condition is created. The adaptive processor will form a null in the direction of the interference because it will detect the power which is passed the filter even though this may only be a small fraction of the total interference power. Clearly the operation of a system like this would require considerable intervention by an operator to ensure that the signal and the interference have suitable spectra. A second technique operates by gating in the time domain and has been applied to the reception of TV signals.[76, 87] In the video modulation of a carrier the amplitude is set to zero during line syncronising pulses, in the UK system, as shown in Fig. 1.4.3. If the signal-to-noise ratio is initially sufficently strong for an oscillator to be locked to these pulses then it may be subsequently used to

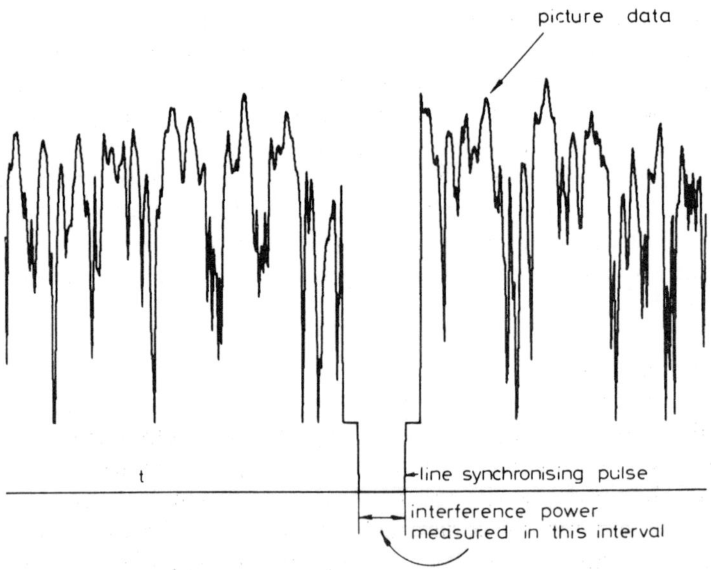

Fig. 1.4.3 *Time of gating of a TV signal to remove desired signal*

open a gate which only allows power measurements to be done during the duration of the synchronising pulses and the power measured is purely that of the interference. Should the interference be another TV transmission there is a tiny probability that the path differences involved may bring the synchronising pulses of the two transmissions into alignment, preventing detection of interference power. This problem must be overcome by putting a time delay in the video signal of the desired transmitter.

These signal-blocking techniques have an advantage even when the array has a look direction constraint because measurement of interference output power becomes very difficult once the array has formed a partial null which reduces the interference strength to a level below that of the desired signal. Power measurements then tend to become noisy and statistically unreliable because of fluctuation of the desired signal power and it is quite difficult to reduce the interference power much more than 10 or 15 dB below the desired signal power. If the signal can be attenuated by 20 dB or so then measurement of interference power is that much more accurate and higher levels of interference cancellation can be achieved.

1.5 The sidelobe canceller

Two-element null steering can be done with dissimilar elements and an extreme case of this occurs when it is required to form nulls in the directional pattern of a parabolic dish antenna. An omnidirectional element is fixed to one edge of the dish and is combined with the main feed through a variable quadrature coefficient as in Fig. 1.5.1. Since null steering is only wanted in the sidelobes of the main antenna the sensitivities of the dish and auxilliary element are comparable as indicated in the directional patterns of Fig. 1.5.2. If it is imagined that both patterns are real initially, then it is easy to see that a suitable attenuation in

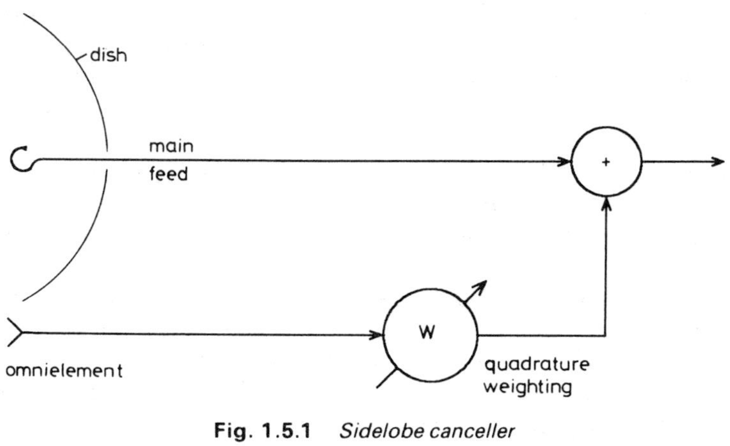

Fig. 1.5.1 *Sidelobe canceller*

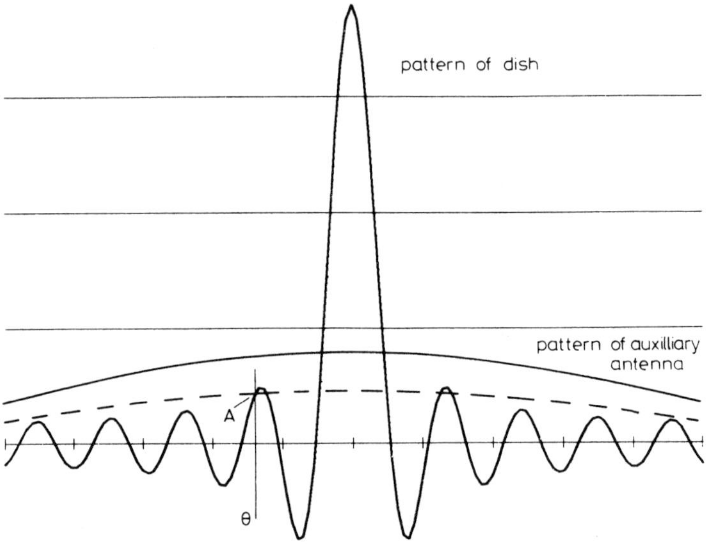

Fig. 1.5.2 *Operation of a sidelobe canceller*

the omnichannel to reduce it to point A will cause a null to be steered to angle θ is the patterns are antiphase at this point. Because of the weak average signal in radars, for which the sidelobe canceller is popular, there is no difficulty in designing feedback loops which will provide automatic cancellation. Because of the separation of the main dish feed and the omnielement laterally there is a rapid change of relative phase between their patterns as the incidence angle varies and it is necessary to change the phase in the omnichannel to achieve complete cancellation. Similar phase changes are caused by imperfections in the dish which cause the phase and polarisation of the pattern sidelobes to be rather random. The adaptive circuit will accommodate such irregularities quite automatically.

The convergence time characteristics of the radar sidelobe canceller are rather stringent because the dish is scanning continuously. Since the interference location remains fixed the cancellation system must form a null which stays fixed in absolute space but rotates relative to the main lobe of the beam and it must be fast enough to track the sidelobe pattern of the dish.

Formation of multiple nulls

1.6 The Davies beamformer[6]

This structure is shown in Fig. 1.6.1. The matrix of $\frac{1}{2}N(N-1)$ phase shifters is connected to a line array with equal interelement spacings of d. Consider the two-element subarray with inputs at A, B and outputting to node E. Its directional pattern is given by

$$E(\theta) = 1 + e^{j(kd\theta + \phi_1)} \tag{1.6.1}$$

where $\theta = \sin(\alpha)$ and it has a null at an angle α_1 such that

$$kd \sin(\alpha_1) + \phi_1 = \pi \tag{1.6.2}$$

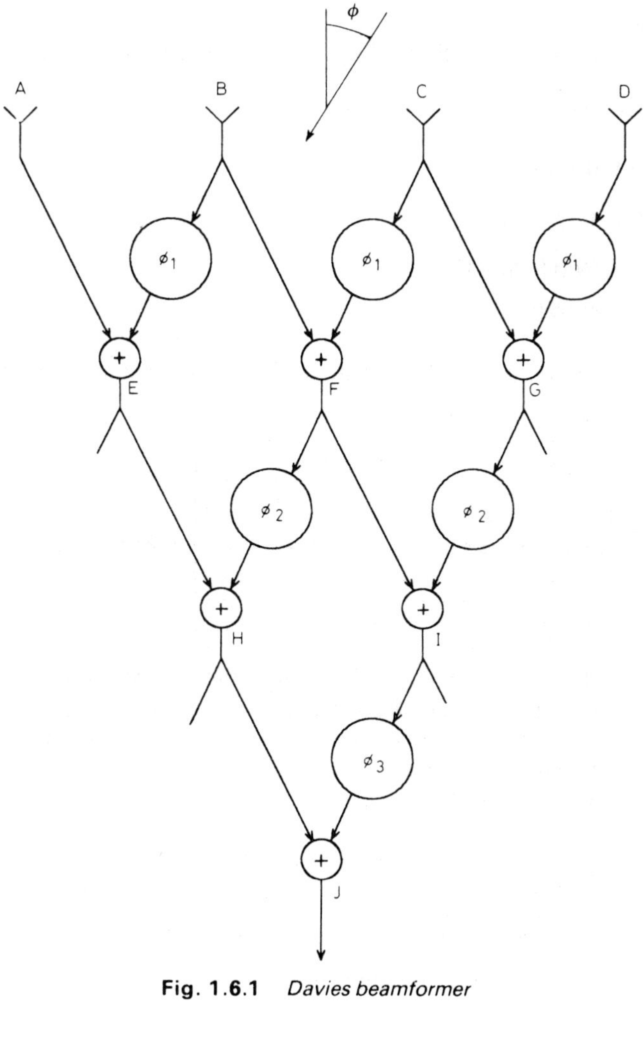

Fig. 1.6.1 *Davies beamformer*

or

$$\sin(\alpha_1) = \frac{\pi - \phi_1}{kd} \tag{1.6.3}$$

It is obvious that all of the two-element subarrays formed by adjacent pairs of antenna elements have the same patterns apart from a relative phase shift between their outputs: $|F(\theta)| = |G(\theta)| = |E(\theta)|$. If a null is steered to some source at θ_1 by varying ϕ_1, then no output due to this source appears at nodes E, F, or G and this null cannot be removed by anything which happens further down in the matrix. If a second source is present at a different angle θ_2, the signals due to it appearing at nodes E, F, and G have the same relative phase as though E, F, and G were connected directly to the antenna elements though the absolute phases may be changed and the amplitudes may be smaller if θ_2 is close to θ_1. By variation of ϕ_2, a null may be trained onto the second source at θ_2. Similarly for a third source at θ_3, the relative phases of its signals at nodes H and I are unchanged and this source may be nulled by variation of ϕ_3.

The resulting directional response of the whole matrix is found as follows. The direction responses at E, F, G is of the form

$$e^{jkdn\theta}[1 + e^{j(kd\theta + \phi_1)}] \tag{1.6.4}$$

where n is the node number from the left (E, F, G = 0, 1, 2). The patterns measured at the nodes H, I are

$$e^{jkdn\theta}[1 + e^{j(kd\theta + \phi_1)}][1 + e^{j(kd\theta + \phi_2)}] \tag{1.6.5}$$

for $n = 0, 1$, and finally at node J we have

$$[1 + e^{j(kd\theta + \phi_1)}][1 + e^{j(kd\theta + \phi_2)}][1 + e^{j(kd\theta + \phi_3)}] \tag{1.6.6}$$

Thus the pattern is factored naturally into three terms each responsible for one null (if $d \leq \frac{1}{2}\lambda$). Fig. 1.6.2 shows a directional pattern for a four-element array with nulls formed at $-60°$, $+60°$, and $+15°$ relative to broadside with $d = \frac{1}{2}\lambda$.

Reduction to N complex coefficients

The matrix of phase shifters of Fig. 1.6.1 may be reduced to a set of N complex coefficients operating directly on the antenna array as in Fig. 1.7.1, and this is done by calculating the network transfer function from each array element to the output. For element No. 1 at A the transfer function is unity, so $w_1 = 1$. For element No. 2 the transfer function to nodes F and I is unity and the output is found by combining the ϕ_1, ϕ_2, and ϕ_3 shifters in parallel to yield

$$w_2 = e^{j\phi_1} + e^{j\phi_2} + e^{j\phi_3} \tag{1.6.7}$$

For element No. 3 the transfer to node G is unity and we obtain

$$w_3 = e^{j\phi_1}e^{j\phi_2} + e^{j\phi_1}e^{j\phi_3} + e^{j\phi_2}e^{j\phi_3} \tag{1.6.8}$$

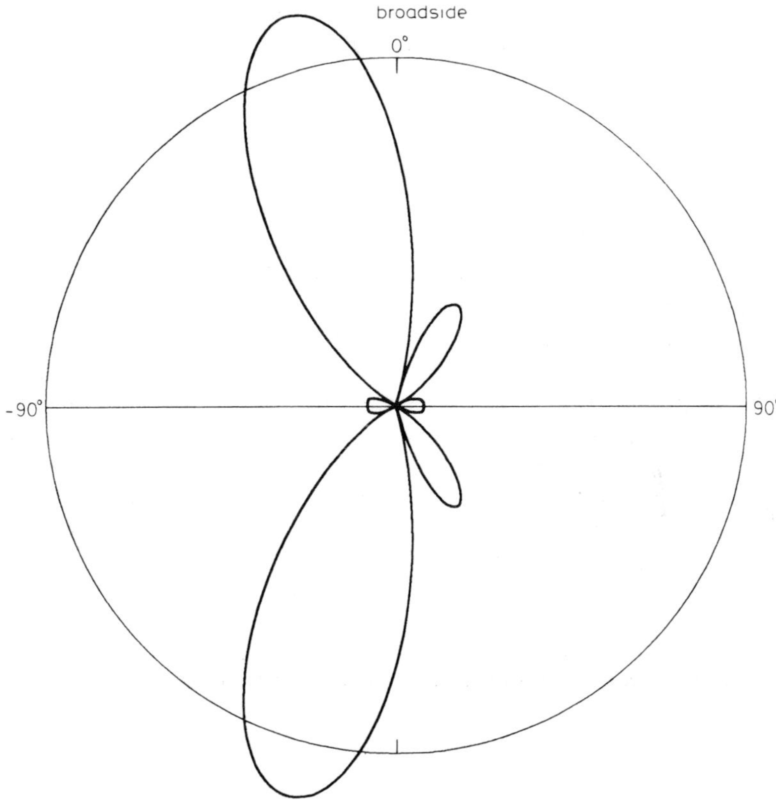

Fig. 1.6.2 *Polar directional pattern formed by Davies beamformer with nulls at 15°, 60°, and −60° relative to broadside*

Finally, for element No. 4 we have

$$w_4 = e^{j\phi_1}e^{j\phi_2}e^{j\phi_3} \tag{1.6.9}$$

These results are somewhat idealised since in practice a hybrid adder would be required at every node and this would introduce some power loss at the edges of the matrix.

If the coefficients calculated by eqns. (1.6.7), (1.6.8), and (1.6.9) including the fixed coefficients $w_1 = 1$ were set up in an array of the type shown in Fig. 1.7.1, the same directional response would be obtained. The hardware would be more economical since there are only $N - 1$ complex variable coefficients instead of the $\frac{1}{2}N(N - 1)$ phase shifters but the ability to steer nulls independently would be lost and each weighting coefficient would have an effect over the whole pattern if varied. The coefficients found by the equations above are the same as the coefficients of the powers of $e^{-jkd\theta}$ which occur when eqn. (1.6.6) is multiplied out and this is an example of the use of Schelkunoff's polynomial which can be used as a design tooi for antennas.[6]

It is possible to make the Davies beamformer adaptive and it does have some advantages because of the independent null steering effect but the design is complicated by the fact that the output power of the antenna is not a simple function of the phase shifts in the rows, i.e. multiple minima may appear, and some difficult stability problems must be solved in a feedback system. Unfortunately the minimisation of output power by variation of each row of phase shifters does not intrinsically cause nulls to be steered to specific sources in a multiple interference situation. For example, if ϕ_1 is varied and the power is minimised at nodes E, F, and G the null formed does not fall on the strongest source but is directed instead to the vicinity of the centre of gravity of the angular configuration of sources. If ϕ_2 is now varied to minimise the power at nodes H and I, the second null is directed close to the centre of gravity of the configuration of sources as modified by the power directional pattern of the first null (noting that the spectrum of sources is effectively repetitive). The process continues for ϕ_3 and further rows of phase shifters. One pass through the rows of phase shifters will not steer perfect nulls at any one source and to do this iterative readjustment of the rows while monitoring the output power at the output node would be required.

Example 1.6.1: For the numerical example above with nulls at 60°, −60°, and 15° the equivalent complex weighting coefficients for Fig. 1.7.1 will be calculated.

We have

$$\phi_2 = -24{\cdot}115°, \qquad \phi_2 = -24{\cdot}155°, \qquad \phi_3 = 133{\cdot}413° \tag{1.6.10}$$

The complex representation of phase shifters with these values is

$$\begin{aligned} e^{j\phi_1} &= 0{\cdot}9127 + j0{\cdot}4086 \\ e^{j\phi_2} &= 0{\cdot}9127 - j0{\cdot}4086 \\ e^{j\phi_3} &= -0{\cdot}6873 + j0{\cdot}7264 \end{aligned} \tag{1.6.11}$$

whence, from eqns. (1.6.7) to (1.6.9), we derive

$$\begin{aligned} w_1 &= 1 \\ w_2 &= e^{j\phi_1} + e^{j\phi_2} + e^{j\phi_3} = 1{\cdot}1382 + j0{\cdot}7264 \\ w_3 &= e^{j(\phi_1+\phi_2)} + e^{j(\phi_2+\phi_3)} + e^{j(\phi_1+\phi_3)} \\ &= -0{\cdot}2545 + j1{\cdot}3260 \\ w_4 &= e^{j(\phi_1+\phi_2+\phi_3)} = -0{\cdot}6873 + j0{\cdot}7264 \end{aligned} \tag{1.6.12}$$

1.7 Multiple null formation with N complex coefficients

Having shown indirectly that an antenna array with N complex weighting coefficients can form $N - 1$ nulls at arbitrary directions, the concept is now investigated more fully using simultaneous equations and linear algebra, the natural language of adaptive antennas. Consider an array of N elements receiving narrow-band signals from M interference sources shown in Fig. 1.7.1. Taking the mth source in isolation gives rise to a signal at the nth element of the array of the form

$$v_{mn}(t) = a_m(t)s_{mn}\, e^{j\omega t} \tag{1.7.1}$$

where $a_m(t)$ is the modulation function of the mth source and the s_{mn} coefficients are unity modulus phasors which represent the phase lag from the source to the nth element. Summing the contributions from all M sources at the nth element gives a resultant of

$$v_n(t) = \sum_{m=1}^{M} a_m(t)s_{mn}\, e^{j\omega t} \tag{1.7.2}$$

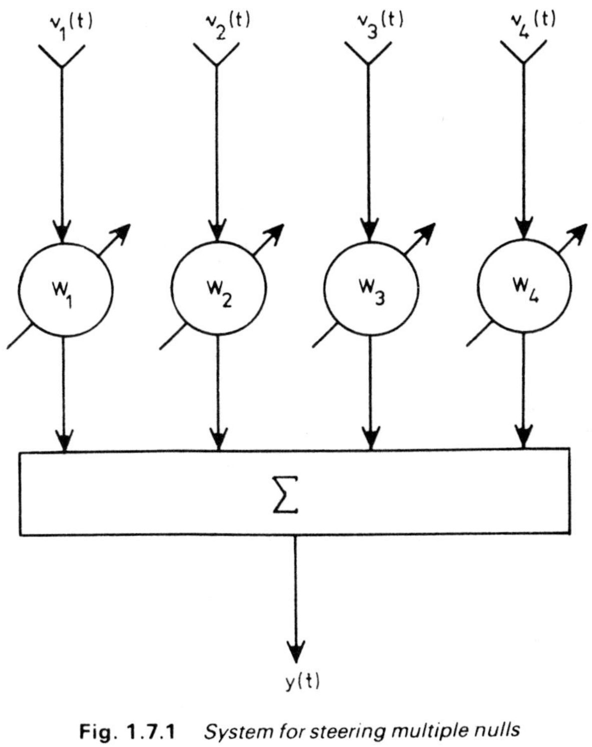

Fig. 1.7.1 *System for steering multiple nulls*

Now consider that the element outputs are weighted by complex coefficients w_n, then added to give an array output signal of

$$y(t) = \sum_{n=1}^{N} v_n(t) w_n = \sum_{n=1}^{N} \sum_{m=1}^{M} a_m(t) s_{mn} w_n e^{j\omega t} \tag{1.7.3}$$

which is input to the receiver. This can be recast into a simultaneous equation form as follows. Consider the mth source only to be energised, with all the rest set to zero, producing an output signal

$$y_m(t) = \sum_{n=1}^{N} a_m(t) s_{mn} w_n e^{j\omega t} = a_m(t) e^{j\omega t} \sum_{n=1}^{N} s_{mn} w_n \tag{1.7.4}$$

The amplitude gain of the array is the multiplier of the source waveform:

$$g_m = \sum_{n=1}^{N} s_{mn} w_n \tag{1.7.5}$$

such that $y(t) = \sum_{m=1}^{M} a_m(t) g_m e^{j\omega t}$. We can write M simultaneous equations based on variation of m in eqn. (1.7.5) as follows:

$$\begin{aligned}
g_1 &= s_{11} w_1 + s_{12} w_2 + s_{13} w_3 + \cdots + s_{1N} w_N \\
g_2 &= s_{21} w_1 + s_{22} w_2 + s_{23} w_3 + \cdots + s_{2N} w_N \\
&\cdots\cdots \\
g_M &= s_{M1} w_1 + s_{M2} w_2 + s_{M3} w_3 + \cdots + s_{MN} w_N
\end{aligned} \tag{1.7.6}$$

Consider first the case that $M = N$ and the matrix of s coefficients is square. A matrix-vector formulation of eqn. (1.7.6) appears as

$$\begin{bmatrix} g_1 \\ g_2 \\ g_3 \\ \cdots \\ g_N \end{bmatrix} = \begin{bmatrix} s_{11} & s_{12} & s_{13} & \cdots & s_{1N} \\ s_{21} & s_{22} & s_{23} & \cdots & s_{2N} \\ s_{31} & s_{32} & s_{33} & \cdots & s_{3N} \\ & & & & \\ s_{N1} & s_{N2} & s_{N3} & \cdots & s_{NN} \end{bmatrix} \begin{bmatrix} w_1 \\ w_2 \\ w_3 \\ \\ w_N \end{bmatrix} \tag{1.7.8}$$

which defines the antenna amplitude gains for the N sources as a function of the N weights w_N. In an adaptive antenna situation, we are more interested in defining the weighting coefficients as a function of the desired gains for the sources g_n. This is done by inverting the matrix of complex coefficients s_{ij} to give the inverse solution

$$\begin{bmatrix} w_1 \\ w_2 \\ w_3 \\ \cdots \\ w_N \end{bmatrix} = \begin{bmatrix} s_{11} & s_{12} & s_{13} & \cdots & s_{1N} \\ s_{21} & s_{22} & s_{23} & \cdots & s_{2N} \\ s_{31} & s_{32} & s_{33} & \cdots & s_{3N} \\ & & & & \\ s_{N1} & s_{N2} & s_{N3} & \cdots & s_{NN} \end{bmatrix}^{-1} \begin{bmatrix} g_1 \\ g_2 \\ g_3 \\ \\ g_N \end{bmatrix} \tag{1.7.9}$$

These equations are directly equivalent to the linear equations of lumped parameter circuits where typically mesh currents and EMFs are related by a similar matrix equation. In fact, the antenna and sources arrangement is nothing more, from a mathematical point of view, than a distributed parameter linear electrical circuit in which couplings are calculated via Maxwell's equations rather than the more specialised lumped parameter equivalents.

Null steering

Suppose that one of the sources, which can be taken as the first quite generally by suitable indexing, radiates the desired transmission and the remainder radiate interference. A solution for the weighting coefficients which selects $g_1 = 1$ and $g_2 = g_3 = \cdots g_N = 0$ should give a good signal to noise ratio, i.e.

$$\begin{bmatrix} g_1 \\ g_2 \\ \cdots \\ g_N \end{bmatrix} = \begin{bmatrix} 1 \\ 0 \\ \cdots \\ 0 \end{bmatrix} \tag{1.7.10}$$

If eqn. (1.7.9) is solved using these values and the resulting coefficients set up in the antenna of Fig. 1.7.1 the resulting directional pattern will have unity gain in the direction of source No. 1 and nulls in the directions of all the other sources. The pattern behaviour in other directions is not defined directly and is of no interest since no sources have been specified to exist there.

If the directions of the sources are known it is possible in principle to calculate the coefficients s_{mn} by geometric considerations only though a weakness of this technique is that it fails to take into account any unexpected variations of the coefficients due to mutual couplings, near-field scattering, etc.

Example 1.7.1: A three-element line array with $\lambda/4$ spacing is required to receive a broadside plane wave with an amplitude gain of 3 and steer nulls in the two end-fire directions. The matrix of coefficients s_{mn} corresponding to eqn. (1.7.8) is found by postulating the existence of sources in the end-fire directions. We have

$$[s_{mn}] = \begin{bmatrix} 1 & 1 & 1 \\ j & 1 & -j \\ -j & 1 & j \end{bmatrix} \tag{1.7.11}$$

Computation of the inverse matrix and postmultiplication by the vector of required amplitude gains results in

$$\begin{bmatrix} w_1 \\ w_2 \\ w_3 \end{bmatrix} = \frac{1}{4j} \begin{bmatrix} 2j & 1-j & -1-j \\ 0 & 2j & 2j \\ 2j & -1-j & 1-j \end{bmatrix} \begin{bmatrix} 3 \\ 0 \\ 0 \end{bmatrix} = \begin{bmatrix} \frac{3}{2} \\ 0 \\ \frac{3}{2} \end{bmatrix} \tag{1.7.12}$$

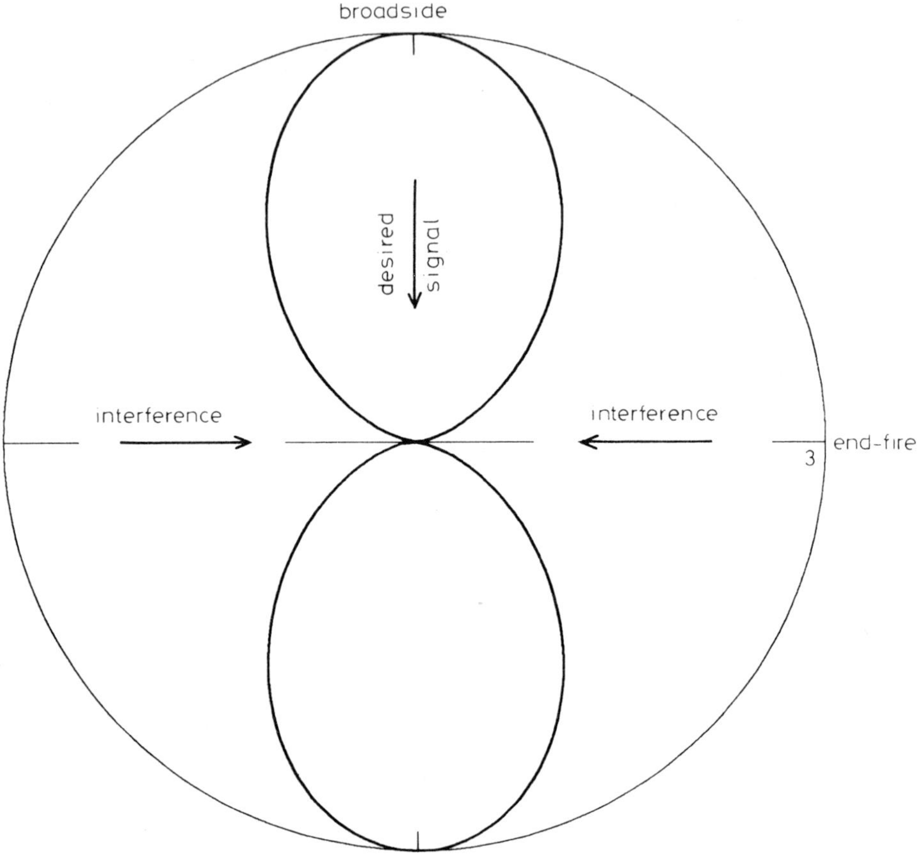

Fig. 1.7.2 *Directional response of 3-element array*

The directional pattern of the antenna using these weights is shown in Fig. 1.7.2 and it is seen to have the required properties.

Example 1.7.2: This example is designed to illustrate some of the potential hazards of null steering. A three-element array with $\lambda/3$ spacing is used and it is required to receive an end-fire signal with an amplitude gain of three and to place nulls at $-45°$ and $+45°$ with respect to end-fire in the horizontal plane. The matrix of signal coefficients for the desired signal and two hypothetical interferences is

$$[s_{mn}] = \begin{bmatrix} \exp(-j\tfrac{2}{3}\pi) & 1 & \exp(j\tfrac{2}{3}\pi) \\ \exp(-j\tfrac{2}{3}\pi\cos 45°) & 1 & \exp(j\tfrac{2}{3}\pi\cos 45°) \\ \exp(-j\tfrac{2}{3}\pi\cos 45°) & 1 & \exp(j\tfrac{2}{3}\pi\cos 45°) \end{bmatrix} \tag{1.7.13}$$

If an attempt is made to invert this matrix on a computer, it will be found to be singular and inspection soon reveals that rows two and three are identical yielding a zero determinant. This does not mean that a solution is not available, rather the technique is wrong. The two interferences are not distinguishable by the array because of the cylindrical symmetry of the pattern about the array axis and the solution for the linear equations is overdetermined. Thus one source will form both nulls and a method for disposing of the surplus row in the matrix must be found. This is most satisfactorily done by the pseudoinverse methods of Section 7.1, but for the moment a solution can be obtained by postulating the existence of an interference at some innocuous bearing which will provide an extra null and enable the equations to be solved. A reasonable choice might be the end fire direction opposite to the signal. Consequently, the modified coefficient matrix is

$$\begin{bmatrix} \exp(-j\tfrac{2}{3}\pi) & 1 & \exp(j\tfrac{2}{3}\pi) \\ \exp(-j\tfrac{2}{3}\pi\cos 45^\circ) & 1 & \exp(j\tfrac{2}{3}\pi\cos 45^\circ) \\ \exp(j\tfrac{2}{3}\pi) & 1 & \exp(-j\tfrac{2}{3}\pi) \end{bmatrix} \tag{1.7.14}$$

whose inverse, now easily computed is

$$[s_{mn}]^{-1} = \begin{bmatrix} -0{\cdot}912 + j0{\cdot}289 & 0{\cdot}848 & 0{\cdot}064 - j0{\cdot}289 \\ -0{\cdot}412 & 0{\cdot}848 & 0{\cdot}564 \\ -0{\cdot}912 - j0{\cdot}289 & 0{\cdot}848 & 0{\cdot}064 + j0{\cdot}289 \end{bmatrix} \tag{1.7.15}$$

and the required weight vector is found as

$$\begin{bmatrix} w_1 \\ w_2 \\ w_3 \end{bmatrix} = [s_{mn}]^{-1} \begin{bmatrix} 3 \\ 0 \\ 0 \end{bmatrix} = \begin{bmatrix} -2{\cdot}736 + j0{\cdot}867 \\ -1{\cdot}236 \\ -2{\cdot}736 - j0{\cdot}867 \end{bmatrix} \tag{1.7.16}$$

The first point to notice about this solution is that two of the weighting coefficients are much greater than one. Since coefficients will normally be implemented by attenuation, if at RF, this implies that the whole vector of coefficients must be scaled down so that the largest in unity; this may have undesired repercussions on the noise performance of the system. Fig. 1.7.3 shows the directional pattern of the array and it has nulls in the correct directions. However, the lobe at end-fire is not the greatest one, and there is a spurious lobe of amplitude 7 at 110° which might have undesirable responses to interference which arises unexpectedly in this direction.

Better performance is obtainable by a little juggling with the direction of the extra null and moving it from end-fire (180°) to around 120° would inhibit the large lobe from appearing. A solution of the coefficients with such a null is shown in Fig. 2.5.5.

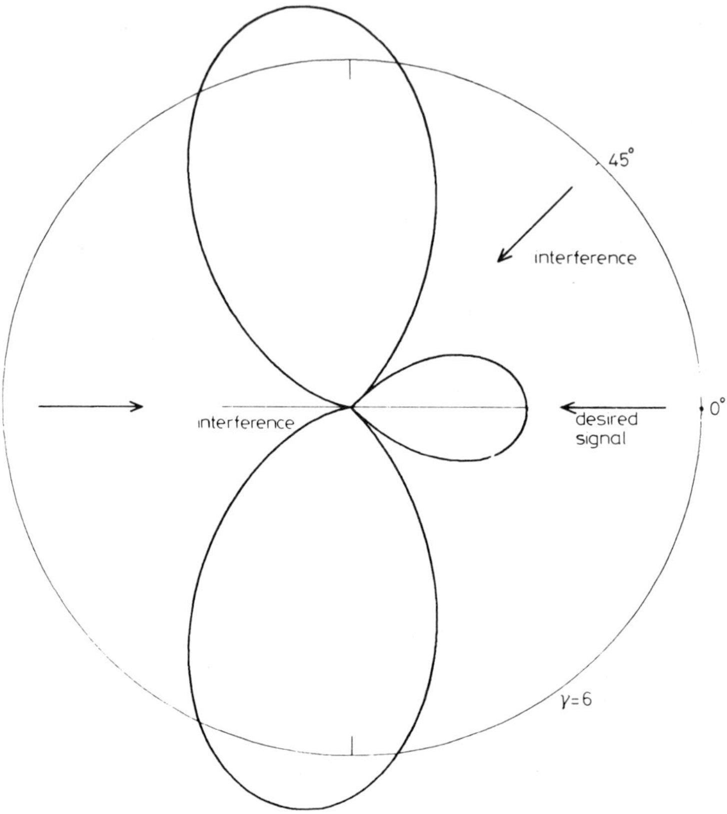

Fig. 1.7.3 *Directional pattern of 3-element array with nulls at 45° and 180°*

Near-field nulls

It is not necessary to assume plane waves in the definition of the source coefficients s_{mn} and an array will steer nulls to local points in space if the coefficients are set up with the appropriate spherical spreading terms. Consider the situation shown in Fig. 1.7.4 where a two-element array with $\frac{1}{4}\lambda$ spacing is required to receive distant source A in the end-fire direction in the presence of interference from source B which lies at $\frac{1}{4}\lambda$ outside the array in the same direction. The source coefficients, allowing for spherical spreading, are

$$[s_{mn}] = \begin{bmatrix} 1 & j \\ -2 & -4j \end{bmatrix} \tag{1.7.17}$$

and the solution for the coefficients is

$$\begin{bmatrix} w_1 \\ w_2 \end{bmatrix} = \frac{1}{2j}\begin{bmatrix} -4j & -j \\ 2 & 1 \end{bmatrix}\begin{bmatrix} 1 \\ 0 \end{bmatrix} = \begin{bmatrix} 2 \\ j \end{bmatrix} \tag{1.7.18}$$

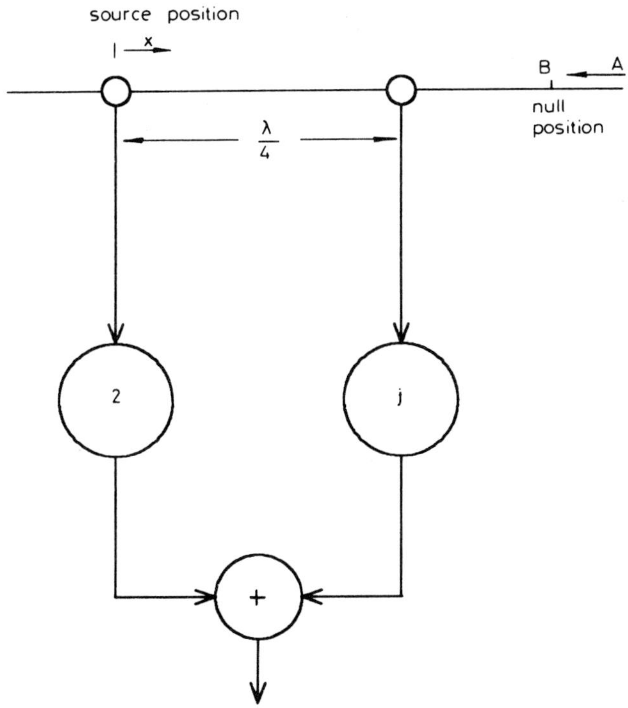

Fig. 1.7.4 *Two-element array for near-field nulling*

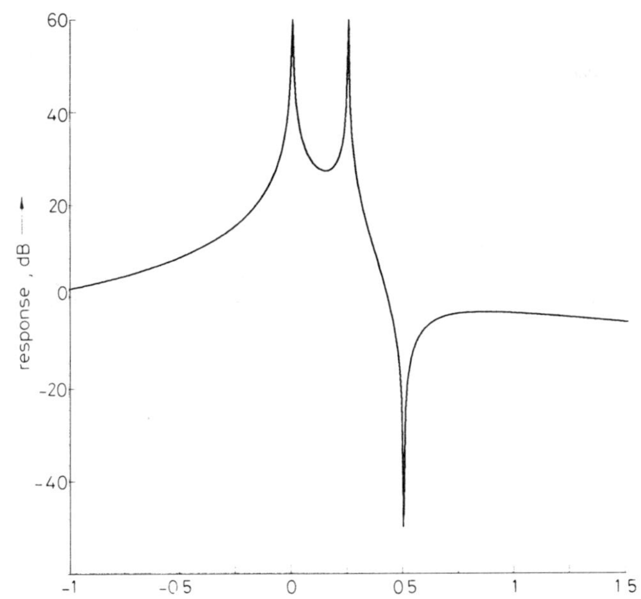

Fig. 1.7.5 *Sensitivity to a source on the x-axis*

Fig. 1.7.5 shows the sensitivity function y of the array for a source on the line joining the elements. The output signal is proportional to

$$y = \frac{2}{x} \exp(-j2\pi x) + \frac{j}{x - \frac{1}{4}} \exp(-j2\pi(x - \tfrac{1}{4})) \qquad (1.7.19)$$

whence

$$|y| = \frac{2}{x} - \frac{1}{x - \frac{1}{4}} \qquad (1.7.20)$$

Chapter 2

Vector and matrix techniques

2.1 Narrow-band signals

Adaptive antenna problems are intrinsically well suited to vector and matrix methods, especially for the narrow-band condition and the appropriate mathematical tools are reviewed in this chapter. The criterion for a signal being narrow-band in the context is that the weighting coefficients can use phase shifters and the nulls which they form remain at constant bearings across the frequency band of interest. The latter criterion is never exactly true so we merely require that the nulls be deep enough across the band.

The condition that the signal is narrow-band is equivalent to requiring that the modulation coherence distance in the direction of propagation should exceed the length of the antenna in the same direction by a large factor. By modulation coherence is meant the modulus of the complex coherence function between the wave field received at two points at different ranges from the source. If the source transmits a pure sine wave then the field is completely coherent at all points but if it transmits a modulated carrier then the field shows a lack of coherence at range differences Δ such that

$$\Delta > c/B \tag{2.1.1}$$

where B is the bandwidth of the modulation and c is the velocity of propagation. Thus the narrow bandwidth approximation is valid if

$$\Delta_a \ll c/B \tag{2.1.2}$$

where the points of the receiving antenna closest and furthest from the source lie within the range interval Δ_a. It is interesting to calculate the various quantities involved for some typical cases and some results are shown in Table 2.1.1 below.

It can be seen that the narrow-band approximation is becoming rather weak for the wideband radar and the sonar applications and in these cases, for good nulling it is necessary to adopt special time domain processing to get deep nulls

Table 2.1.1: *Ratio of array length to modulation coherence distance for some typical systems*

Application	Bandwidth	Array length	$\Delta_a B/c$
Radar	1 MHz	5 m	0·016
Radar	10 MHz	5 m	0·16
Sonar	1 kHz	1 m	0·6
VHF speech link	12 kHz	2 m	8×10^{-5}
HF speech link	6 kHz	100 m	2×10^{-3}

over a broad band. First-order finite bandwidth effects are discussed in Section 5.3 and time domain processing in Section 7.10.

2.2 Representations for narrow-band signals[25, 125]

Complex representation: A source which excites the transmitting antenna with the current time function

$$I(t) = m(t)e^{j\omega t} \tag{2.2.1}$$

where $m(t)$ is the complex baseband modulation function and ω the carrier frequency will produce an E field at a range R, in the plane of polarisation, equal to

$$E(R_i, t) = Km(t - R_i/c)\frac{1}{R_i}e^{j\omega(t - R_i/c)} \tag{2.2.2}$$

where R_i is the radial range from the transmitter to the ith receiving array element and K is a constant depending on directivity of the transmitter and other factors which remain constant in the receiving array. In the closely related scalar acoustic case the excitation function is a piston velocity and the measured field is a pressure.

Under the narrow-band approximation, the modulation function $m(t - R_i/c)$ is constant across the array at a given time instant and we obtain

$$E(R_i, t) = Km(t - \bar{T})\frac{1}{R_i}e^{-jkR_i}e^{j\omega t} \tag{2.2.3}$$

in which $k = 2\pi/\lambda = \omega/c$ is the wavenumber and $\bar{T}$ is the average propagation time delay between the source and the receiving array. This equation is separable into a product of time and space variables and in narrow-band arrays the null steering and pattern optimisation function is concerned solely with the space variation of the field.

The space vector of a given source is defined as a vector whose elements are the space-dependent quantities in eqn. 2.2.3

$$\mathbf{S} \triangleq \begin{bmatrix} R_1^{-1} e^{-jkR_1} \\ R_2^{-1} e^{-jkR_2} \\ \cdots \\ R_N^{-1} e^{-jkR_N} \end{bmatrix} \tag{2.2.4}$$

In most cases the receiving array is located so far from the source that the spherical spreading terms R_i^{-1} are not significantly different at the various elements of the receiving array and in this case it is convenient to omit these factors and to define the source vector as

$$\mathbf{S} \triangleq \begin{bmatrix} e^{-jkR_1} \\ e^{-jkR_2} \\ \cdots \\ e^{-jkR_N} \end{bmatrix} \tag{2.2.5}$$

in which the vector elements each have unit modulus.

Example 2.2.1 : A 3-element receiving array receives a signal from a source located at 30° relative to broadside and the interelement spacing is $\frac{1}{2}\lambda$. The source space vector is

$$\mathbf{S} = \begin{bmatrix} -j \\ 1 \\ j \end{bmatrix} e^{-jkR_2} \tag{2.2.6}$$

the first element in the vector is farthest from the source and R_2 is the range from the source to the centre element.

In many cases, the exact range from the source to the array is not of interest (except in the case of multipath propagation phenomena) and the common factor e^{-jkR} which defines the range to the phase centre of the array can be dropped. Thus for a line array of N elements receiving a signal from bearing α relative to broadside the space vector is

$$\mathbf{S}(\alpha) = \begin{bmatrix} 1 \\ e^{jkd \sin \alpha} \\ e^{j2kd \sin \alpha} \\ \cdots \\ e^{j(N-1)kd \sin \alpha} \end{bmatrix} \tag{2.2.7}$$

where the reference point is taken as element No. 1 and phases are measured relative to this. Often it is more convenient to define the centre of the line array as the phase centre. The space vector for a circular array is given in eqn. (5.3.10(a)).

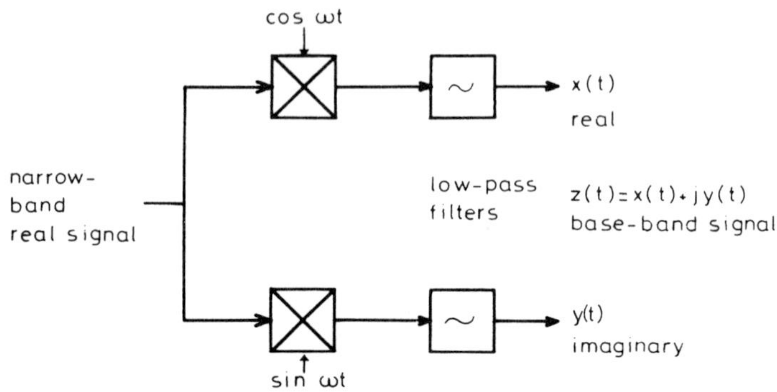

Fig. 2.2.1 *Generation of baseband complex signal by quadrature modulators*

In place of eqn. (2.2.3) we now write simply

$$\mathbf{X}(t) = m(t - \bar{T})\mathbf{S} \tag{2.2.8}$$

as the vector representing the complex signals received at the array elements from one source.

Application of complex representation : The complex signal representation has two main areas of application. First, for numerical calculations, simulations, and analytical work where the cumbersome real variable expressions (described below) are a source of considerable difficulty and secondly for hardware implementations of base-band processing where complex variables allow resolution of the potential ambiguity of frequency offsets from the local oscillator frequency. In this connection they are most familiar in the video processing of radar signals for Doppler analysis. Fig. 2.2.1 shows the down-conversion to baseband of a narrow-band real signal to a quadrature real signal pair which can be described by complex numbers.[25]

Weighting of complex signals : Complex weighting is obligatory for base band signals. Fig. 2.2.2 shows a complex multiplier which produces a phase shift and scaling equivalent to

$$\tilde{z}(t) = ae^{j\phi}z(t) \tag{2.2.9}$$

where

$$z(t) = x(t) + jy(t),\ x,\ y \text{ real} \tag{2.2.10}$$

Setting

$$ae^{j\phi} = u + jv \tag{2.2.11}$$

implementation of the product eqn. 2.2.9 in hardware requires four real multipliers to calculate

$$\tilde{z}(t) = ux(t) - vy(t) + j[vx(t) + uy(t)] \tag{2.2.12}$$

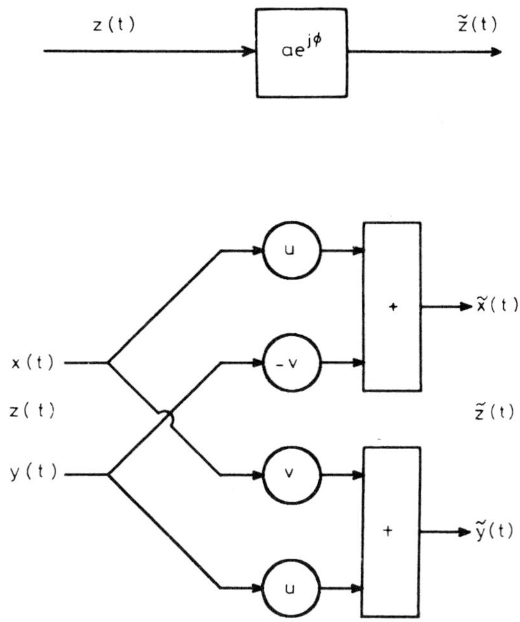

Fig. 2.2.2 *Complex multiplication of baseband signals*

Fig. 2.2.3 is a representation of an antenna with quadrature down-converters and complex weighting coefficients. The output signal of the antenna is complex too. Sometimes the output pair is specifically kept separate, for example in four-phase modulation systems where there are independent modulations of the in-phase and quadrature components or in radar video processing where the pair resolves frequency ambiguity. For optimisation purposes we only wish to know the output power of the antenna and this is the squared modulus of the complex output.

Real signal representation : Sometimes it is desirable to analyse the operation of an antenna with real signals. This might occur if the detailed operation of a hardware implemented real-signal processor is under investigation. The real signal in the medium is of the form

$$\begin{aligned} x_R(t) &= \mathrm{Re}\,\{[m_c(t) + jm_s(t)]e^{j\omega t}\} \\ &= m_c(t)\cos\omega t - m_s(t)\sin\omega t \end{aligned} \tag{2.2.14}$$

where m_c and m_s are independent narrow-band modulations on the in-phase and quadrature components of the carrier. Given that the elements of the array have differential delays T_i from the source, or some receiving array related reference point, then the observable signal at the nth element has the form

$$x_p^n(t) = m_c(t)\cos\omega(t - T_n) - m_s(t)\sin\omega(t - T_n) \tag{2.2.15}$$

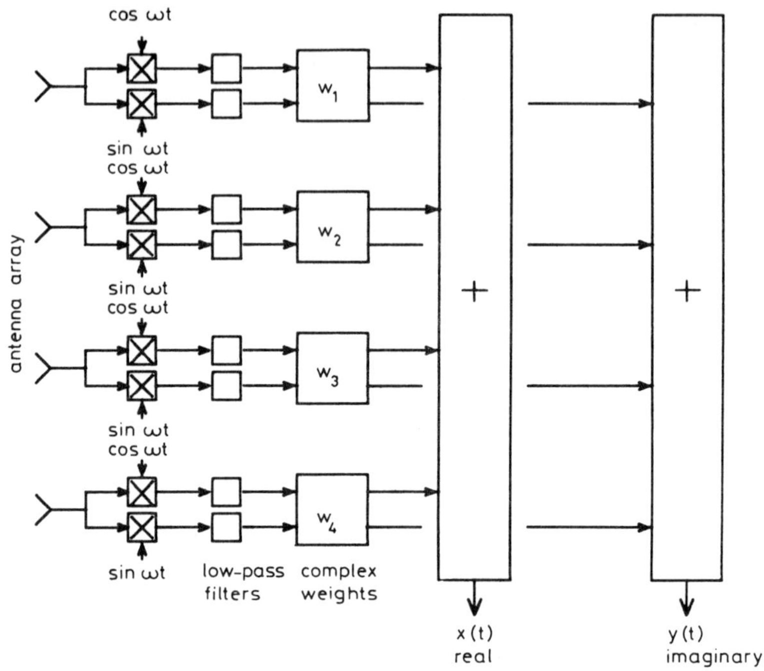

Fig. 2.2.3 *Complete antenna with complex weightings at baseband*

and the locally derived quadrature signal formed by a 90° phase lag or the equivalent quadrature hybrid circuit is

$$x_q^n(t) = m_c(t) \sin \omega(t - T_n) + m_s(t) \cos \omega(t - T_n) \tag{2.2.16}$$

These two signals are present at the points indicated in Fig. 2.2.4.

A vector representation of eqns. (2.2.15 and 2.2.16 is not yet worthwhile because everything is time varying. By expanding the cosine and sine terms they may, however, be written in the form

$$\begin{aligned} x_p^n(t) &= [m_c(t) \cos(-\omega T_n) - m_s(t) \sin(-\omega T_n)] \cos \omega t \\ &\quad - [m_c(t) \sin(-\omega T_n) + m_s(t) \cos(-\omega T_n)] \sin \omega t \\ x_q^n(t) &= [m_c(t) \cos(-\omega T_n) - m_s(t) \sin(-\omega T_n)] \sin \omega t \\ &\quad + [m_c(t) \sin(-\omega T_n) + m_s(t) \cos(-\omega T_n)] \cos \omega t \end{aligned} \tag{2.2.17}$$

Defining the vectors

$$\begin{aligned} \mathbf{C} &= [\cos(-\omega T_1), \cos(-\omega T_2), \ldots, \cos(-\omega T_N)]^t \\ \mathbf{S} &= [\sin(-\omega T_1), \sin(-\omega T_2), \ldots, \sin(-\omega T_N)]^t \end{aligned} \tag{2.2.18}$$

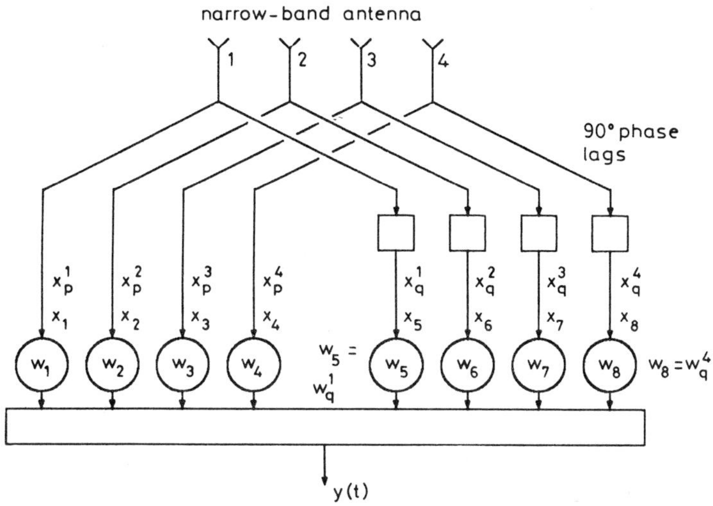

Fig. 2.2.4 *Generation of a 2N-vector of quadrature antenna signals*

substitution into eqns. (2.2.17) yields the vectors

$$\begin{aligned}\mathbf{X}_p(t) &= [m_c(t)\mathbf{C} - m_s(t)\mathbf{S}] \cos \omega t - [m_c(t)\mathbf{S} + m_s(t)\mathbf{C}] \sin \omega t \\ \mathbf{X}_q(t) &= [m_c(t)\mathbf{C} - m_s(t)\mathbf{S}] \sin \omega t + [m_c(t)\mathbf{S} + m_s(t)\mathbf{C}] \cos \omega t\end{aligned} \tag{2.2.19}$$

The final 2 N-element real vector $\mathbf{X}_R(t)$ is then formed as the column

$$\mathbf{X}_R(t) = \begin{bmatrix}\mathbf{X}_p(t)\\ \mathbf{X}_q(t)\end{bmatrix} \tag{2.2.20}$$

Weighting of real signals: There are some differences in the real signal case which require careful treatment if the real signals are to be equivalent to the complex ones at every point. Fig. 1.3.2 showed a quadrature weighting coefficient with in-phase and quadrature weightings of w_p, w_q. With input signal as defined in eqn. (2.2.14) the output is

$$\begin{aligned}y_R(t) &= w_p[m_c(t) \cos \omega t - m_s(t) \sin \omega t] \\ &\qquad + w_q[m_c(t) \sin \omega t + m_s(t) \cos \omega t] \\ &= [w_p m_c(t) + w_q m_s(t)] \cos \omega t \\ &\qquad - [w_p m_s(t) - w_q m_c(t)] \sin \omega t\end{aligned} \tag{2.2.21}$$

This result can be equated with the complex case, eqn. (2.2.12) if we put

$$m_c(t) = x(t) \qquad m_s(t) = y(t) \tag{2.2.22}$$

$$w_p = u \qquad w_q = -v \tag{2.2.23}$$

The multiplier of cos ωt in eqn. (2.2.21) is equivalent to the real part of $\tilde{z}(t)$ in eqn. (2.2.12) and that of sin ωt is equivalent to the imaginary part. If, instead of using a phase lag to generate the local quadrature signals, a phase lead had been used the sign of $x_q(t)$ would have been reversed and it would not have been necessary to invert the sign of w_q in eqn. (12.2.23). It appears that the complex signals can be fully compatible with the real signals or the weightings can be fully compatible but not both at the same time. The conventions used above will be used consistently in the remainder of the book.

Multiple signal sources: When the antenna receives signals from more than one source, the element output voltages are a linear superposition of the individual signals. However, the antenna is going to offer different amplitude gains to different sources and the source vectors must be kept separate. We have, in the complex representation, for M sources

$$\begin{aligned}\mathbf{X}(t) &= m_1(t)\mathbf{S}_1 + m_2(t)\mathbf{S}_2 + \cdots + m_M(t)\mathbf{S}_M \\ &= \sum_{i=1}^{M} m_i(t)\mathbf{S}_i\end{aligned} \tag{2.2.24}$$

A compact representation for eqn. (2.2.24) which is sometimes useful is obtained by writing the $\mathbf{S}_i$ vectors into the columns of a matrix:

$$\mathbf{Q} \triangleq [\mathbf{S}_1 | \mathbf{S}_2 | \mathbf{S}_3 | \cdots | \mathbf{S}_M] \tag{2.2.25}$$

which is the transpose of the matrix used in eqn. (1.7.8) if $M = N$. Defining the M-vector $\mathbf{A}(t)$ of modulations:

$$\mathbf{A}(t) \triangleq \begin{bmatrix} m_1(t) \\ m_2(t) \\ \cdots \\ m_M(t) \end{bmatrix} \tag{2.2.26}$$

the composite array signal vector is given by

$$\mathbf{X}(t) = \mathbf{Q}\mathbf{A}(t) \tag{2.2.27}$$

the vector equivalent of eqn. (1.7.2).

2.3 Vector inner products

These are ubiquitous in adaptive algorithms. For real vectors **A** and **B** with elements A_i, B_i the definition of the inner product is

$$d = \mathbf{A}^t\mathbf{B} = \sum_{i=1}^{N} A_i B_i \tag{2.3.1}$$

sometimes known as the scalar product or dot product. For complex vectors there are two possible forms, both useful,

(i) $$d = \mathbf{A}^t\mathbf{B} = \mathbf{B}^t\mathbf{A} \tag{2.3.2}$$

and the Hermitian inner product

(ii) $$d = \mathbf{A}^H\mathbf{B} = \sum_{i=1}^{N} A_i^* B_i = (\mathbf{A}^t\mathbf{B}^*)^* \tag{2.3.3}$$

Everywhere in this book the convention is that $(\,.\,)^t$ is a simple transpose without conjugation, $(\,.\,)^H$ is an Hermitian or conjugate transpose, and $(\,.\,)^*$ is a simple conjugate. Notations elsewhere are very diverse.

Examples of use

Array output : For an array with element voltages defined by complex N-vector $\mathbf{X}(t)$ and with complex weighting vector $\mathbf{W}$, the output signal is

$$y(t) = \mathbf{X}(t)^t\mathbf{W} = \mathbf{W}^t\mathbf{X}(t) \tag{2.3.4}$$

This representation is in closest correspondence with the physical situation and increasing the argument or either **X** or **W** corresponds to a time advance of the signal. In the literature the Hermitian inner product is sometimes used to get $y(t)$ and this has the advantage that the resulting equations conform more closely to the accepted usage in mathematical texts.

If the data $\mathbf{X}(t)$ is generated by a single source as in eqn. (2.2.8) we can write

$$y(t) = m(t - \bar{T})\mathbf{S}_0^t\mathbf{W} \doteq m(t - \bar{T})g \tag{2.3.5}$$

where complex scalar g is the amplitude gain of the antenna for the particular source, relating the summed output voltage to that at a single reference element.

When a real signal representation is used as in eqn. (2.2.20) we have a $2N$ element real vector

$$\mathbf{W}_R \triangleq \begin{bmatrix} \mathbf{W}_p \\ \mathbf{W}_q \end{bmatrix} \tag{2.3.6}$$

and the output signal $y_R(t)$ is given by

$$\begin{aligned} y_R(t) &= \mathbf{W}_R^t\mathbf{X}_R(t) = \mathbf{W}_p^t\mathbf{X}_p(t) + \mathbf{W}_q^t\mathbf{X}_q(t) \\ &= [m_c(t)\cos\omega t - m_s(t)\sin\omega t][\mathbf{W}_p^t\mathbf{C} + \mathbf{W}_q^t\mathbf{S}] \\ &\quad - [m_c(t)\sin\omega t + m_s(t)\cos\omega t][\mathbf{W}_p^t\mathbf{S} - \mathbf{W}_q^t\mathbf{C}] \end{aligned} \tag{2.3.7}$$

with the equivalences $\mathbf{W}_p = \mathrm{Re}(\mathbf{W})$, $\mathbf{W}_q = -\mathrm{Im}(\mathbf{W})$, $\mathbf{C} = \mathrm{Re}(\mathbf{S}_0)$, $\mathbf{S} = \mathrm{Im}(\mathbf{S}_0)$.

Vector norms: The norm of a vector is usually taken as the Euclidian length L which, for a real vector $\mathbf{A}$ with elements A_i, is

$$L = \left[\sum_{i=1}^{N} A_i^2\right]^{1/2} = (\mathbf{A}^t\mathbf{A})^{1/2} = \|\mathbf{A}\| \tag{2.3.8}$$

For three-element real vectors, the length corresponds to physical length in three dimensions if the elements are thought of as defining vector components on the x, y, and z axes. In the complex case we have

$$L = \left[\sum_{i=1}^{N} A_i^* A_i\right]^{1/2} = \left[\sum_{i=1}^{N} |A_i|^2\right]^{1/2} = \|\mathbf{A}\| = (\mathbf{A}^H\mathbf{A})^{1/2} = (\mathbf{A}^t\mathbf{A}^*)^{1/2} \tag{2.3.9}$$

where the conjugation is obligatory. These norms have another interpretation in connection with time series. If the elements of $\mathbf{A}$, either in the real or complex case are the values in a zero mean sequence of length N then $L/N^{1/2}$ is the standard deviation of the sequence and L^2/N is the variance.

2.4 Angles between vectors[97]

For a real 3-vector, if the elements are thought of as spatial components in the standard form:

$$\begin{aligned} \mathbf{A} &\triangleq A_1\mathbf{i} + A_2\mathbf{j} + A_3\mathbf{k} \\ \mathbf{B} &\triangleq B_1\mathbf{i} + B_2\mathbf{j} + B_3\mathbf{k} \end{aligned} \tag{2.4.1}$$

where $\mathbf{i}$, $\mathbf{j}$, and $\mathbf{k}$ are unit vectors in the x, y, and z directions, then the angle θ between $\mathbf{A}$ and $\mathbf{B}$ is defined by

$$\cos(\theta) = \frac{\mathbf{A}\cdot\mathbf{B}}{(\mathbf{A}\cdot\mathbf{A}\,.\,\mathbf{B}\cdot\mathbf{B})^{1/2}} = \frac{A_1B_1 + A_2B_2 + A_3B_3}{(\mathbf{A}\cdot\mathbf{A}\,.\,\mathbf{B}\cdot\mathbf{B})^{1/2}} \tag{2.4.2}$$

The generalisation to N dimensions involves simply replacing the dot products by inner products; for the real case

$$\cos(\theta) = \mathbf{A}^t\mathbf{B}/(\mathbf{A}^t\mathbf{A}\mathbf{B}^t\mathbf{B})^{1/2} \tag{2.4.3}$$

In the complex case we can set

$$\cos(\theta)e^{j\phi} = \mathbf{A}^t\mathbf{B}/(\|\mathbf{A}\| \cdot \|\mathbf{B}\|) \tag{2.4.4}$$

$$\cos(\theta)e^{j\phi} = \mathbf{A}^H\mathbf{B}/(\|\mathbf{A}\| \cdot \|\mathbf{B}\|) \tag{2.4.5}$$

or

$$\cos(\theta) = \mathrm{Re}\,(\mathbf{A}^H\mathbf{B})/(\|\mathbf{A}\| \cdot \|\mathbf{B}\|) \tag{2.4.6}$$

and some degree of physical insight into the problem at hand is necessary to select the most appropriate definition.

The angle between vectors, defined as eqn. (2.4.3) or (2.4.5), is equivalent to the definition of the cross correlation between the two sequences formed by the elements of the vectors.

Application to directional pattern determination: An example of the use of eqn. (2.4.5) is in the calculation of antenna beam patterns. We postulate an incoming unit amplitude test signal arriving from direction α, as defined by eqn. (2.2.5), yielding

$$\mathbf{X}(t) = e^{j\omega t}\mathbf{S}(\alpha) \tag{2.4.7}$$

and, if the antenna has weighting vector $\mathbf{W}$, using eqn. (2.3.5) we have an amplitude gain of

$$g(\alpha) = \mathbf{S}(\alpha)^t\mathbf{W} \tag{2.4.8}$$

If the weighting coefficients have been set up to point a beam in a certain direction α_0 by phase compensation with zero attenuation we have

$$\mathbf{W} = \mathbf{W}_c = \mathbf{S}(\alpha_0)^* \tag{2.4.9}$$

and

$$\|\mathbf{W}\| = \|\mathbf{S}(\alpha_0)\| = \|\mathbf{S}(\alpha_0)\|^* = N^{1/2} \tag{2.4.10}$$

$\mathbf{W}_c$ will be defined as a conventional weighting vector. The equation

$$\rho(\alpha) = \frac{\mathbf{S}(\alpha)^t\mathbf{W}_c}{\|\mathbf{S}(\alpha)\| \cdot \|\mathbf{W}_c\|} = \frac{\mathbf{S}(\alpha)^t\mathbf{S}(\alpha_0)^*}{N} \tag{2.4.11}$$

defines a normalised beam pattern whose maximum amplitude is $1 + j0$.

In the case of a uniformly spaced line array the vector $\mathbf{S}(\alpha)$ is defined by eqn. (2.2.7) and it is a standard problem in array theory to show that

$$g(\alpha) = \sum_{n=0}^{N-1} \exp jnkd\phi = \exp(\tfrac{1}{2}jkNd\phi)\,\frac{\sin(\tfrac{1}{2}Nkd\phi)}{\sin(\tfrac{1}{2}kd\phi)} \tag{2.4.12}$$

and

$$\rho(\alpha) = \frac{1}{N}\,g(\alpha) \tag{2.4.13}$$

where $\phi = \sin(\alpha) - \sin(\alpha_0)$. By a judicious choice of phase centre, the argument of the complex exponential in eqn. (2.4.12) can be made zero and the antenna pattern is then purely real.

Cauchy–Schwartz inequality: In the real vector case, eqn. (2.4.3), $\cos(\theta)$ attains its maximum value of unity when the vectors **A** and **B** are parallel, either having the same elements or one being a scaled version of the other. Multiplying throughout by $\|\mathbf{A}\| \, . \, \|\mathbf{B}\|$ yields

$$\mathbf{A}^t\mathbf{B} = \|\mathbf{A}\| \, . \, \|\mathbf{B}\| \cos(\theta) \leq \|\mathbf{A}\| \, . \, \|\mathbf{B}\| \tag{2.4.14}$$

an inequality known as the Cauchy inequality, the discrete equivalent of the Schwartz inequality.

The complex vector equivalents are

$$|\mathbf{A}^t\mathbf{B}| \leq \|\mathbf{A}\| \, . \, \|\mathbf{B}\|$$

$$|\mathbf{A}^H\mathbf{B}| = |\mathbf{A}^t\mathbf{B}^*| \leq \|\mathbf{A}\| \, . \, \|\mathbf{B}\| \tag{2.4.15}$$

Nulls and orthogonality

In the antenna case, the cosine of the angle between the weighting vector **W** and the signal vector **S** due to a single source is directly related to the directional gain and becomes equal to it if the latter is normalised to unity maximum. If there is a null in a beam pattern at some bearing α it means that a source placed there gives rise to no output from the weighted antenna

$$\mathbf{W}^t\mathbf{S}(\alpha) = 0 \tag{2.4.16}$$

Vectors which have this property are orthogonal and the concept finds great application in optimal antennas. For example suppose we have array data represented by eqn. (2.2.24), a superposition of signals, but are only interested in one of them, say $m_1(t)$. If a weighting vector can be found such that $\mathbf{W}^t\mathbf{S}_2 = \mathbf{W}^t\mathbf{S}_3 = \cdots = \mathbf{W}^t\mathbf{S}_M = 0$, while $\mathbf{W}^t\mathbf{S}_1 = 1$, then

$$\mathbf{W}^t\mathbf{X}(t) = m_1(t)\mathbf{W}^t\mathbf{S}_1 = m_1(t) \tag{2.4.17}$$

and perfect rejection of the undesired sources is achieved. This idea is the same one used in eqn. (1.7.10). Adaptive antennas get very close to this condition and a theoretical discussion of the direct derivation of such vectors is given in the following Section. A real or complex vector of N elements can be made orthogonal to $N - 1$ other vectors and still have a finite inner product with the Nth which would be the desired signal. Thus the performance limit of a null steering antenna is $N - 1$ nulls which in principle may be arbitrarily placed.

In the complex case, the orthogonality relation eqn. (2.4.16) does not imply that $\mathbf{W}^H\mathbf{S}(\alpha) = 0$ since $\mathbf{W}^t$ and $\mathbf{W}^H$ are quite independent vectors.

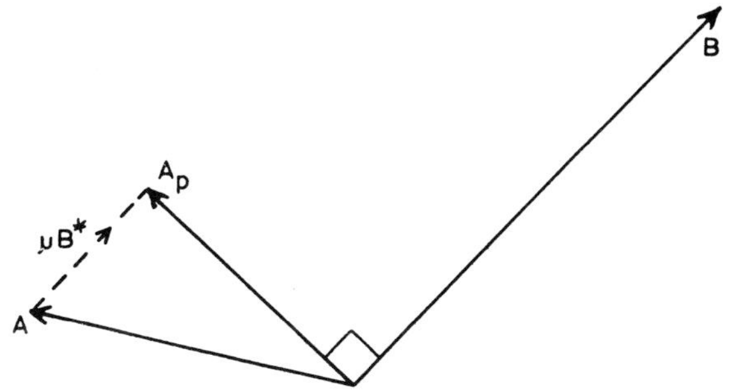

Fig. 2.5.1 *Projection of a vector*

2.5 Orthogonality achieved by projections

An efficient mathematical method for achieving orthogonality between two vectors is the process of projection. In Fig. 2.5.1, two vectors in a space of arbitarily high dimensionality are shown in the plane which contains them both. In order to make vector **A** orthogonal to **B** it is necessary to rotate it to a right-angled orientation and the preferred technique is to add to A a vector parallel to **B*** so that orthogonality is achieved the position of the new vector being indicated. This projected vector has the form

$$\mathbf{A}_P = \mathbf{A} + \mu\mathbf{B}^* \tag{2.5.1}$$

and since the end result of the process is $\mathbf{B}^t\mathbf{A}_P = 0$ the scalar variable μ is found from

$$\mathbf{B}^t\mathbf{A}_P = \mathbf{B}^t\mathbf{A} + \mu\mathbf{B}^t\mathbf{B}^* = 0$$

whence

$$\mu = -\frac{\mathbf{B}^t\mathbf{A}}{\mathbf{B}^t\mathbf{B}^*} \tag{2.5.2}$$

Substituting eqn. (2.5.2) into (2.5.1) gives rise to the first step of the Gram–Schmidt formula,

$$\mathbf{A}_P = \mathbf{A} - \frac{\mathbf{B}^*(\mathbf{B}^t\mathbf{A})}{\mathbf{B}^t\mathbf{B}^*} = \left(\mathbf{I} - \frac{\mathbf{B}^*\mathbf{B}^t}{\mathbf{B}^t\mathbf{B}^*}\right)\mathbf{A} = \mathbf{PA} \tag{2.5.3}$$

The matrix $\mathbf{P} = \mathbf{I} - (\mathbf{B}^*\mathbf{B}^t/\mathbf{B}^t\mathbf{B}^*)$ is a projection matrix and operates on an arbitary vector converting it to a form orthogonal to **B**. Geometrically, the process finds the shadow of the **A** vector on the plane orthogonal to **B** when it is illuminated from a light source along the axis of **B**, hence the term projection.

Beam pattern interpretation

Projection operations applied to weighting vectors of antennas have a very direct interpretation in terms of beam patterns and knowledge of this helps greatly in the understanding of adaptive antenna solutions. In eqn. (2.5.3) we had the result

$$\mathbf{W}_p = \mathbf{W} - \mu\mathbf{S}^* \tag{2.5.4}$$

where the complex variable μ is chosen such that $\mathbf{S}^t\mathbf{W}_p = 0$, producing orthogonality. Suppose $\mathbf{W}$ is the initial weighting vector of an antenna, $\mathbf{S}$ the space vector due to a source which it is to null, and $\mathbf{W}_p$ is the weighting vector having such a null. Each of the three vectors has a directional pattern $g(\alpha)$ corresponding to the beam pattern which it would produce if used as a weighting vector. From eqn. (2.4.8),

$$g_W(\alpha) = \mathbf{S}(\alpha)^t\mathbf{W} = \sum_{i=1}^{N} \mathbf{W}_i e^{-jkx_i\theta} \tag{2.5.5}$$

if the antenna is collinear where $\theta = \sin(\alpha)$, $k = 2\pi/\lambda$, x_i is the coordinate of the ith element and α is bearing relative to broadside. Substituting the directional patterns g_{W_p}, g_W and g_{S^*} into eqn. (2.5.4) yields

$$g_{\mathbf{W}_p}(\alpha) = g_{\mathbf{W}}(\alpha) - \mu g_{\mathbf{S}^*}(\alpha) \tag{2.5.6}$$

and it can be seen that the directional pattern of the vector $\mathbf{W}_p$ is equal to the initial pattern $g_{\mathbf{W}}$ less a pattern $g_{\mathbf{S}^*}$ which is that due to the conjugate source vector. As the latter is a plane wave with a phase gradient its directional pattern looks like that of a phased array with a lobe directed toward the source. The beam pattern $g_{\mathbf{S}^*}(\alpha)$ is known as the retro-beam[48] and its main lobe as the retro-lobe.

Example 2.5.1: Let $\mathbf{W}$ be the broadside steered cophasal weighting vector for a five-element array, $\mathbf{W} = (1, 1, 1, 1, 1)^t$ and $\mathbf{S}$ correspond to a source at $-12°$ relative to broadside:

$$\mathbf{S} = \begin{bmatrix} 0{\cdot}2614 + j0{\cdot}9652 \\ 0{\cdot}7942 + j0{\cdot}6077 \\ 1{\cdot}0000 + j0{\cdot}0000 \\ 0{\cdot}7942 - j0{\cdot}6077 \\ 0{\cdot}2614 - j0{\cdot}9652 \end{bmatrix} \tag{2.5.7}$$

The directional patterns of these two vectors are shown in Fig. 2.5.2. Projection of $\mathbf{W}$ into the plane perpendicular to $\mathbf{S}$ yields

$$\mathbf{W}_p = \mathbf{W} - 0{\cdot}622\mathbf{S}^* = \begin{bmatrix} 0{\cdot}8374 + j0{\cdot}6006 \\ 0{\cdot}5059 + j0{\cdot}3781 \\ 0{\cdot}3778 + j0{\cdot}0000 \\ 0{\cdot}5059 - j0{\cdot}3781 \\ 0{\cdot}8374 - j0{\cdot}6006 \end{bmatrix} \tag{2.5.8}$$

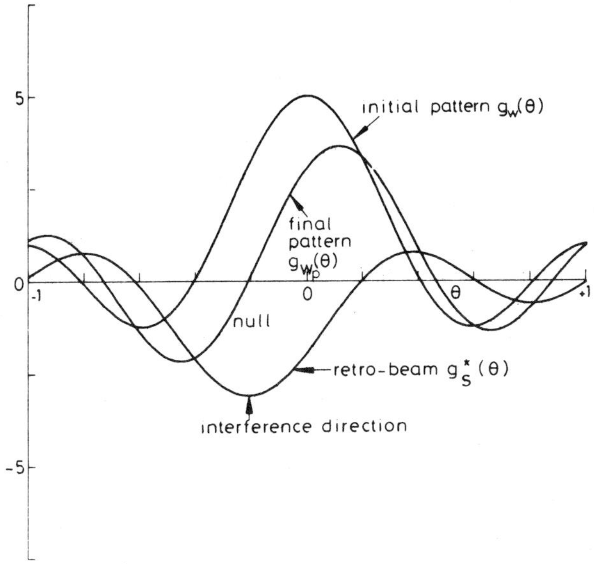

Fig. 2.5.2 *Directional pattern of a broadside steered array, patterns of beam steered to interference source (retrobeam), and sum of the two having a null at the interference direction*

whose directional pattern $g_{\mathbf{W}_p}(\theta)$, shown in Fig. 2.5.2, is seen to have a null in the direction of the interference.

Noncollinear arrays: The same concept applies to arbitarily spatially disposed arrays of elements though derivation of the space vectors is more difficult. In every case, for maximum final amplitude gain for the desired source, the initial weighting vector $\mathbf{W}$ is chosen as the conjugate of the space vector of the incoming desired signal.

Multidimensional projections

It is frequently required to achieve a result of the form of eqn. (2.4.17) where a vector is found which is orthogonal to several other vectors $\mathbf{B}_i$, $i = 1, \ldots, M$. This cannot be done directly by successive projections of type (2.5.3) which leads to failure. To show this consider that there are two vectors $\mathbf{B}_1$ and $\mathbf{B}_2$ and it required to make vector $\mathbf{A}$ orthogonal to both. The first projection gives

$$\mathbf{A}(1) = \mathbf{P}_1 \mathbf{A} = \left(\mathbf{I} - \frac{\mathbf{B}_1^* \mathbf{B}_1^t}{\mathbf{B}_1^t \mathbf{B}_1^*}\right) \mathbf{A} \tag{2.5.9}$$

and application of the second to $\mathbf{A}(1)$ yields

$$\mathbf{A}(2) = \mathbf{P}_2 \mathbf{A}(1) = \left(\mathbf{I} - \frac{\mathbf{B}_2^* \mathbf{B}_2^t}{\mathbf{B}_2^t \mathbf{B}_2^*}\right) \mathbf{A}(1) = \mathbf{A}(1) - \frac{\mathbf{B}_2^* \mathbf{B}_2^t \mathbf{A}(1)}{\mathbf{B}_2^t \mathbf{B}_2^*} \tag{2.5.10}$$

The orthogonality of $\mathbf{A}(2)$ and $\mathbf{B}_1$ is tested:

$$\mathbf{B}_1^t \mathbf{A}(2) = \mathbf{B}_1^t \mathbf{A}(1) - \frac{\mathbf{B}_1^t \mathbf{B}_2^* \mathbf{B}_2^t \mathbf{A}(1)}{\mathbf{B}_2^t \mathbf{B}_2^*} \tag{2.5.11}$$

Now $\mathbf{B}_1^t \mathbf{A}(1) = 0$ by virtue of the projection (2.5.9) but in the second term of eqn. (2.5.11) neither $\mathbf{B}_1^t \mathbf{B}_2^*$ or $\mathbf{B}_2^t \mathbf{A}(1)$ are necessarily zero, therefore $\mathbf{B}_2^t \mathbf{A}(2) \neq 0$ in general and the method has not succeeded.

In eqn. (2.5.11), if $\mathbf{B}_1^t \mathbf{B}_2^*$ were zero then the second term would vanish and $\mathbf{A}(2)$ would be orthogonal to both $\mathbf{B}_1$ and $\mathbf{B}_2$. This is an example of the rule that successive projections by a set of orthogonal vectors are commutative and as the final vector $\mathbf{A}(M)$ in such a process is orthogonal to the last, $\mathbf{B}_M$, then it must be orthogonal to them all since any projection may be taken as the last. Thus we wish to convert the set of vectors $\mathbf{B}_i$ into an orthogonal set before beginning operations on $\mathbf{A}$. This itself is done by projection. The first vector $\mathbf{B}_1$ is used as the first direction in a new orthogonal coordinate system $\mathbf{X}_i$, $i = 1, \ldots, M$ and the first projection uses $\mathbf{X}_1$

$$\mathbf{X}_1^* = \mathbf{B}_1^*$$

$$\mathbf{P}_1 = \mathbf{I} - \frac{\mathbf{X}_1^* \mathbf{X}_1^t}{\mathbf{X}_1^t \mathbf{X}_1^*} \tag{2.5.12}$$

The second vector $\mathbf{B}_2$ is now projected to become orthogonal to $\mathbf{X}_1$ and at the same time forms the second coordinate direction and defines a new projection:

$$\mathbf{X}_2^* = \mathbf{P}_1 \mathbf{B}_2^* = \left(\mathbf{I} - \frac{\mathbf{X}_1^* \mathbf{X}_1^t}{\mathbf{X}_1^t \mathbf{X}_1^*}\right) \mathbf{B}_2^*$$

$$\mathbf{P}_2 = \mathbf{I} - \frac{\mathbf{X}_2^* \mathbf{X}_2^t}{\mathbf{X}_2^t \mathbf{X}_2^*} \tag{2.5.13}$$

Vector $\mathbf{B}_3$ is projected to become orthogonal to both $\mathbf{X}_1$ and $\mathbf{X}_2$ (and hence orthogonal to $\mathbf{B}_1$ and $\mathbf{B}_2$) and forms the third co-ordinate $\mathbf{X}_3$

$$\mathbf{X}_3^* = \mathbf{P}_2 \mathbf{P}_1 \mathbf{B}_3^* = \mathbf{P}_1 \mathbf{P}_2 \mathbf{B}_3^*$$

$$\mathbf{P}_3 = \mathbf{I} - \frac{\mathbf{X}_3^* \mathbf{X}_3^t}{\mathbf{X}_3^t \mathbf{X}_3^*} \tag{2.5.14}$$

and so on until finally

$$\mathbf{X}_M^* = \mathbf{P}_{M-1} \mathbf{P}_{M-2} \cdots \mathbf{P}_2 \mathbf{P}_1 \mathbf{B}_M^*$$

$$\mathbf{P}_M = \mathbf{I} - \frac{\mathbf{X}_M^* \mathbf{X}_M^t}{\mathbf{X}_M^t \mathbf{X}_M^*} \tag{2.5.15}$$

We now have a set of orthogonal directions $\mathbf{X}_i$ which span the same subspace as the original set of vectors $\mathbf{B}_i$, i.e. any $\mathbf{B}_i$ is a linear combination of the $\mathbf{X}_i$ and vice versa, and if vector $\mathbf{A}$ is projected into a direction which is orthogonal to every

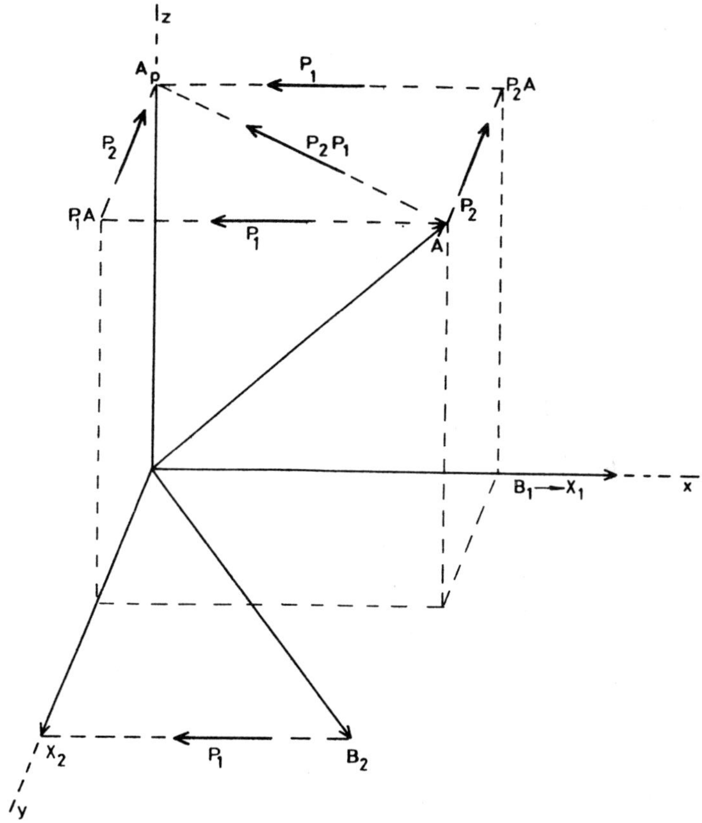

Fig. 2.5.3 *Double projection*

$\mathbf{X}_i$ then it is also orthogonal to every $\mathbf{B}_i$, the original requirement. The set of projections defined by these $\mathbf{X}_i$ directions can be used in any order to project vector $\mathbf{A}$.

$$\mathbf{A}_p = \mathbf{P}_M \mathbf{P}_{M-1} \cdots \mathbf{P}_2 \mathbf{P}_1 \mathbf{A} \tag{2.5.16}$$

The process is illustrated for the simplest case in Fig. 2.5.3 in which vector $\mathbf{A}$ is projected into a subspace orthogonal to $\mathbf{B}_1$ and $\mathbf{B}_2$ which lie in the xy plane. It is clear that $\mathbf{A}$ must finish up along the z axis. The first co-ordinate $\mathbf{X}_1$ is parallel to $\mathbf{B}_1$ along the x axis while the projection of $\mathbf{B}_2$ to become orthogonal to $\mathbf{X}_1$ moves it to the y axis $\mathbf{X}_2$. Projection of $\mathbf{A}$ by $\mathbf{X}_1$ and $\mathbf{X}_2$ in any order, as shown, moves $\mathbf{A}$ onto the z axis.

The new vector $\mathbf{A}_p$ does not have any particular magnitude and this has to be adjusted by scaling since direction changes are inadmissable at this point.

$$\tilde{\mathbf{A}}_p = k\mathbf{A}_p \tag{2.5.17}$$

An interesting numerical equivalence to the projections 2.5.16 is obtainable by using the representation

$$\mathbf{Q} = [\mathbf{B}_1 \,|\, \mathbf{B}_2 \,|\cdots|\, \mathbf{B}_M] \tag{2.5.18}$$

as in eqn. (2.2.25) for the vectors. The matrix

$$\tilde{\mathbf{P}} = \mathbf{Q}^*(\mathbf{Q}^t\mathbf{Q}^*)^{-1}\mathbf{Q}^t \tag{2.5.19}$$

is a projection matrix which projects any vector into the subspace spanned by the M vectors $\mathbf{B}_n$ by setting to zero any components which lie outside it. Conversely, the complementary projection

$$\mathbf{P} = \mathbf{I} - \tilde{\mathbf{P}} \tag{2.5.20}$$

has the effect of projecting any vector onto the subspace orthogonal to the vectors $\mathbf{B}_n$ and it can be shown that $\mathbf{P} = \mathbf{P}_1\,\mathbf{P}_2 \cdots \mathbf{P}_M$. This relationship is used in Section 5.2.

Application to multiple null formation: The Gram-Schmidt procedure generates a sequence of orthogonal vectors, or coordinates, $\mathbf{X}_n$ such that X_n is orthogonal to $\mathbf{B}_{n-1}, \mathbf{B}_{n-2}, \ldots, \mathbf{B}_1$. When the projections $\mathbf{P}_M\,\mathbf{P}_{M-1} \cdots \mathbf{P}_1$ operate on $\mathbf{A}$ in eqn. (2.5.16) they can be thought of a generating a new coordinate $\mathbf{X}_{M+1} = \mathbf{A}_p$. Thus $\mathbf{A}_p$ is orthogonal to $\mathbf{B}_M, \mathbf{B}_{M-1}, \ldots, \mathbf{B}_1$:

$$\mathbf{B}_i^t\,\mathbf{A}_p = 0, \qquad i = 1, 2, \ldots, M \tag{2.5.21}$$

However,

$$\mathbf{A}^H\mathbf{A}_p \neq 0 \tag{2.5.22}$$

except when $\mathbf{A}$ is a linear combination of the $\mathbf{B}_i^*$. From eqn. (2.5.21), $\mathbf{A}_p$ is seen to be a suitable candidate for a weighting vector in an array which requires nulls in its directional pattern in the directions of sources which give rise to space vectors $\mathbf{B}_1, \mathbf{B}_2, \ldots, \mathbf{B}_M$ and, if the desired source has space vector $\mathbf{S}$, we set

$$\mathbf{A} = \mathbf{S}^* \tag{2.5.23}$$

and, from eqn. (2.5.22),

$$\mathbf{S}^t\mathbf{A}_p \neq 0 \tag{2.5.24}$$

showing that there is normally some residual amplitude gain for the desired signal. The qualifier following eqn. (2.5.22) will, in practice, be dormant unless one of the undesired sources lies in the same direction as the desired signal or on a grating lobe of the cophasal beam pattern for the desired signal.

Figure 2.5.4 shows an example of an array steering nulls at three jamming sources at 18°, 22°, −53°. The array has nine elements spaced at $\lambda/2$ and the desired signal is broadside. The directional patterns of the interference space vectors $\mathbf{S}_1^*$, $\mathbf{S}_2^*$, and $\mathbf{S}_3^*$ are shown on the left of the figure and of the desired signal space vector pattern at the bottom, this pattern is the same as that of a cophasal beam directed at the desired signal and it is clear that the jammers

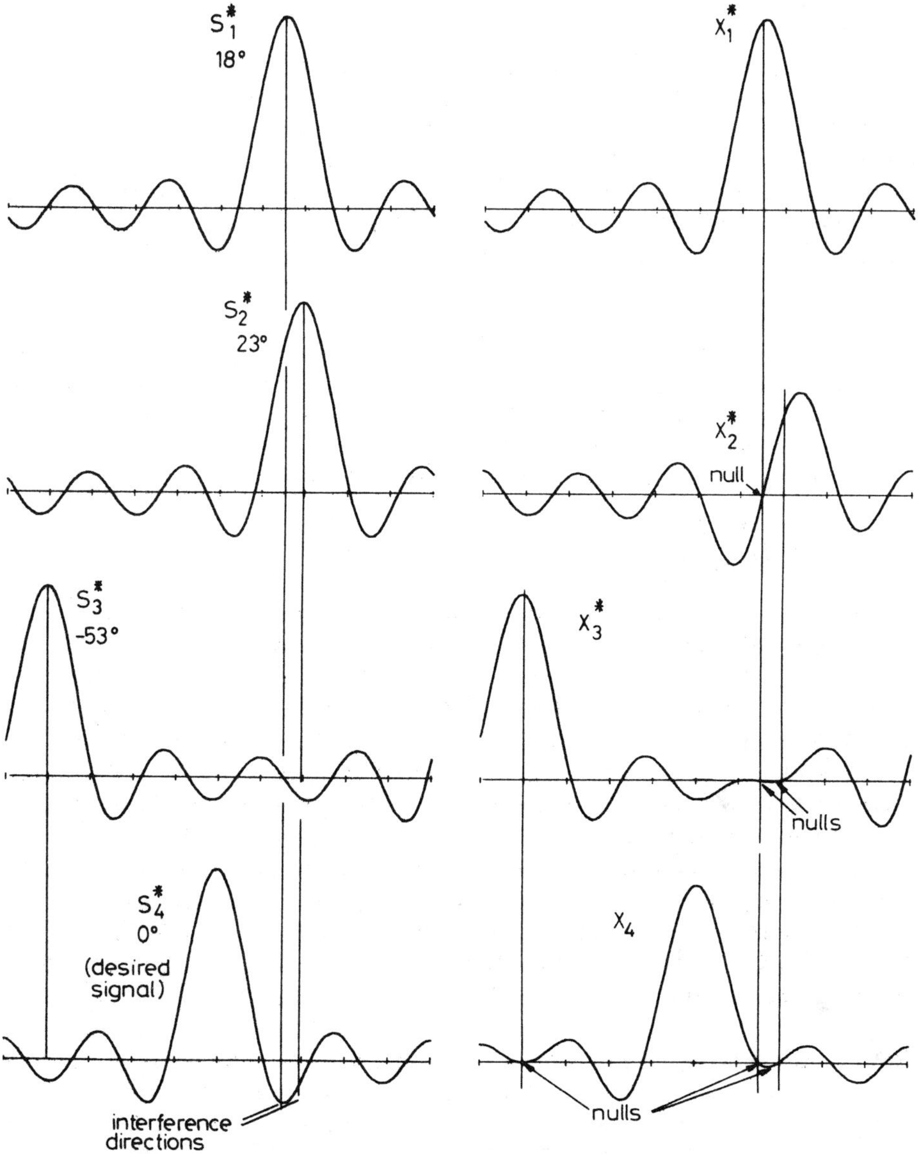

Fig. 2.5.4 *Directional pattern derived by Gram-Schmidt orthogonalisation*

Directional patterns of the conjugate space vectors of the 3 interferences $\mathbf{S}_1^*$, $\mathbf{S}_2^*$, and $\mathbf{S}_3^*$ are shown. The pattern of $\mathbf{X}_2^*$ has a null at $\mathbf{S}_1$, that of $\mathbf{X}_3^*$ has a null at $\mathbf{S}_1$ and $\mathbf{S}_2$, while $\mathbf{X}_4$ has nulls at all three sources have some response in the cophasal pattern. If the interference sources were much stronger than the desired signal there would be a potential loss of useful reception of the desired signal.

On the right are the directional patterns of the conjugated co-ordinate vectors $\mathbf{X}_i^*$. $\mathbf{X}_1^*$ is the same as $\mathbf{S}_1^*$, $\mathbf{X}_2^*$ has a pattern with a null directed at the 18° jammer S_1, $\mathbf{X}_3^*$ has nulls at both the 18° source S_1 and the 22° source S_2, and finally $\mathbf{X}_4$ has nulls at the 18°, 22°, and −53° sources. The overall change in the pattern going from $\mathbf{S}_4^*$ to $\mathbf{X}_4$ is remarkably small and the change in the main lobe region is insignificant. However, if a main lobe jammer were present it would distort the main lobe considerably as was the case in Fig. 2.5.2.

Minimum norm property of projections

The use of projections to achieve orthogonality is more than just a convenient mathematical procedure; it is also a minimum norm solution in the following sense. If a vector is required with a certain amplitude gain g for a desired signal (with vector $\mathbf{S}_1$ say) and zero response for up to $N-1$ other sources with vectors $\mathbf{S}_m$, $m = 2, \ldots, M$ then the vector defined by eqns. (2.5.16) and (2.5.17),

$$\mathbf{S}_p = \frac{g\mathbf{P}_M \mathbf{P}_{M-1} \cdots \mathbf{P}_2 \mathbf{S}_1^*}{\mathbf{S}_1^t \mathbf{P}_M \mathbf{P}_{M-1} \cdots \mathbf{P}_2 \mathbf{S}_1^*} \tag{2.5.25}$$

which has been normalised by multiplying by scalar g is the vector having these properties with the smallest norm, a statement proved in Appendix III. In engineering terms $\mathbf{S}_p$ will have the smallest mean square coefficient values and will lead to minimum loss of desired signal response if the coefficients are achieved by attenuation and to smallest sensitivity to errors in construction.

Given these properties it is not surprising that optimal solutions (i.e. maximum SNR conditions) can come very close to the projection solutions in some circumstances. However, proof of this is rather difficult (see Section 5.2). The similarity of the solutions is useful in the sense that the projection solutions are mathematically tractible and can be used to study the optimal solutions.

It is interesting to reconsider an earlier example (eqn. 1.7.2) in the light of this to see if a better solution is obtainable. It was found necessary to use two nulls in order to obtain a full rank matrix there and doing this uses all the degrees of freedom of the antenna and the solution is unique. Therefore, if two nulls are formed, the method of projections will give the same solution as the matrix inversion technique. However, since only one null is actually necessary to solve the problem we will use a projection method to see if the result is better. Thus we wish to project the desired signal vector $\mathbf{S}_1^* = (e^{j2\pi/3}, 1, e^{-j2\pi/3})^t$ into a subspace orthogonal to the interference vector $\mathbf{S}_2 = (\exp(-j2\pi \cos(45°)/3, 1, \exp(j2\pi \cos(45°)/3)^t$. The projection matrix from 2.5.3 is

$$\mathbf{P} = \frac{\mathbf{S}_2^* \mathbf{S}_2^t}{\mathbf{S}_2^t \mathbf{S}_2^*} = \begin{bmatrix} \frac{2}{3} & -\frac{1}{3}e^{j\pi\sqrt{2}/3} & -\frac{1}{3}e^{j\pi\sqrt{8}/3} \\ -\frac{1}{3}e^{-j\pi\sqrt{2}/3} & \frac{2}{3} & -\frac{1}{3}e^{j\pi\sqrt{2}/3} \\ -\frac{1}{3}e^{-j\pi\sqrt{8}/3} & -\frac{1}{3}e^{-j\pi\sqrt{2}/3} & \frac{2}{3} \end{bmatrix} \tag{2.5.26}$$

and the vector obtained by projecting $\mathbf{S}_1^*$ is

$$\mathbf{W} = \mathbf{PS}_1^* = (0{\cdot}282 - j0{\cdot}506, \quad -0{\cdot}061 + j0{\cdot}105, \quad 0{\cdot}297 - j0{\cdot}497)^t \qquad (2.5.27)$$

This vector does not have a gain of three in the end-fire direction where the desired signal lies and it is scaled (as in eqn. (2.5.25)) to yield finally,

$$\mathbf{W} = \frac{3\mathbf{PS}_1^*}{\mathbf{S}_1^t\,\mathbf{PS}_1^*} = \begin{bmatrix} 1{\cdot}234 - j2{\cdot}215 \\ -0{\cdot}266 + j0{\cdot}461 \\ 1{\cdot}301 - j2{\cdot}176 \end{bmatrix} \qquad (2.5.28)$$

Figure 2.5.5 shows the directional response of this vector used as a weighting vector. It is clear that the projection has formed two nulls; the second, not specified in the problem, occurs at about 135° relative to the end-fire direction. It was observed earlier that steering a null in this direction would probably lead to a better solution than that found in example 1.7.2, and this is indeed the case though the appearance of extra nulls is not a general feature of the projection method. The broadside lobe is smaller in the projection solution having an amplitude of about 4·6 as against 7 in the matrix inversion method and the norm of the vector in eqn. (2.5.28) is 3·62 as compared to 4·24 for eqn. (1.7.16), so that in these two important aspects the projection solution is much superior. It is not much better when the maximum modulus in the weighting vector is

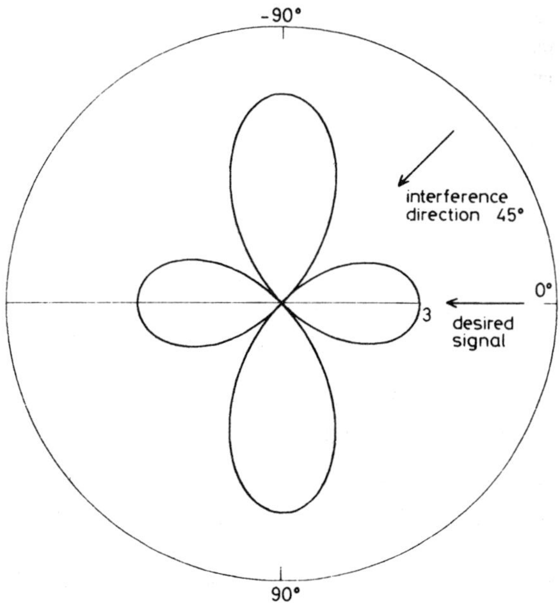

Fig. 2.5.5 *Directional response for weighting vector obtained by projection*

compared yielding 2·54 instead of 2·87 so that if the coefficients were implemented using attenuators there would be much the same loss in signal level when the largest coefficient modulus was scaled down to unity, the projection solution being only about 1 dB better.

2.6 Output powers, covariance matrices, and quadratic forms

The determination of output powers in an adaptive array is a task of great importance since the aim is eventually to optimise the ratio of signal output power to noise and interference output power.

Real signals

In the real signal case, typified by Fig. 2.2.4 the array output signal is given by

$$y(t) = \mathbf{W}^t\mathbf{X}(t) \tag{2.6.1}$$

and the mean output power is

$$\begin{aligned}\overline{y(t)^2} &= \overline{\mathbf{W}^t\mathbf{X}(t)\mathbf{X}(t)^t\mathbf{W}}\\ &= \mathbf{W}^t\overline{\mathbf{X}(t)\mathbf{X}(t)^t}\mathbf{W}\\ &= \mathbf{W}^t\mathbf{R}\mathbf{W}\end{aligned} \tag{2.6.2}$$

where the covariance matrix $\mathbf{R}$ is defined, element by element, for the zero mean signals, as

$$R_{ij} \triangleq \begin{bmatrix} \overline{x_1x_1} & \overline{x_1x_2} & \overline{x_1x_3} & \cdots & \overline{x_1x_{2N}} \\ \overline{x_2x_1} & \overline{x_2x_2} & x_2x_3 & \cdots & x_2x_{2N} \\ \overline{x_3x_1} & \overline{x_3x_2} & \cdots & \cdots & \cdots \\ \cdots & \cdots & \cdots & \cdots & \cdots \\ \overline{x_{2N}x_1} & \overline{x_{2N}x_2} & \overline{x_{2N}x_3} & \cdots & \overline{x_{2N}x_{2N}} \end{bmatrix} = \overline{x_ix_j} \tag{2.6.3}$$

By inspection $\mathbf{R}$ is symmetric, i.e. $\mathbf{R}_{ij} = \mathbf{R}_{ji}$ since these are covariances of the same pair of signals. If the matrix is partitioned as follows:

$$\mathbf{R} = \begin{bmatrix} \mathbf{A} & \mathbf{B} \\ \mathbf{B}^t & \mathbf{A} \end{bmatrix} = \begin{bmatrix} \mathbf{A} & \mathbf{B} \\ -\mathbf{B} & \mathbf{A} \end{bmatrix} \tag{2.6.4}$$

then each submatrix $\mathbf{A}$ is symmetric but the submatrices $\mathbf{B}$ are skew-symmetric, i.e. $\mathbf{B}^t = -\mathbf{B}$ (Appendix A1.1). This occurs because for the term on the opposite side of the appropriate quarter diagonal the same pair of signals are correlated but with a total of 180° of phase shift introduced in one channel so inverting the correlation (actually +90° is introduced in one channel and −90° in the other).

If the weighting vector $\mathbf{W}$ is written in the partitioned form

$$\mathbf{W} \triangleq \begin{bmatrix} \mathbf{W}_p \\ \mathbf{W}_q \end{bmatrix} \tag{2.6.5}$$

as in eqn. (2.2.20) we can rewrite eqn. (2.6.2) in the form

$$\begin{aligned} \mathbf{W}^t\mathbf{R}\mathbf{W} &= \begin{bmatrix} \mathbf{W}_p \\ \mathbf{W}_q \end{bmatrix}^t \begin{bmatrix} \mathbf{A} & \mathbf{B} \\ \mathbf{B}^t & \mathbf{A} \end{bmatrix} \begin{bmatrix} \mathbf{W}_p \\ \mathbf{W}_q \end{bmatrix} \\ &= \mathbf{W}_p^t\mathbf{A}\mathbf{W}_p + \mathbf{W}_q^t\mathbf{A}\mathbf{W}_q + \mathbf{W}_q^t\mathbf{B}^t\mathbf{W}_p + \mathbf{W}_p^t\mathbf{B}\mathbf{W}_q \\ &= \mathbf{W}_p^t\mathbf{A}\mathbf{W}_p + \mathbf{W}_q^t\mathbf{A}\mathbf{W}_q + 2\mathbf{W}_p^t\mathbf{B}\mathbf{W}_q \end{aligned} \tag{2.6.6}$$

For the case of a collinear equally spaced array receiving exclusively far-field signals plus uncorrelated noise the submatrix $\mathbf{A}$ is, in principle, Toeplitz and $\mathbf{B}$ is skew symmetric Toeplitz, i.e. the coefficients down a given diagonal have the same values. However, this assumption cannot be relied upon because of channel gain mismatches, near-field scattering into the array aperture, sampling errors and so on.

Complex signals : The complex variable case is closely analagous, yielding an output power of

$$\overline{|y(t)|^2} = \mathbf{W}^H\overline{\mathbf{X}(t)^*\mathbf{X}(t)^t}\mathbf{W} = \mathbf{W}^H\mathbf{R}\mathbf{W} \tag{2.6.7}$$

where N by N covariance matrix $\mathbf{R}$ is defined coefficient by coefficient as

$$\mathbf{R}_{ij} = \overline{x_i^*(t)x_j(t)} \tag{2.6.8}$$

and it is easily shown from this equation that $\mathbf{R}$ is Hermitian, i.e.

$$\mathbf{R}^H = \mathbf{R}, \qquad \mathbf{R}^t = \mathbf{R}^* \tag{2.6.9}$$

Defining $\mathbf{W} = \mathbf{W}_r + j\mathbf{W}_j$ and $\mathbf{R} = \mathbf{R}_r + j\mathbf{R}_j$ as in Appendix 2, we can rewrite eqn. (2.6.7) as

$$\mathbf{W}^H\mathbf{R}\mathbf{W} = \mathbf{W}_r^t\mathbf{R}_r\mathbf{W}_r + \mathbf{W}_j^t\mathbf{R}_r\mathbf{W}_j - 2\mathbf{W}_r^t\mathbf{R}_j\mathbf{W}_j \tag{2.6.10}$$

which is closely analagous to eqn. (2.6.6). In all cases $\mathbf{R}$ is positive semidefinite and for any complex vector $\mathbf{V}$ we have

$$\mathbf{V}^H\mathbf{R}\mathbf{V} \geq 0 \tag{2.6.11}$$

This is also clear from physical principles since $\mathbf{V}^H\mathbf{R}\mathbf{V}$ can be interpreted as an output power as in eqn. (2.6.7).

Relation between real and complex covariances[125]*:* For arbitrary input data, if the array elements are ordered as suggested in Figs. 2.2.4 and 2.2.3 for the real and complex cases, the real covariance matrix eqn. (2.6.3) is in direct correspondence with the complex one, eqn. (2.6.8). If $\mathbf{R}_C$ corresponds to the latter and

$$\mathbf{R}_C = \mathbf{R}_r + j\mathbf{R}_j \tag{2.6.12}$$

where $\mathbf{R}_r$ is real symmetric and $\mathbf{R}_j$ is real skew symmetric, then

$$\mathbf{R}_R = \begin{bmatrix} \mathbf{R}_r & \mathbf{R}_j \\ \mathbf{R}_j^t & \mathbf{R}_r \end{bmatrix} \tag{2.6.13}$$

Similarly, if $\mathbf{Z}^H\mathbf{R}_C\mathbf{Z}$ is an Hermitian form over $\mathbf{R}_C$ and $\mathbf{Z} = \mathbf{X} + j\mathbf{Y}$, then the equivalent real quadratic form is

$$\begin{bmatrix} \mathbf{X} \\ -\mathbf{Y} \end{bmatrix}^t \begin{bmatrix} \mathbf{R}_r & \mathbf{R}_j \\ \mathbf{R}_j^t & \mathbf{R}_r \end{bmatrix} \begin{bmatrix} \mathbf{X} \\ -\mathbf{Y} \end{bmatrix} \tag{2.6.14}$$

Thus all real and complex representations are in direct correspondence. The sign change in $\mathbf{Y}$ is discussed in Section 2.2.

If the signal splitting in Fig. 2.2.4 conserves power, as is the case with hybrid circuits, the magnitude of $\mathbf{R}_R$ in eqn. (2.6.13) should be halved.

Estimation of the covariance matrix

It is shown later that estimation of covariance matrices on-line is an essential part of adaptive array processing. The maximum likelihood estimate of $\mathbf{R}$ appears to be the most useful and is just a direct average:

$$\hat{\mathbf{R}} = \frac{1}{T}\int_0^T \mathbf{X}(t)^*\mathbf{X}(t)^t\,dt \tag{2.6.15}$$

in the complex case or the same for the real case without the conjugate. For sampled data the integral is replaced by a discrete summation

$$\hat{\mathbf{R}} = \frac{1}{K}\sum_{k=1}^{K} \mathbf{X}(t_k)^*\mathbf{X}(t_k)^t \tag{2.6.16}$$

in which the data is instantaneously sampled at the times t_k, $k = 1, 2, \ldots, K$.

For the discrete case with independent sampling the estimate of $\mathbf{R}$ follows the Wishart distribution, discussed by Capon.[96]

Theoretical discrete source covariance matrices

Single signals: For the real case, the covariance matrix $\overline{\mathbf{X}(t)\mathbf{X}(t)^t}$ is found by discarding all terms whose correlation falls to zero over one period of the carrier, namely cos ωt sin ωt products. Using representation eqn. (2.2.20) if $m_c(t)$ and $m_s(t)$ are assumed uncorrelated and the signal power is

$$\pi = \overline{m_c(t)^2 + m_s(t)^2}$$

then

$$\mathbf{R} = \pi \begin{bmatrix} \mathbf{CC}^t + \mathbf{SS}^t & \mathbf{CS}^t - \mathbf{SC}^t \\ -\mathbf{CS}^t + \mathbf{SC}^t & \mathbf{CC}^t + \mathbf{SS}^t \end{bmatrix} \tag{2.6.17}$$

which in turn is equal to

$$\pi \left\{ \begin{bmatrix} \mathbf{C} \\ \mathbf{S} \end{bmatrix} \begin{bmatrix} \mathbf{C} \\ \mathbf{S} \end{bmatrix}^t + \begin{bmatrix} -\mathbf{S} \\ \mathbf{C} \end{bmatrix} \begin{bmatrix} -\mathbf{S} \\ \mathbf{C} \end{bmatrix}^t \right\} \tag{2.6.18}$$

which is a sum of rank one dyads. Thus in the real case each source can contribute two to the rank of **R**.

In the complex case the signal is represented as $\mathbf{X}(t) = m(t)\mathbf{S}$ as in eqn. (2.2.8) where both m and **S** are complex and

$$\mathbf{R} = \overline{\mathbf{X}(t)^*\mathbf{X}(t)^t} = \overline{|m(t)|^2}\mathbf{S}^*\mathbf{S}^t \tag{2.6.19}$$

which is only of rank one. Thus each source contributes only one to the rank of **R**.

Multiple uncorrelated sources: Using eqn. (2.2.24) for the multiple sources we obtain, in the complex case,

$$\mathbf{R} = E \left\{ \begin{array}{c} |m_1(t)|^2\mathbf{S}_1^*\mathbf{S}_1^t + |m_2(t)|^2\mathbf{S}_2^*\mathbf{S}_2^t + \cdots + |m_M(t)|^2\mathbf{S}_M^*\mathbf{S}_M^t \\ + 2\,\mathrm{Re}\,[m_1^*(t)m_2(t)\mathbf{S}_1^*\mathbf{S}_2^t + m_2^*(t)m_3(t)\mathbf{S}_2^*\mathbf{S}_3^t + \cdots] \\ + 2\,\mathrm{Re}\,[m_1^*(t)m_3(t)\mathbf{S}_1^*\mathbf{S}_3^t + m_2^*(t)m_4(t)\mathbf{S}_2^*\mathbf{S}_4^t + \cdots] \\ + \cdots \\ + 2\,\mathrm{Re}\,[m_1^*(t)m_M(t)\mathbf{S}_1^*\mathbf{S}_M^t] \end{array} \right\} \tag{2.6.20}$$

If the modulation functions $m_i(t)$ are coherent then the covariance matrix is very complicated, involving all possible pairs of signal products. The general two-source case is discussed in more detail in the following Section. However, if the modulation functions $m_i(t)$ are uncorrelated, then only the first line in eqn. (2.6.20) remains and it reduces to

$$\mathbf{R} = E\left\{ \sum_{i=1}^{M} |m_i(t)|^2 \mathbf{S}_i^* \mathbf{S}_i^t \right\} = \sum_{i=1}^{M} \pi_i \mathbf{S}_i^* \mathbf{S}_i^t \tag{2.6.21}$$

where the π_is are a measure of the source powers. If the space vectors $\mathbf{S}_i$ are normalised to have unity modulus elements, then π_i can be interpreted as the channel power of the ith source.

Since the total covariance matrix is the sum of the covariance matrices of the individual sources its rank can be estimated. Since Rank $\mathbf{R} \leq \sum$ Rank $(\mathbf{A}_i)$ if $\mathbf{R} = \sum \mathbf{A}_i$ (Appendix A1.2) and in this case each $\mathbf{A}_i$ is a rank one dyad we have the result

$$\text{Rank}\,(\mathbf{R}) \leq M \tag{2.6.22}$$

in which the equality will normally prevail if $M \leq N$.

2.7 Covariance matrices for two discrete sources[68, 62]

This is an important extension of the single-signal result as the performance of optimal antennas can be much worse with certain two-source configurations than single-source results might indicate. The general M-source covariance matrix is indicated in eqn. (2.6.20) and for the case $M = 2$ we will discuss the forms of the inverse of $\mathbf{R}$, and its eigenvalues, when the sources are correlated. We have

$$\mathbf{X}(t) = m_1(t)\mathbf{S}_1 + m_2(t)\mathbf{S}_2 \tag{2.7.1}$$

and

$$\begin{aligned}\mathbf{R} = \overline{\mathbf{X}^*\mathbf{X}'} &= \overline{[m_1(t)\mathbf{S}_1 + m_2(t)\mathbf{S}_2]^*[m_1(t)\mathbf{S}_1 + m_2(t)\mathbf{S}_2]'}\\ &= \overline{|m_1|^2}\mathbf{S}_1^*\mathbf{S}_1' + \overline{|m_2|^2}\mathbf{S}_2^*\mathbf{S}_2' + \overline{m_1^* m_2}\mathbf{S}_1^*\mathbf{S}_2'\\ &\quad + \overline{m_1 m_2^*}\mathbf{S}_2^*\mathbf{S}_1'\\ &= \pi_1\mathbf{S}_1^*\mathbf{S}_1' + \pi_2\mathbf{S}_2^*\mathbf{S}_2' + \psi\mathbf{S}_1^*\mathbf{S}_2' + \psi^*\mathbf{S}_2^*\mathbf{S}_1'\end{aligned} \tag{2.7.2}$$

where ψ is the complex covariance of m_1 and m_2. Defining

$$\mathbf{Q} = [\mathbf{S}_1 | \mathbf{S}_2] \quad \text{and} \quad \mathbf{P} = \begin{bmatrix}\pi_1 & \psi\\ \psi^* & \pi_2\end{bmatrix} \tag{2.7.3}$$

we have

$$\mathbf{R} = \mathbf{Q}^*\mathbf{P}\mathbf{Q}' \tag{2.7.4}$$

If there is some uncorrelated noise present whose covariance matrix (eqn. 2.8.3) is $\sigma^2\mathbf{I}$ then adding this to $\mathbf{R}$ produces a full rank matrix $\mathbf{M}$ which is invertible.

$$\mathbf{M} = \sigma^2\mathbf{I} + \mathbf{R} = \sigma^2\mathbf{I} + \mathbf{Q}^*\mathbf{P}\mathbf{Q}' \tag{2.7.5}$$

From a matrix inversion formula, Appendix A1.2, the inverse of $\mathbf{M}$ is

$$\mathbf{M}^{-1} = \sigma^{-2}\mathbf{I} - \sigma^{-4}\mathbf{Q}^*[\mathbf{P}^{-1} + \sigma^{-2}\mathbf{Q}'\mathbf{Q}^*]^{-1}\mathbf{Q}' \tag{2.7.6}$$

Since the inverse in this form is only of order two, explicit results are easily obtained. Kwok and Brandon[62] and Morgan[68] have described such results.

Eigenvalue determination: It is useful to determine the eigenstructure of $\mathbf{R}$ and as it is Hermitian it has the general representation (A1.2)

$$\mathbf{R} = \lambda_1\mathbf{U}_1^*\mathbf{U}_1' + \lambda_2\mathbf{U}_2^*\mathbf{U}_2' \tag{2.7.7}$$

where $\mathbf{U}_1$ and $\mathbf{U}_2$ are orthonormal and λ_1 and λ_2 are real. We can think of eqn. (2.7.7) as a representation of the covariance matrix produced by a new pair of sources with space vectors $\mathbf{U}_1$ and $\mathbf{U}_2$ and with powers λ_1 and λ_2. There is no way of distinguishing, from measurements made at the antenna whether the source vectors are the **U**s of eqn. (2.7.7) or the **S**s of 2.7.1 except that the **U** vectors will not correspond to plane waves in general. Thus, eqn. (2.7.2) may be the unique plane-wave representation.

The eigenvalues of $\mathbf{R}$ are found from the equation

$$\det(\mathbf{R} - \lambda\mathbf{I}) = 0 \tag{2.7.8}$$

but this equation is not easy to apply directly if $\mathbf{R}$ is of high order. The labour is reduced by noting that either $\mathbf{U}_1$ is a linear combination of $\mathbf{S}_1$ and $\mathbf{S}_2$

$$\mathbf{V}^* = a\mathbf{S}_1^* + b\mathbf{S}_2^* \qquad V = U_1 \text{ or } U_2 \tag{2.7.9}$$

and from eqn. (2.7.2) we have

$$\mathbf{RV}^* = \mathbf{S}_1^*(\pi_1 aN + \pi_1 b\phi + a\psi\phi^* + b\psi N) + \mathbf{S}_2^*(\pi_2 a\phi + \pi_2 bN + \psi^* aN + \psi^* b\phi) \tag{2.7.10}$$

where

$$\mathbf{S}_1^t\mathbf{S}_1^* = \mathbf{S}_2^t\mathbf{S}_2^* = N, \qquad \overline{m_1^* m_2} = \psi, \qquad \mathbf{S}_1^t\mathbf{S}_2^* = N\phi \tag{2.7.11}$$

ϕ will be recognised as the cosine of the angle between $\mathbf{S}_1$ and $\mathbf{S}_2$. If $\mathbf{V}^*$ is an eigenvector of $\mathbf{R}$ then

$$\mathbf{RV}^* = \lambda\mathbf{V}^* = \lambda a\mathbf{S}_1^* + \lambda b\mathbf{S}_2^* \tag{2.7.12}$$

and substituting eqn. (2.7.10) into this yields

$$\begin{aligned}(\pi_1 N + \psi N\phi^*)a + (\pi_1 N\phi + \psi N)b &= \lambda a\\ (\pi_2 N\phi^* + \psi^* N)a + (\pi_2 N + \psi^* N\phi)b &= \lambda b\end{aligned} \tag{2.7.13}$$

which is a new eigenvalue equation with roots determined by

$$\det\begin{bmatrix}\pi_1 N + \psi N\phi^* - \lambda & \pi_1 N\phi + \psi N\\ \pi_2 N\phi^* + \psi^* N & \pi_2 N + \psi^* N\phi - \lambda\end{bmatrix} = 0 \tag{2.7.14}$$

which expansion shows to have roots of

$$\tfrac{1}{2}N[\pi_1 + \pi_2 + (\pi_1\pi_2)^{1/2} 2\,\mathrm{Re}\,(\rho\phi^*)] \times\left[1 \pm \sqrt{1 - \frac{4\pi_1\pi_2(1 - |\phi|^2)(1 - |\rho|^2)}{[\pi_1 + \pi_2 + 2(\pi_1\pi_2)^{1/2}\,\mathrm{Re}\,(\rho\phi^*)]^2}}\right] \tag{2.7.15}$$

where ρ is the correlation coefficient of the modulations $m_1(t)$ and $m_2(t)$:

$$\rho = \overline{m_1^*(t)m_2(t)}/(\pi_1\pi_2)^{1/2} = \psi/(\pi_1\pi_2)^{1/2} \tag{2.7.16}$$

Uncorrelated modulations: This is the normal condition when the transmitters are separated in space and is the case usually treated.[61, 68] Setting $\rho = 0$ in eqn. (2.7.16) yields

$$\lambda_1, \lambda_2 = \tfrac{1}{2}N(\pi_1 + \pi_2)\left\{1 \pm \sqrt{1 - \frac{4\pi_1\pi_2(1 - |\phi|^2)}{(\pi_1 + \pi_2)^2}}\right\} \tag{2.7.17}$$

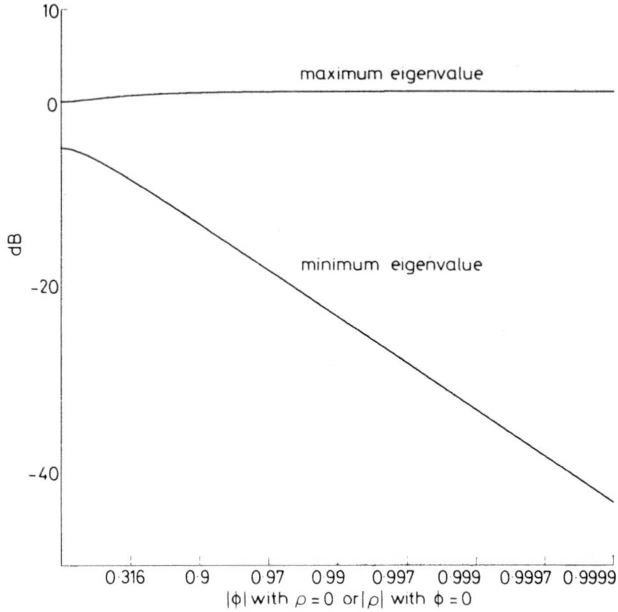

Fig. 2.7.1 *Eigenvalues of the covariance matrix of a close or correlated transmitters. Intrinsic source powers 0 and – 5* dB

Figure 2.7.1 shows how λ_1 and λ_2 vary as a function of ϕ for $\pi_1 = 0$ dB, $\pi_2 = -5$ dB. The ordinate is actually linear in log $(1 - |\phi|^2)$. If the space vectors are orthogonal then

$$\lambda_1, \lambda_2 = N\pi_1, N\pi_2 \tag{2.7.18}$$

It can be shown that the eigenvectors of **R** are parallel to the space vectors $\mathbf{S}_1$ and $\mathbf{S}_2$ for this condition. As the sources move close together their space vectors get increasingly parallel and $|\phi|$ approaches unity. As seen from the figure, the smaller eigenvalue becomes small and vanishes in the limit $|\phi| = 1$ when the sources coalesce into one and at the same time the rank of **R** falls from 2 to 1.

If we set

$$|\phi|^2 = 1 - \epsilon^2 \tag{2.7.19}$$

then it is easily shown that for small ϵ^2

$$\lambda_1 \simeq N(\pi_1 + \pi_2), \qquad \lambda_2 \simeq \frac{N\pi_1\pi_2\epsilon^2}{\pi_1 + \pi_2} \tag{2.7.20}$$

For a collinear array we have, from eqn. (2.4.12),

$$\phi = \frac{\sin(\frac{1}{2}Nkd\theta)}{N\sin(\frac{1}{2}kd\theta)} \underset{\theta \to 0}{\simeq} 1 - \frac{1}{24}(kd\theta)^2(N^2 - 1) \tag{2.7.21}$$

where θ is the difference of the source bearing sines and

$$\lambda_2 \simeq \frac{N(N^2-1)\pi_1\pi_2(kd\theta)^2}{24(\pi_1+\pi_2)} \tag{2.7.22}$$

is seen to increase as the square of the source separations for small ϵ^2.

Correlated Sources: Although correlation between different transmitters is unlikely it is possible to receive the same transmission from two or more directions as a result of multipath propagation and if the relative time delays between the different paths is small enough the correlation coefficient can approach unity. Such multipath will normally arise from the vehicle on which the array is mounted. Orthogonality of the space vectors is assumed and substituting $\phi = 0$ in eqn. (2.7.15) yields

$$\lambda_1, \lambda_2 = \tfrac{1}{2}N(\pi_1+\pi_2)\left\{1 \pm \sqrt{1 - \frac{4\pi_1\pi_2(1-|\rho|^2)}{(\pi_1+\pi_2)^2}}\right\} \tag{2.7.23}$$

As this has the same form as eqn. (2.7.17) with ρ replacing ϕ, Fig. 2.7.1 can be used to show variation of the eigenvalues with $|\rho|^2$. Also setting $|\rho|^2 = 1 - \epsilon^2$ yields the analogous results

$$\lambda_1 \simeq N(\pi_1+\pi_2), \qquad \lambda_2 \simeq \frac{N\pi_1\pi_2\epsilon^2}{\pi_1+\pi_2} \tag{2.7.24}$$

The value of ρ can be found from the power spectrum of the modulation function and this is discussed in Section 5.5.

Closely spaced sources with high correlation: Setting $|\rho|$ and $|\phi|$ simultaneously close to unity in eqn. (2.7.15) yields eigenvalues approximated by

$$\lambda_1 \simeq N(\pi_1^{1/2}+\pi_2^{1/2})^2, \qquad \lambda_2 \simeq \frac{2N\pi_1\pi_2\epsilon_\phi^2\epsilon_\rho^2}{(\pi_1^{1/2}+\pi_2^{1/2})^2} \tag{2.7.25}$$

2.8 Incoherent noise fields

Discrete noise sources may not be the only significant sources of interference in an adaptive antenna. Depending upon the application receiver noise, man-made interference from electrical machinery, even sky noise may be present.

Uncorrelated noise: Uncorrelated noise is normally considered to arise in preamplifiers connected to the antenna or sensor array and is due to thermal noise or device noise. In the absence of cross-coupling between preamplifiers which might be due to stray inductive or capacitive effects such noise will be totally incoherent between the channels and taking two channels $x_i(t)$ and $x_j(t)$ then we have in the real signal case

$$\overline{x_i(t)x_j(t)} = \sigma^2\delta_{ij} \qquad i, j = 1, \ldots, 2N \tag{2.8.1}$$

where σ^2 is the variance of the noise at the measuring point and $\delta_{ij} = 0$ if $i \neq j$ and 1 if $i = j$. Even if the signals x_i and x_j are derived from the same channel via a quadrature network they will be uncorrelated given some basic assumptions on the frequency spectra of the noise. For the complex case we have

$$\overline{x_i(t)^* x_j(t)} = \sigma^2 \delta_{ij} \qquad i, j = 1, \ldots, N \tag{2.8.2}$$

Substituting eqn. (2.8.1) into (2.6.3) (which refers to Fig. 2.2.4) gives a covariance matrix of the form

$$\mathbf{R}_u = \sigma^2 \mathbf{I} \tag{2.8.3}$$

where $\mathbf{I}$ is an identity matrix of dimension $2N$ by $2N$. In the complex case, substitution of eqn. (2.8.2) into (2.6.8) also gives an identity matrix but only of dimension N by N.

The output power due to uncorrelated noise is found by substituting eqn. (2.8.3) into (2.6.2) or (2.6.7) yielding

$$\overline{y(t)^2} = \mathbf{W}^t \mathbf{R}_u \mathbf{W} = \sigma^2 \mathbf{W}^t \mathbf{W}$$

or

$$\overline{|y(t)|^2} = \mathbf{W}^H \mathbf{R}_u \mathbf{W} = \sigma^2 \mathbf{W}^H \mathbf{W} \tag{2.8.3a}$$

for the real and complex cases, respectively.

Spherically isotropic noise:[14, 98] This is often considered as a model for sky noise. Placing a large number of uncorrelated sources on the inside of a spherical shell at a large radius gives rise to a correlation between the signals received at a pair of elements, spaced by d in any direction, of the form, shown in Fig. 2.8.1,

$$\psi(d) = \sigma^2 \frac{\sin(kd)}{kd} \tag{2.8.4}$$

where $k = 2\pi/\lambda$. As this function is frequency dependent we restrict attention to narrow band signals for which wavelength λ is well defined. The correlation is purely real in the complex case, and in the real signal case there is no correlation between in-phase signals and quadrature signals. Thus the covariance matrix is of the form

$$\mathbf{R}_{s(ij)} = \psi(|r_i - r_j|) \tag{2.8.5}$$

where r_i is the three-dimensional coordinate of the ith receiver. The correlation due to spherically isotropic noise decays very quickly with increasing receiver spacing and can usually be neglected for spacings of more than $\frac{1}{3}\lambda$, the covariance matrix approximating to that of uncorrelated noise. The optimal antenna can, in principle, offer better signal-to-noise ratio when receiving a plane wave on closely spaced elements in the presence of spherically isotropic noise than can a conventional antenna but the gains only amount to a few decibels[9, 13, 14] if uncorrelated noise is present too.

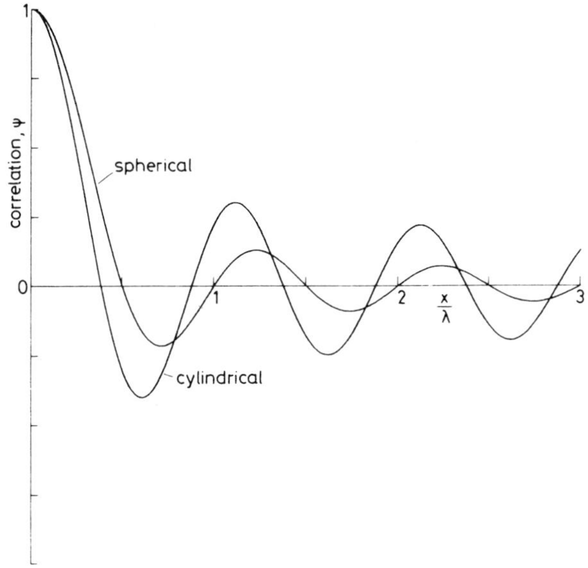

Fig. 2.8.1 *Spatial correlation of cylindrically and spherically isotropic noise fields as a function of separation in wavelengths*

Cylindrically isotropic noise:[98] If a large number of uncorrelated sources are placed in a ring at a large range from the receiving antenna, the correlation between two elements spaced by d in the plane of the ring takes the form shown in Fig. 2.8.1,

$$\psi(d) = \sigma^2 J_0(kd) \tag{2.8.6}$$

where J_0 is the Bessel function of the first kind and zero order. Separation of the elements in the direction of the axis of the ring may be ignored. Again the correlation is purely real or 'in-phase'. Since $J_0(x)$ decays roughly as $|x|^{-1/2}$ for large $|x|$, the noise field is more coherent than for the spherical case.

Cylindrically isotropic noise is a good model for source distributions which lie in a plane such as that due to the ocean surface, or possibly man-made noise at HF in an industrial area.

Total array interference and noise output:[9, 14] Let the covariance matrices due to uncorrelated, spherically isotropic and cylindrically isotropic noise be $\mathbf{R}_u$, $\mathbf{R}_s$, and $\mathbf{R}_c$, respectively. The total noise output from an antenna with weighting coefficients $\mathbf{W}$ is then

$$\begin{aligned} p_{\text{tot}} &= W^H\{\mathbf{R}_u + \mathbf{R}_s + \mathbf{R}_c\}\mathbf{W} \\ &= \mathbf{W}^H\mathbf{R}_u\mathbf{W} + \mathbf{W}^H\mathbf{R}_s\mathbf{W} + \mathbf{W}^H\mathbf{R}_c\mathbf{W} \end{aligned} \tag{2.8.7}$$

provided all the sources of noise are uncorrelated. When there are discrete noise sources present too their covariance matrix eqn. (2.6.21) is added in:

$$p_{tot} = \mathbf{W}^H\{\mathbf{R}_u + \mathbf{R}_s + \mathbf{R}_c + \sum_{i=1}^{M} \pi_i \mathbf{S}_i^* \mathbf{S}_i^t\}\mathbf{W} = \mathbf{W}^H\mathbf{M}\mathbf{W} \tag{2.8.8}$$

Since the signal output power is of the form

$$p_{sig} = \pi_s |\mathbf{W}^t\mathbf{S}_0|^2 = \pi_s \mathbf{W}^H\mathbf{S}_0^* \mathbf{S}_0^t \mathbf{W} \tag{2.8.9}$$

where $\mathbf{S}_0$ is the space vector of the desired source and π_s its power the output signal-to-noise and interference ratio (SNIR) is

$$Q = \frac{\pi_s \mathbf{W}^H\mathbf{S}_0^* \mathbf{S}_0^t \mathbf{W}}{\mathbf{W}^H\{\mathbf{R}_u + \mathbf{R}_s + \mathbf{R}_c + \sum_{i=1}^{M} \pi_i \mathbf{S}_i^* \mathbf{S}_i^t\}\mathbf{W}} = \frac{\mathbf{W}^H\mathbf{R}_s\mathbf{W}}{\mathbf{W}^H\mathbf{M}\mathbf{W}} \tag{2.8.10}$$

and the function of an optimal antenna which maximises signal to noise ratio is to maximise Q by variation of the coefficients in $\mathbf{W}$, a topic discussed in the next chapter.

Modelling of array signals:[30] In simulations it is frequently necessary to model signals which are specified only by their covariance matrices like the noise fields above. This can be conveniently done by factorisation of the covariance matrix. Let $\mathbf{M}$ be the total noise covariance matrix (e.g. as in the denominator of eqn. (2.8.10)) and let $\mathbf{M}$ be factored into the product

$$\mathbf{M} = \mathbf{A}^*\mathbf{A}^t \tag{2.8.11}$$

This factorisation is not unique and a variety of methods may be used to obtain it such as the Cholesky method.[128] Let $\mathbf{X}$ be a vector of N uncorrelated Gaussian samples and perform the transformation

$$\mathbf{Y} = \mathbf{A}\mathbf{X} \tag{2.8.12}$$

Then the covariance matrix of $\mathbf{Y}$ is

$$\mathbf{M}_y = E[\mathbf{Y}^*\mathbf{Y}^t] = \mathbf{A}^*E[\mathbf{X}^*\mathbf{X}^t]\mathbf{A}^t = \mathbf{A}^*\mathbf{A}^t = \mathbf{M} \tag{2.8.13}$$

and the vector $\mathbf{Y}$ is a set of samples which are suitable for simulation of an array processor or for theoretical purposes.

Chapter 3

Optimal antennas

3.1 Meaning of optimality

The antenna patterns discussed previously were designed for null steering and complete cancellation of one or more interference sources. This could be fairly close to the optimum condition but suspicion remains that it may be possible to obtain even better performance. For example, one of the interferences may be quite weak, only a few dB above the receiver noise floor, and it is not clear if steering a null at such a source will improve performance by removing this interference or whether the change in the weighting coefficients required to do it will create an undesirable condition with large coefficients which, as has been seen in example 1.7.2, could reduce the amplitude gain for the desired transmission and actually reduce overall output SNR. Perhaps it might be optimum in such conditions to steer a partial null at such a source or at least ensure that no large sidelobes are created in its direction when stronger sources are nulled. The method of projections offers no help in such cases as it gives an all or nothing condition; either a source is completely cancelled or it is ignored. Another difficult problem is what to do in circumstances where the predominant interference is not due to discrete sources but is set up by isotropic noise in the medium, a typical case being medium noise in a passive sonar, which is highly correlated from element to element in the array. Proper weighting of the element outputs in such a case can offer a significant improvement in signal to noise performance. The final purpose of pursuing an optimal solution is that the projection method is not very compatible with adaptive processing since it requires estimation of the source directions before computation of the weighting vector and is a rather indirect technique. It will be shown that the nature of the optimal solution is very conducive to the development of adaptive algorithms which have a remarkable simplicity.

The theory of antennas which are optimised in the output SNR sense has been well understood for a long time[2,4,9,17,20] and the required weighting vector can be derived from the statistics of the noise field. It is also possible to maximise a probability of detection of a weak signal, e.g. a radar echo, in the presence of interference and this is a worth while exercise.

3.2 Eigenvalue solution for maximum SNIR

This technique has been widely discussed mostly in connection with optimal rather than adaptive antennas.[4, 8, 9, 13, 20] The output signal-to-noise plus interference power ratio (SNIR) is the ratio

$$Q = \frac{\text{mean signal output}}{\text{mean noise-plus-interference output}} \tag{3.2.1}$$

which has been quantified as a ratio of Hermitian forms in eqn. (2.8.10):

$$Q = \frac{\pi_s \mathbf{W}^H \mathbf{S}^* \mathbf{S}^t \mathbf{W}}{\mathbf{W}^H \mathbf{M} \mathbf{W}} \tag{3.2.2}$$

The maximum value of Q as $\mathbf{W}$ is varied is a stationary point and is a well known eigenvalue problem. The solution also has a physical interpretation and this is derived below.

If the noise-plus-interference covariance matrix $\mathbf{M}$ is positive definite it can be factored into the product of two equal Hermitian matrices

$$\mathbf{M} = \mathbf{A}\mathbf{A} = \mathbf{A}^2$$

$$\mathbf{M}^{-1} = \mathbf{A}^{-1}\mathbf{A}^{-1} = \mathbf{A}^{-2} \tag{3.2.3}$$

and this factorisation, obtained from the eigenstructure of $\mathbf{M}$ is unique (see Appendix A1.2). The positive definiteness of $\mathbf{M}$ can be ensured by adding a small amount of uncorrelated noise to the data which augments its diagonal coefficients and eigenvalues by σ^2. Matrix $\mathbf{A}$ can be used to transform $\mathbf{W}$ into a new vector $\mathbf{V}$ and vice versa,

$$\mathbf{V} = \mathbf{A}\mathbf{W} \qquad \mathbf{W} = \mathbf{A}^{-1}\mathbf{V} = (\mathbf{A}^{-1})^H\mathbf{V} \tag{3.2.4}$$

Substituting eqn. (3.2.4) into (3.2.2) yields

$$\frac{Q}{\pi_s} = \frac{\mathbf{V}^H \mathbf{A}^{-1} \mathbf{S}^* \mathbf{S}^t \mathbf{A}^{-1} \mathbf{V}}{\mathbf{V}^H \mathbf{V}} = \frac{\mathbf{V}^H \mathbf{B} \mathbf{V}}{\mathbf{V}^H \mathbf{V}} \tag{3.2.5}$$

where

$$\mathbf{B} = \mathbf{A}^{-1}\mathbf{S}^*\mathbf{S}^t\mathbf{A}^{-1} \tag{3.2.6}$$

Maximisation of Q has now been reduced to the simplest eigenvalue problem, that of maximising eqn. (3.2.5), whose solution is

$$\mathbf{V} = \mathbf{U}_{\max}$$

$$\mathbf{B}\mathbf{U}_{\max} = \lambda_{\max}\mathbf{U}_{\max} \tag{3.2.7}$$

where λ_{max} is the largest eigenvalue of **B** and $\mathbf{U}_{max}$ is the associated eigenvector. Substitution of eqn. (3.2.7) into (3.2.5) yields

$$\frac{Q_{max}}{\pi_s} = \lambda_{max} \tag{3.2.8}$$

and the equivalent weighting vector in array space is

$$\mathbf{W}_{opt} = \mathbf{A}^{-1}\mathbf{U}_{max} \tag{3.2.9}$$

from eqn. (3.2.4). If this vector is used as the coefficient vector, as in Fig. 2.2.3, the output SNIR is maximum.

The eigenvalue $\mathbf{U}_{max}$ is found very easily since **B** is a rank one matrix,

$$\mathbf{B} = \mathbf{C}^*\mathbf{C}^t, \qquad \mathbf{C}^* = \mathbf{A}^{-1}\mathbf{S}^* \tag{3.2.10}$$

and $\mathbf{U}_{max}$ is the only eigenvector with a non-zero eigenvalue. By inspection,

$$\mathbf{B}\mathbf{U}_{max} = \mathbf{C}^*\mathbf{C}^t\mathbf{U}_{max} = \lambda\mathbf{U}_{max} \tag{3.2.11}$$

where

$$\lambda = \mathbf{C}^t\mathbf{C}^* \tag{3.2.12}$$

and substitution of this result into eqn. (3.2.9), using also (3.2.10), yields

$$\mathbf{W}_{opt} = \mathbf{A}^{-1}\mathbf{A}^{-1}\mathbf{S}^* = \mathbf{M}^{-1}\mathbf{S}^* \tag{3.2.13}$$

whereas from eqns. (3.2.8), (3.2.12), and (3.2.10)

$$Q_{max}/\pi_s = \mathbf{S}^t\mathbf{A}^{-2}\mathbf{S}^* = \mathbf{S}^t\mathbf{M}^{-1}\mathbf{S}^* \tag{3.2.14}$$

Incoherent signal case: Sometimes the desired signal cannot be represented accurately by a dyad of the form (2.8.9). Examples of this are (i) when there is multipath propagation such that signal components arrive from a number of directions (ii) the signal traverses an inhomogenous medium as in tropospheric scatter communication (iii) the signal has a fairly broad bandwidth (this problem is discussed in Section 5.3). We require now to maximise the ratio of quadratic forms

$$Q = \frac{\mathbf{W}^t\mathbf{R}_s\mathbf{W}^*}{\mathbf{W}^t\mathbf{M}\mathbf{W}^*} \tag{3.2.15}$$

where $\mathbf{R}_s$ is a matrix of rank greater than unity. The procedure is the same as that for the coherent signal up to eqn. (3.2.9) but there is no simplification possible after this point and matrix B must be analysed numerically for its eigenstructure.

Prewhitening interpretation of optimality[3]

The eigenvector analysis above can be given a physically oriented interpretation. Construct a linear transformation unit **T** or generalised beamformer with

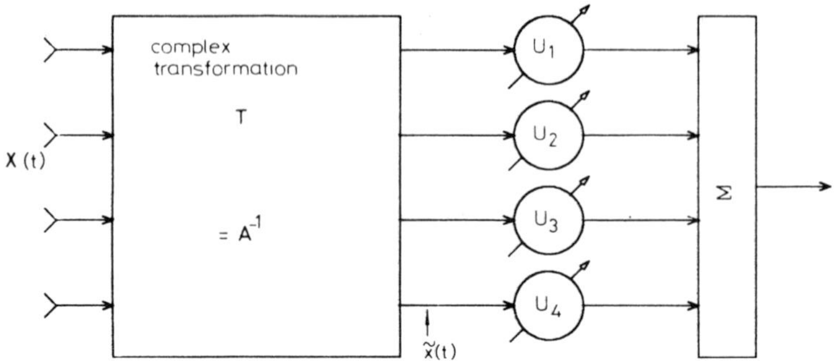

Fig. 3.2.1 *Prewhitening transformation*

N input and N output ports such that

$$\tilde{x}_i = \sum_{j=1}^{N} T_{ij} x_j \tag{3.2.16}$$

where matrix $\mathbf{T}$ with elements T_{ij} defines the transformation. The narrow band antenna element voltage vector

$$\mathbf{X}(t) = \mathbf{X}_s(t) + \mathbf{X}_n(t) \tag{3.2.17}$$

is connected to its input ports as shown in Fig. 3.2.1. $\mathbf{X}_s$ is the desired signal part of the element voltages and $\mathbf{X}_n$ the noise plus interference part. If the N^2 elements of $\mathbf{T}$ are suitably chosen the interference outputs $(\tilde{x}_n)_i(t)$ can be made mutually uncorrelated and of unit power in every channel so that the covariance matrix of $\tilde{\mathbf{X}}_n(t)$ is an identity matrix:

$$\tilde{\mathbf{R}}_n = \overline{\tilde{\mathbf{X}}_n^*(t)\tilde{\mathbf{X}}_n(t)^t} = \mathbf{I} \tag{3.2.18}$$

If the input noise-only covariance matrix $\mathbf{M}$ is positive definite, there is a unique Hermitian matrix $\mathbf{T}$ which achieves the output decorrelation and it is obtained from the eigenvector decomposition of $\mathbf{M}$ (Appendix A1.2) by taking the positive square roots of the eigenvalue reciprocals. It is the same matrix as $\mathbf{A}^{-1}$ obtained previously.

The transformation $\mathbf{T}$ also affects the desired signal. In the case of a coherent signal, eqn. (3.2.10), the desired signal vector at the transformation output is

$$\tilde{\mathbf{S}} = \mathbf{TS} = (\mathbf{A}^{-1})^t\mathbf{S} \tag{3.2.19}$$

and as the time variation is

$$\tilde{\mathbf{X}}_s(t) = m_s(t)(\mathbf{A}^{-1})^t\mathbf{S} \tag{3.2.20}$$

if a weighting vector $\mathbf{U}$ is used on the transformed output before summing the channels the desired signal output power is

$$\tilde{p}_s = \overline{|\tilde{\mathbf{X}}_s(t)^t\mathbf{U}|^2} = \pi_s \mathbf{U}^H\mathbf{A}^{-1}\mathbf{S}^*\mathbf{S}^t\mathbf{A}^{-1}\mathbf{U} \tag{3.2.21}$$

where $\pi_s = \overline{|m_s(t)|^2}$, and the noise output power is

$$\tilde{p}_n = \mathbf{U}^H\mathbf{I}\mathbf{U} = \mathbf{U}^H\mathbf{U} \tag{3.2.22}$$

using the noise covariance matrix of eqn. (3.2.18). The output signal-to-noise power ratio is

$$\tilde{Q} = \frac{\tilde{p}_s}{\tilde{p}_n} = \pi_s \frac{\mathbf{U}^H\mathbf{A}^{-1}\mathbf{S}^*\mathbf{S}^t\mathbf{A}^{-1}\mathbf{U}}{\mathbf{U}^H\mathbf{U}} \tag{3.2.23}$$

which agrees with eqn. (3.2.5).

It is a well known axiom that the optimum processing for a signal vector in uncorrelated noise is a matched filter and this is the conjugate of $\tilde{\mathbf{S}}$:

$$\mathbf{U}_{opt} = \tilde{\mathbf{S}}^* = \mathbf{A}^{-1}\mathbf{S}^* \tag{3.2.24}$$

Using this result in eqn. (3.2.23) gives the optimum signal-to-noise ratio

$$\tilde{Q}_{max} = \pi_s \frac{\mathbf{S}^t\mathbf{A}^{-2}\mathbf{S}^*\mathbf{S}^t\mathbf{A}^{-2}\mathbf{S}^*}{\mathbf{S}^t\mathbf{A}^{-2}\mathbf{S}^*} \tag{3.2.25}$$

$$= \pi_s \mathbf{S}^t\mathbf{M}^{-1}\mathbf{S}^* \tag{3.2.26}$$

which agrees with eqn. (3.2.14).

For the incoherent signal case eqn. (3.2.15), the desired signal covariance matrix at the transformation output ports is

$$\tilde{\mathbf{R}}_s = \mathbf{A}^{-1}\mathbf{R}_s\mathbf{A}^{-1} = \mathbf{B} \tag{3.2.27}$$

Optimum output signal-to-noise power ratio is found by using a weighting vector $\mathbf{U}$ with a unit norm, $\mathbf{U}^H\mathbf{U} = 1$, since such a vector will produce an output of constant noise power as in eqn. (3.2.22), and varying it under this constraint so as to maximise the signal output power

$$p_s = \mathbf{U}^t\mathbf{B}\mathbf{U}^*, \qquad \|\mathbf{U}\| = 1 \tag{3.2.28}$$

which is equivalent to the definition of the largest eigenvalue to $\mathbf{B}$ as in eqn. (3.2.7). Maximisation of signal output power corresponds to maximisation of total output power in this case: $p_{total} = 1 + p_s$.

Graphical interpretation

For a simple two-dimensional real variable case, $\mathbf{W} \triangleq (w_1, w_2)^t$, the contours of output signal-and-noise power form two independent sets of ellipses in $\mathbf{W}$-space as shown in Fig. 3.2.2(*a*) and the optimisation problem is to find the direction $\mathbf{W}_{opt}$ such that the ratio is a maximum, a highly nonlinear problem even in two dimensions. The effect of the transformation $\mathbf{A}^{-1}$ on the input data vector is to circularise the ellipses of the noise output power contour in $\mathbf{U}$-space as in Fig. 3.2.2(*b*) and as defined mathematically in eqn. (3.2.22). Selection of the vector $\mathbf{U}_{opt}$ which maximises the signal-to-noise power ratio is obvious in Fig. 3.2.2(*b*) as it corresponds to the direction of maximum signal power for a given radius,

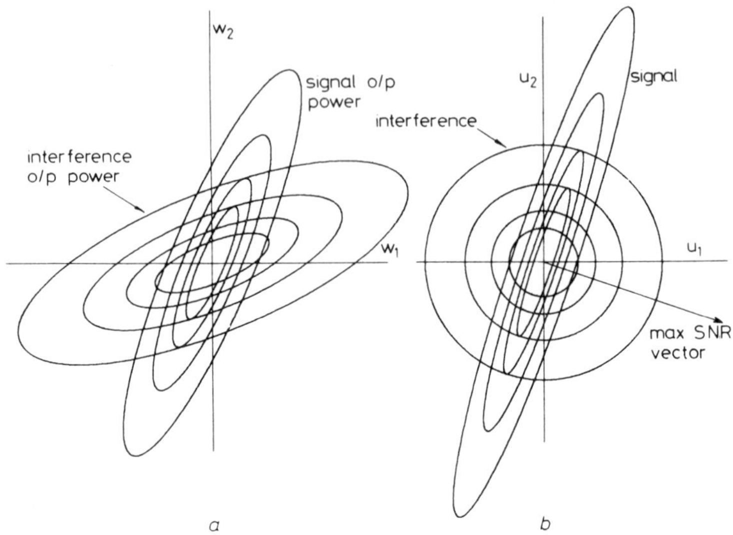

Fig. 3.2.2 Contours
a Signal and interference output power contours in array space
b Output power contours after data transformation

i.e. the direction of the minor ellipse axis. The relation of the ellipse dimensions to the eigenvalues of the covariance matrix is discussed in Appendix A1.2.

The transformation of the contours from **W**-space to **U**-space can be thought of as a scaling of the coordinates along the axes of the noise output power ellipse. In the case of the dyad signal covariance matrix eqn. (3.2.21), the signal ellipse becomes infinitely extended on its major axis due to $N - 1$ null eigenvalues and the ellipse is a parabolic trough or cylinder whose equation is

$$|\mathbf{W}^t\mathbf{S}|^2 = \text{constant}$$

or

$$|\mathbf{U}^t\tilde{\mathbf{S}}|^2 = \text{constant} \tag{3.2.33}$$

for the array data and transformed data respectively. These contours are shown in Fig. 3.2.3.

The transformation unit **T** can be implemented as a hardware device and White[80] has discussed its design as an adaptive preprocessor.

3.3 Least mean square error criterion[84]

The adaptive antenna can be considered as a filter which has variable parameters. If the desired antenna output can be defined then the antenna parameters can be varied until the mean square error between the desired output

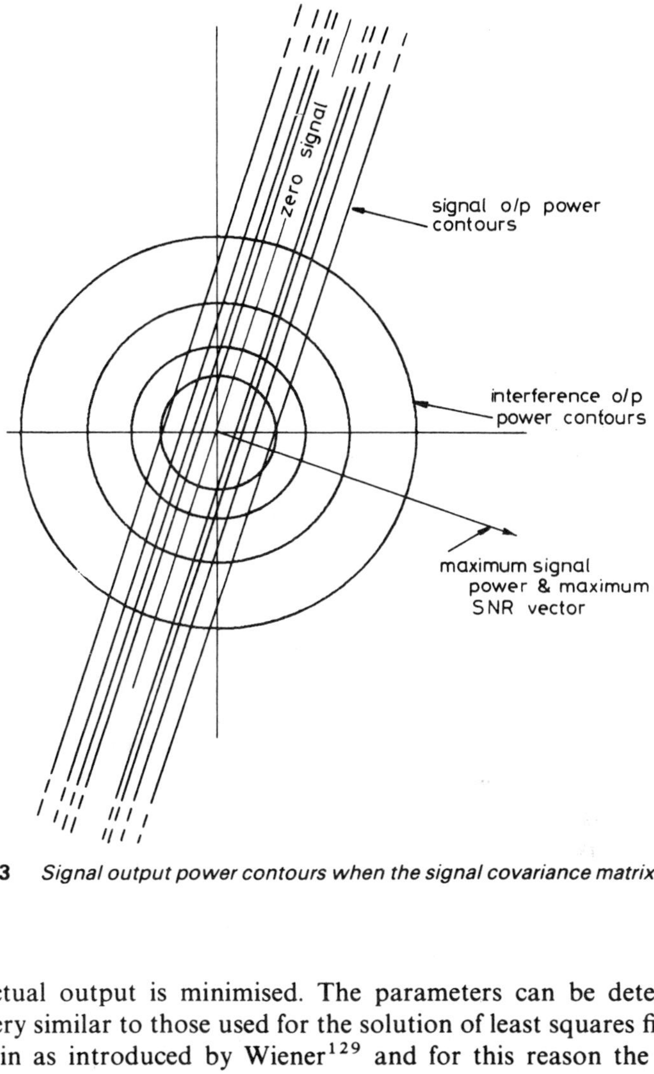

Fig. 3.2.3 *Signal output power contours when the signal covariance matrix is a dyad*

and the actual output is minimised. The parameters can be determined by methods very similar to those used for the solution of least squares filters in the time domain as introduced by Wiener[129] and for this reason the parameter solution is known as the Wiener-Hopf solution.

Figure 3.3.1 shows the variable narrow-band antenna, the reference signal, and finally the error signal which, for the real signal representation case, is

$$e(t) = d(t) - \mathbf{W}^t\mathbf{X}(t) \tag{3.3.1}$$

and the mean square error, or cost function, p is

$$p = \overline{e(t)^2} = \overline{[d(t) - \mathbf{W}^t\mathbf{X}(t)]^2} \tag{3.3.2}$$

which can be written in the form

$$p = \overline{d(t)^2} - 2\mathbf{W}^t\overline{\mathbf{X}(t)\, d(t)} + \mathbf{W}^t\mathbf{R}\mathbf{W} \tag{3.3.3}$$

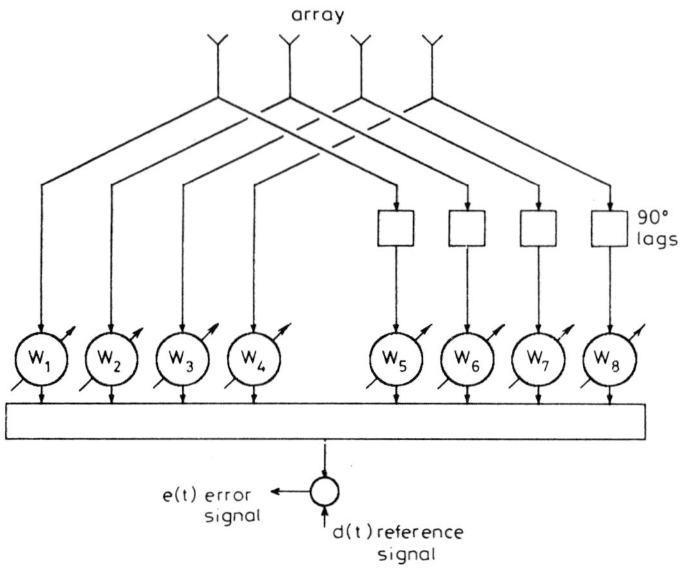

Fig. 3.3.1 *Real-signal implementation of the LMS algorithm*

Since p is a quadratic function of the weighting coefficients **W** it can be shown that there is a unique set of coefficients which minimises p, provided **R** is of full rank, and this solution can be found by determining the coefficient values for which the partial differential coefficients of p with respect to each are zero. The correlation $\overline{\mathbf{X}(t)\ d(t)}$ is written as the invariant vector **S** and the derivative of p with respect to the nth element of **W** (derived in Appendix 2) is

$$\delta p/\delta \mathbf{W}^n = -2\mathbf{S}^n + 2(\mathbf{RW})^n \tag{3.3.4}$$

where $(\ )^n$ is the nth coefficient of a vector ( ). Setting $\delta p/\delta \mathbf{W}^n$ to zero yields

$$(\mathbf{RW})^r = \mathbf{S}^n \tag{3.3.5}$$

and setting the coefficients simultaneously to zero results in

$$\mathbf{RW} = \mathbf{S} \tag{3.3.6}$$

Given that **R** has an inverse the weight vector **W** can be solved by premultiplying throughout by $\mathbf{R}^{-1}$:

$$\mathbf{W} = \mathbf{R}^{-1}\mathbf{S} \tag{3.3.7}$$

This solution is the Weiner-Hopf solution referred to earlier.[130]

Complex signals: For the case of complex signals and weights, (Fig. 3.3.2), the derivatives are somewhat more difficult to find. In place of eqn. (3.3.3) we have

$$p = \overline{|d(t)|^2} - 2\ \mathrm{Re}\ (\mathbf{W}^t\overline{\mathbf{X}(t)\ d(t)^*}) + \mathbf{W}^H\mathbf{RW} \tag{3.3.8}$$

$$= \overline{|d(t)|^2} - 2\ \mathrm{Re}\ (\mathbf{W}^t\mathbf{S}) + \mathbf{W}^H\mathbf{RW} \tag{3.3.9}$$

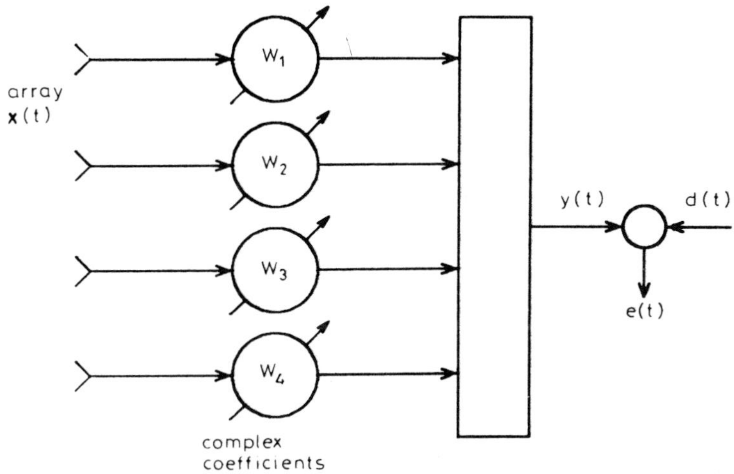

Fig. 3.3.2 *LMS criterion, complex implementation*

Defining $\mathbf{W} = \mathbf{W}_r + j\mathbf{W}_j$ and $\mathbf{S} = \mathbf{S}_r + j\mathbf{S}_j$ eqn. (3.3.9) becomes

$$p = \overline{|d(t)|^2} - 2\mathbf{W}_r^t\mathbf{S}_r + 2\mathbf{W}_j^t\mathbf{S}_j + \mathbf{W}^H\mathbf{R}\mathbf{W} \tag{3.3.10}$$

The quadratic form $\mathbf{W}^H\mathbf{R}\mathbf{W}$ is discussed in Appendix 2 and is shown to have partial derivatives of

$$\partial(\mathbf{W}^H\mathbf{R}\mathbf{W})/\partial\mathbf{W}_r^n = 2\ \mathrm{Re}\ (\mathbf{R}\mathbf{W})^n$$

$$\partial(\mathbf{W}^H\mathbf{R}\mathbf{W})/\partial\mathbf{W}_j^n = 2\ \mathrm{Im}\ (\mathbf{R}\mathbf{W})^n \tag{3.3.11}$$

Using these results we have

$$\partial p/\partial\mathbf{W}_r^n = -2\mathbf{S}_r^n + 2\ \mathrm{Re}\ (\mathbf{R}\mathbf{W})^n$$

$$\partial p/\partial\mathbf{W}_j^n = 2\mathbf{S}_j^n + 2\ \mathrm{Im}\ (\mathbf{R}\mathbf{W})^n \tag{3.3.12}$$

Equating all partial derivatives to zero yields

$$\mathrm{Re}\ (\mathbf{R}\mathbf{W}) = \mathbf{S}_r$$

$$\mathrm{Im}\ (\mathbf{R}\mathbf{W}) = -\mathbf{S}_j \tag{3.3.13}$$

and adding eqn. [3.3.13(a)] to j times [3.3.13(b)] gives the final complex form

$$\mathbf{R}\mathbf{W} = \mathrm{Re}\ (\mathbf{R}\mathbf{W}) + j\ \mathrm{Im}\ (\mathbf{R}\mathbf{W}) = \mathbf{S}_r - j\mathbf{S}_j = \mathbf{S}^* \tag{3.3.14}$$

and if $\mathbf{R}$ has an inverse the solution for $\mathbf{W}$ is

$$\mathbf{W}_{\mathrm{opt}} = \mathbf{R}^{-1}\mathbf{S}^* \tag{3.3.15}$$

A physical meaning can now be attached to the vector $\mathbf{S} = \overline{\mathbf{X}(t)\,d(t)^*}$.

The reference signal is chosen to correlate very highly with the desired signal waveform and, equally important, to be uncorrelated with any interference signals. Suppose the signals from the wanted source present at the array are of the form

$$\mathbf{X}_s(t) = m(t)\mathbf{S}_w\, e^{j\omega t}$$

as in eqn. (2.2.3) then a suitable reference signal would be $m(t)e^{j\omega t}$. The cross correlation $\mathbf{X}(t)\ d(t)^*$ reduces to

$$\overline{m(t)e^{j\omega t}\mathbf{S}_w(m(t)e^{j\omega t})^*} = \overline{|m(t)|^2}\mathbf{S}_w \tag{3.3.16}$$

thus **S** is simply a scaled replica of the desired signal space vector $\mathbf{S}_w$. It can now be seen that the result of eqn. (3.3.15) is identical to the result obtained earlier, eqn. 3.2.13, and the Wiener-Hopf solution is identical to the maximum signal-to-noise ratio solution for the signal of the assumed type, namely a narrow-band emission from a single source in the absence of multipath effects and other complications. The nature of this optimal solution is discussed in Chapter 5. It is worth drawing attention to the fact that covariance **R** above is a signal-bearing covariance matrix and variation of the amplitude of the incoming desired signal causes changes in it. This point is also discussed in Chapter 5 where it is shown that the direction of the optimal weighting vector is invariant under power changes of the desired signal.

Hypothetical implementation: Fig. 3.3.3 shows a direct implementation of the LMS optimal solution where the only changes are that the averaged covariance matrix and correlation vector are replaced by estimated values obtained on-line. This type of solution is the fastest converging one in the sense that given a

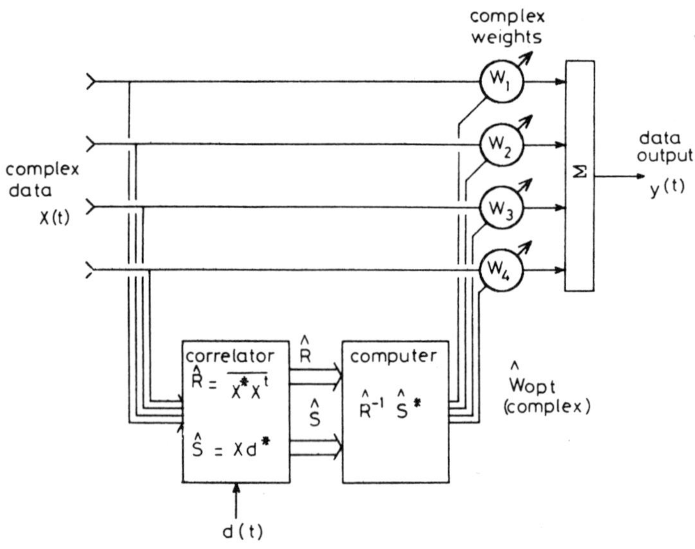

Fig. 3.3.3 *Direct implementation of a reference system*

limited amount of time to estimate an optimal weighting vector direct estimation appears to give best results (see Section 4.13). However, the system has a drawback which is that the control of the weighting coefficients is not monitored by the algorithm and since practical coefficients are often not very accurate some loss of performance is inevitable. Therefore this 'open-loop' system is best suited for a digital implementation where such errors are under tighter control. Gradient methods, discussed in the next chapter, incorporate the weighting coefficients into a feedback loop and continually correct control anomalies.

Provision of a reference signal: The desired transmitter, if co-operative, can send a predictable 'pilot' signal superposed on the data stream which is used at the receiver as a reference signal $d(t)$. This reference correlates only with the pilot component of the transmitter modulation, not with the main data stream, which causes an attenuation of the reference vector $\mathbf{S}_w$ in eqn. 3.3.16 but this neither changes the direction of the estimated optimal weight vector nor the final data output SNIR. Clearly the pilot signal must be detectable by the receiver before adaptation against strong interference can commence yet, just at this time, the antenna offers no SNR improvement for the pilot. The way out of this impasse is to use time-domain filtering to provide sufficient anti-jam protection for the pilot signal, quite feasible as it need occupy little bandwidth, and thus the pilot signal receiver can be connected to one array element to operate in a path external to the adaptive array.

The simplest narrow-band pilot is the carrier of an AM transmission but this is hardly jamming-proof and spectrum spreading is an advisable feature of the pilot signal. A very narrow bandwidth pilot receiver will give a high anti-jam margin but will introduce an extended cold-start pilot aquisition delay and Doppler shift problems may emerge. A more complex reference reconstruction technique is described in Section 4.3.

3.4 Steered systems[50]

It was seen in the last section that the optimal solution for least mean square output error is of the form

$$\mathbf{W}_{opt} = \lambda \mathbf{R}^{-1}\mathbf{S}^* \tag{3.4.1}$$

where $\mathbf{S}$ is the space vector of the desired signal generated at the receiver by a correlation process. If $\mathbf{S}$ is known already then this correlation can be dispensed with, a typical case being when the direction of the desired signal is known and $\mathbf{S}$ can be calculated from geometric considerations as in eqn. (2.2.5). This type of approach is essential if the desired source is non-co-operative, e.g. in radio monitoring or passive sonar applications when no pilot signals are transmitted or can be decoded.

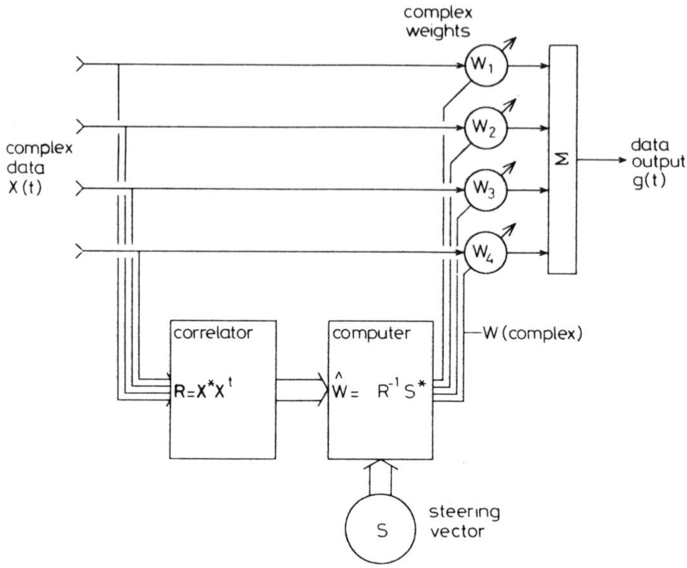

Fig. 3.4.1 *'Steered' LMS algorithm*

The scaling factor λ in eqn. (3.4.1) is arbitrary and will be chosen so that the weighting coefficients occupy some convenient dynamic range; this will usually be with the largest coefficient having a modulus of unity or 0 dB attenuation. The receiver output will depend only slightly upon λ because of its AGC loop which will compensate any input variations in the usual way.

Figure 3.4.1 shows a block diagram of a steered system in which it is postulated that a digital computer will compute **S** given bearing and elevation co-ordinates and physical parameters of the antenna array and will then proceed to calculate the optimal solution when the data covariance matrix becomes available from the correlator unit. The steered system suffers greatly from sensitivity to error in the estimation of **S** and is probably unusable for desired signal to noise ratios exceeding 20 or 30 dB unless some form of correction for steering error is incorporated into the system. The sensitivity increases with input SNR. Methods of correction are discussed in Chapter 6.

Linearly constrained output power minimisation

Real signal representation: The optimal steered solution is obtainable without the need to invoke a reference signal. Suppose the weighting vector in Fig. 3.3.1 is constrained such that the in-phase antenna response, defined in eqn. (2.3.7), is fixed at a value of g:

$$g_r = \mathbf{W}_p^t\mathbf{C} + \mathbf{W}_q^t\mathbf{S} = \mathbf{W}^t\mathbf{D} = g \tag{3.4.2}$$

then from eqn. (3.2.2) the output in-phase signal-to-noise ratio is maximised at the maximum of

$$\frac{\mathbf{W}^t\mathbf{D}\mathbf{D}^t\mathbf{W}}{\mathbf{W}^t\mathbf{M}\mathbf{W}} = \frac{g^2}{\mathbf{W}^t\mathbf{M}\mathbf{W}} \tag{3.4.3}$$

which clearly corresponds to a minimum of the interference output power $\mathbf{W}^t\mathbf{M}\mathbf{W}$. If the interference plus noise covariance matrix $\mathbf{M}$ is positive definite the quadratic function $\mathbf{W}^t\mathbf{M}\mathbf{W}$ has a unique minimum under the constraint of eqn. (3.4.2) and this can be found, using the method of Lagrange, by minimisation of the augmented cost function

$$p = \mathbf{W}^t\mathbf{M}\mathbf{W} + \lambda(\mathbf{W}^t\mathbf{D} - g) \tag{3.4.4}$$

The derivative of p with respect to each coefficient $\mathbf{W}^n$ is (Appendix 2)

$$\delta p/\delta \mathbf{W}^n = 2(\mathbf{M}\mathbf{W})^n + \lambda \mathbf{D}^n \tag{3.4.5}$$

and equating this to zero coefficient by coefficient yields

$$(\mathbf{M}\mathbf{W})^n = -\tfrac{1}{2}\lambda \mathbf{D}^n$$

or

$$\mathbf{M}\mathbf{W} = -\tfrac{1}{2}\lambda \mathbf{D} \tag{3.4.6}$$

With the positive definite assumption, eqn. (3.4.6) can be solved to give

$$\mathbf{W} = -\tfrac{1}{2}\lambda \mathbf{M}^{-1}\mathbf{D} \tag{3.4.7}$$

The initial objective of satisfying eqn. (3.4.2) is now achieved by varying λ, and substituting eqn. (3.4.7) into (3.4.2) gives

$$\mathbf{W}^t\mathbf{D} = -\tfrac{1}{2}\lambda \mathbf{D}^t\mathbf{M}^{-1}\mathbf{D} = g \tag{3.4.8}$$

which has the solution $\lambda = -2g/(\mathbf{D}^t\mathbf{M}^{-1}\mathbf{D})$, giving finally the constrained optimal solution with a defined desired signal in-phase amplitude gain:

$$\mathbf{W}_{\text{opt}} = \frac{g\mathbf{M}^{-1}\mathbf{D}}{\mathbf{D}^t\mathbf{M}^{-1}\mathbf{D}} \tag{3.4.9}$$

The quadrature part of the antenna gain for the desired signal corresponding to phase shift,

$$g_q = \mathbf{W}_p^t\mathbf{S} - \mathbf{W}_q^t\mathbf{C} \tag{3.4.10}$$

[see eqn. (2.3.7)] must be shown to vanish in the optimal solution eqn. 3.4.9. The interference covariance matrix and its inverse [eqn. (2.6.4)],

$$\mathbf{M} \triangleq \begin{bmatrix} \mathbf{A} & \mathbf{B} \\ \mathbf{B}^t & \mathbf{A} \end{bmatrix} \qquad \mathbf{M}^{-1} = \begin{bmatrix} \mathbf{U} & \mathbf{V} \\ \mathbf{V}^t & \mathbf{U} \end{bmatrix} \tag{3.4.11}$$

are each symmetric as a whole and, whereas the submatrices **A** and **U** are also symmetric, the matrices **B** and **V** are skew symmetric (Appendix A1.1). Substituting the optimal weighting vector $\mathbf{W}_{opt}$ into (3.4.10) yields

$$g_q = \begin{bmatrix} \mathbf{S} \\ -\mathbf{C} \end{bmatrix}^t \mathbf{W}_{opt} = -\frac{\lambda}{2}\begin{bmatrix} \mathbf{S} \\ -\mathbf{C} \end{bmatrix}^t \begin{bmatrix} \mathbf{U} & \mathbf{V} \\ \mathbf{V}^t & \mathbf{U} \end{bmatrix}\begin{bmatrix} \mathbf{C} \\ \mathbf{S} \end{bmatrix} \tag{3.4.12}$$

$$\frac{-2g_q}{\lambda} = \mathbf{S}^t\mathbf{U}\mathbf{C} + \mathbf{S}^t\mathbf{V}\mathbf{S} - \mathbf{C}^t\mathbf{V}^t\mathbf{C} - \mathbf{C}^t\mathbf{U}\mathbf{S} = \mathbf{S}^t\mathbf{V}\mathbf{S} - \mathbf{C}^t\mathbf{V}\mathbf{C} \tag{3.4.13}$$

Since a real quadratic form over a skew symmetric matrix is zero (Appendix A1.1) we have $g_q = 0$ and it is concluded that in the real variable case it is sufficient to constrain only the in-phase part of the antenna gain for the desired signal. This conclusion is rather theoretical and in practice in the presence of hardware irregularities and coefficient control problems it might be advisable to constraint the quadrature response also if maintaining a fixed phase response for the desired signal during adaptation is critical.

Complex case

In the complex case, a similar line can be followed regarding the quadrature part of the antenna response. However, an opportunity will be taken to simultaneously constrain the real and imaginary parts with two Lagrange multipliers. Thus if $\mathbf{W}^t\mathbf{S} = g + j0$ then

$$\mathrm{Re}\,(\mathbf{W}^t\mathbf{S}) = g, \qquad \mathrm{Im}\,(\mathbf{W}^t\mathbf{S}) = 0 \tag{3.4.14}$$

A similar procedure is followed by Owsley.[107] We create the augmented cost function

$$p = \mathbf{W}^H\mathbf{M}\mathbf{W} + \lambda_r(\mathbf{W}^t\mathbf{S} + \mathbf{W}^H\mathbf{S}^* - 2g) + \lambda_j(\mathbf{W}^t\mathbf{S} - \mathbf{W}^H\mathbf{S}^*) \tag{3.4.15}$$

and defining

$$\mathbf{W} \triangleq \mathbf{W}_r + j\mathbf{W}_j, \qquad \mathbf{S} \triangleq \mathbf{S}_r + j\mathbf{S}_j \tag{3.4.16}$$

have

$$p = \mathbf{W}^H\mathbf{M}\mathbf{W} + 2\lambda_r(\mathbf{W}_r^t\mathbf{S}_r - \mathbf{W}_j^t\mathbf{S}_j - g) + 2\lambda_j(\mathbf{W}_r^t\mathbf{S}_j + \mathbf{W}_j^t\mathbf{S}_r) \tag{3.4.17}$$

The differentials of $\mathbf{W}^H\mathbf{M}\mathbf{W}$ are discussed in Appendix 2, and for the other terms we have

$$\frac{\partial\,\mathrm{Re}\,(\mathbf{W}^t\mathbf{S})}{\partial\mathbf{W}_r} = \mathbf{S}_r, \qquad \frac{\partial\,\mathrm{Re}\,(\mathbf{W}^t\mathbf{S})}{\partial\mathbf{W}_j} = -\mathbf{S}_j$$

$$\frac{\partial\,\mathrm{Im}\,(\mathbf{W}^t\mathbf{S})}{\partial\mathbf{W}_r} = \mathbf{S}_j, \qquad \frac{\partial\,\mathrm{Im}\,(\mathbf{W}^t\mathbf{S})}{\partial\mathbf{W}_j} = \mathbf{S}_r \tag{3.4.18}$$

Thus the differentials of the real and imaginary parts of p with respect to the kth element of $\mathbf{W}$ are

$$\frac{\partial p}{\partial \mathbf{W}_r^k} = 2\ \mathrm{Re}\ (\mathbf{MW})^k + 2\lambda_r \mathbf{S}_r^k + 2\lambda_j \mathbf{S}_j^k$$

$$\frac{\partial p}{\partial \mathbf{W}_j^k} = 2\ \mathrm{Im}\ (\mathbf{MW})^k - 2\lambda_r \mathbf{S}_j^k + 2\lambda_j \mathbf{S}_r^k \tag{3.4.19}$$

and equating both to zero and adding to form a complex vector yields

$$\mathbf{MW} = -\lambda_r \mathbf{S}^* - j\lambda_j \mathbf{S}^* \tag{3.4.20}$$

which has the solution

$$\mathbf{W}_{\mathrm{opt}} = -\mathbf{M}^{-1}(\lambda_r \mathbf{S}^* + j\lambda_j \mathbf{S}^*) \tag{3.4.21}$$

We now calculate the multipliers λ_r and λ_j using the original constraint (3.4.14)

$$\mathbf{S}^t\mathbf{W}_{\mathrm{opt}} = g + j0 = -\lambda_r \mathbf{S}^t\mathbf{M}^{-1}\mathbf{S}^* - j\lambda_j \mathbf{S}^t\mathbf{M}^{-1}\mathbf{S}^* \tag{3.4.22}$$

Noting that an Hermitian form is real, the real and imaginary parts of eqn. (3.4.22) remain separate and the solutions are simply

$$\lambda_r = -\frac{g}{\mathbf{S}^t\mathbf{M}^{-1}\mathbf{S}^*} \qquad \lambda_j = 0 \tag{3.4.23}$$

Substituting back into eqn. (3.4.21) yields the final optimal vector

$$\mathbf{W}_{\mathrm{opt}} = \frac{g\mathbf{M}^{-1}\mathbf{S}^*}{\mathbf{S}^t\mathbf{M}^{-1}\mathbf{S}^*} \tag{3.4.24}$$

Minimum output power level: The actual minimum output power is obtained by substituting eqn. (3.4.24) back into (3.4.15) to get

$$p_{\min} = \frac{g^2}{\mathbf{S}^t\mathbf{M}^{-1}\mathbf{S}^*} \tag{3.4.25}$$

but this, of course, is only the interference output power. The signal output power is easily found from the look direction gain g. The presence of the desired signal does not change the optimal solution, this being proved below.

Fig. 3.4.2 shows a representation of the output power minimisation in an array with 3 real weighting coefficients for which the output power contours in $\mathbf{W}$-space form a set of ellipsoids centred at the origin where $p = 0$. The intersection of the ellipsoids with the constraint 'surface' $\mathbf{D}^t\mathbf{W} = g$ forms a new set of ellipsoids in $N - 1$ dimensional space whose centre and eccentricity is a function of vector $\mathbf{D}$ and the covariance matrix $\mathbf{M}$. The linearly constrained solution must lie on the surface at all times and the minimum output power point clearly lies at the centre of the $N - 1$ dimensional ellipsoids. This point is the one given by eqn. (3.4.9) or (3.4.30). The gradient at this point is normal to the surface and the

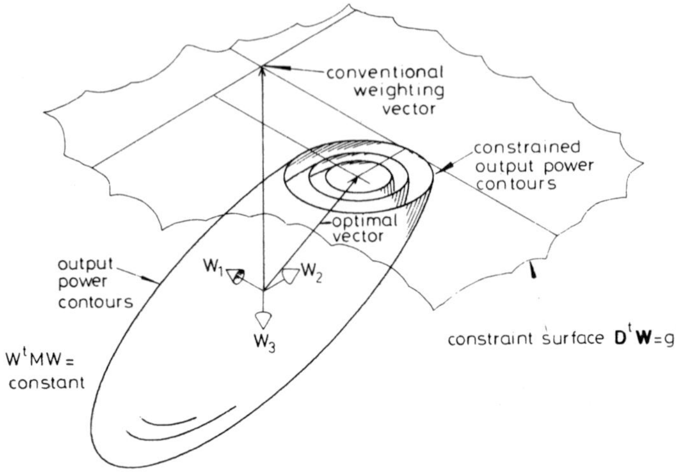

Fig. 3.4.2 *Three-dimensional example of linearly constrained output power minimisation*

surface forms a tangent to one of the N-dimensional ellipsoids with appropriate output power level.

Effect of desired signal on optimal constrained solution: The LMS solutions (3.3.7) and (3.3.15) look similar to the minimum output power solutions (3.4.9) and (3.4.24) but there is a difference in the desired signal which is assumed to be present in the covariance matrix **R** of the LMS case and absent in the matrix **M** of the constrained solution. It is important to know whether this has any effect on the optimal weighting vector otherwise we have an ambiguity. Consider the complex solution eqn. (3.4.30) (for simplicity) for the constrained antenna and add some desired signal of power π_s

$$\mathbf{R} = \mathbf{M} + \pi_s \mathbf{S}^* \mathbf{S}^t \tag{3.4.26}$$

From a matrix inversion formula (Appendix A1.2) we have

$$\mathbf{R}^{-1} = \mathbf{M}^{-1} - \frac{\pi_s \mathbf{M}^{-1} \mathbf{S}^* \mathbf{S}^t \mathbf{M}^{-1}}{1 + \pi_s \mathbf{S}^t \mathbf{M}^{-1} \mathbf{S}^*} \tag{3.4.27}$$

and

$$\mathbf{R}^{-1}\mathbf{S}^* = \mathbf{M}^{-1}\mathbf{S}^* \left[1 - \frac{\pi_s \mathbf{S}^t \mathbf{M}^{-1} \mathbf{S}^*}{1 + \pi_s \mathbf{S}^t \mathbf{M}^{-1} \mathbf{S}^*} \right] \tag{3.4.28}$$

$$= \frac{\mathbf{M}^{-1}\mathbf{S}^*}{1 + \pi_s \mathbf{S}^t \mathbf{M}^{-1} \mathbf{S}^*} \tag{3.4.29}$$

Thus the effect of adding the desired signal to **M** is to scale the optimal vector $\mathbf{R}^{-1}\mathbf{S}^*$ without changing its direction. The effect of this scaling is removed by the normalisation in eqn. (3.4.23) and it is concluded that the level of desired signal has no effect on the optimal solution.

Related effects occur in the LMS algorithm in which, if the output SNIR is good, the antenna output matches the reference amplitude. If the desired signal level is increased at the antenna input the weighting coefficients will fall to retain a constant output amplitude. This can be a source of difficulty in the LMS system where the weighting coefficients may run out of dynamic range if they get too small or too large. Either the reference amplitude must be adjusted to keep the weights operating in a convenient range or the antenna output must be scaled before subtracting the reference signal.

The invariance of the direction of the optimal weighting vector under changes in desired signal power is a property often taken for granted. However, it is not always true for suboptimal systems and great care in defining 'optimality' for reduced degrees of freedom (Chapter 7) is required.

3.5 Maximisation of probability of detection

An independent line of thought in the development of an optimal weight vector solution is to consider the detection of an echo signal by an array in the presence of interference, then vary the weights so as to maximise the probability of detection.[1,2]

The desired signal has the space-time representation

$$\mathbf{S}(t) = s(t)\mathbf{S}_d \tag{3.5.1}$$

and the interference is $\mathbf{N}(t)$. It is postulated that the array receives either noise only, $\mathbf{N}(t)$, or signal plus noise $\mathbf{S}(t) + \mathbf{N}(t)$ and must choose either one hypothesis or the other based upon the received vector

$$\mathbf{X}(t) = \mathbf{N}(t)$$

or

$$\mathbf{X}(t) = \mathbf{S}(t) + \mathbf{N}(t) \tag{3.5.2}$$

A statistic[118] for the detection of the signal is to calculate the *a posteriori* probability ratio or likelihood ratio

$$L = \frac{p(\mathbf{X} \mid \mathbf{S} + \mathbf{N})}{p(\mathbf{X} \mid \mathbf{N})} \tag{3.5.3}$$

where $p(\mathbf{X} \mid \mathbf{S} + \mathbf{N})$ is the probability density at $\mathbf{X}$ given signal plus noise and $p(\mathbf{X} \mid \mathbf{N})$ is the probability density at $\mathbf{X}$ given noise only. In eqn. (3.5.3) we deal only with probability distributions of samples of $\mathbf{X}$, $\mathbf{N}$, and $\mathbf{S}$ and this implies that they are each stationary.

If the vectors **S** and **N** are assumed to have a complex multidimensional normal form,[73] then if vector **V** represents **S** or **N** as appropriate,

$$p(\mathbf{V}) = \frac{\pi^{-N}}{\det(\mathbf{R}_v)} \exp\left[-\mathbf{V}^t\mathbf{R}_v^{-1}\mathbf{V}^*\right] \tag{3.5.4}$$

where $\mathbf{R}_v$ is the covariance of $\mathbf{V}$: $E(\mathbf{V}^*\mathbf{V}^t)$.

Substitution into eqn. (3.5.3) yields

$$L = \frac{\det(\mathbf{M}) \exp\left[-\mathbf{X}^t(\mathbf{R}_s + \mathbf{M})^{-1}\mathbf{X}^*\right]}{\det(\mathbf{M} + \mathbf{R}_s) \exp\left[-\mathbf{X}^t\mathbf{M}^{-1}\mathbf{X}^*\right]} \tag{3.5.5}$$

in which $\mathbf{R}_s$ is the covariance matrix of the desired signal and **M** that of the noise plus interference. Taking logs on both sides results in

$$\log(L) = K + \mathbf{X}^t\mathbf{M}^{-1}\mathbf{X}^* - \mathbf{X}^t(\mathbf{R}_s + \mathbf{M})^{-1}\mathbf{X}^* \tag{3.5.6}$$

where K is a constant independent of **X**. For the case of a signal represented by a dyadic covariance matrix,

$$\mathbf{R}_s = \pi_s \mathbf{S}^*\mathbf{S}^t \tag{3.5.7}$$

we have

$$(\mathbf{R}_s + \mathbf{M})^{-1} = \mathbf{M}^{-1} - \frac{\pi_s \mathbf{M}^{-1}\mathbf{S}^*\mathbf{S}^t\mathbf{M}^{-1}}{1 + \pi_s \mathbf{S}^t\mathbf{M}^{-1}\mathbf{S}^*} \tag{3.5.8}$$

and using this in equation (3.5.6) gives

$$\log(L) = K + \frac{\pi_s \mathbf{X}^t\mathbf{M}^{-1}\mathbf{S}^*\mathbf{S}^t\mathbf{M}^{-1}\mathbf{X}^*}{1 + \pi_s \mathbf{S}^t\mathbf{M}^{-1}\mathbf{S}^*} \tag{3.5.9}$$

which is of the form

$$\log(L) = K + K_1 |\mathbf{X}^t\mathbf{W}|^2 \tag{3.5.10}$$

where $\mathbf{W} = \mathbf{M}^{-1}\mathbf{S}^*$. Thus the best detection statistic reduces to sampling the output power of an array with weighting vector **W**, and comparing to a threshold; a result which conforms to other optimality criteria based on output SNIR. However, as the present result is only obtained using Gaussian probability densities, it is somewhat restrictive in comparison.

3.6 Direct maximisation of SNR

Starting with the expression for output signal-to-noise power ratio, eqn. (2.8.10), we can maximise the SNR directly by finding a stationary point of the quotient which follows the same general form as for a scalar quotient. Using results in Appendix A2 we obtain

$$dQ/d\mathbf{W}_r^m = \frac{\mathbf{W}^H\mathbf{M}\mathbf{W} \,.\, 2\,\mathrm{Re}\,(\mathbf{R}_s\mathbf{W})^m - \mathbf{W}^H\mathbf{R}_s\mathbf{W} \,.\, 2\,\mathrm{Re}\,(\mathbf{M}\mathbf{W})^m}{(\mathbf{W}^H\mathbf{M}\mathbf{W})^2}$$

and

$$dQ/d\mathbf{W}_j^m = \frac{\mathbf{W}^H\mathbf{M}\mathbf{W} \,.\, 2\,\mathrm{Im}\,(\mathbf{R}_s\mathbf{W})^m - \mathbf{W}^H\mathbf{R}_s\mathbf{W} \,.\, 2\,\mathrm{Im}\,(\mathbf{M}\mathbf{W})^m}{(\mathbf{W}^H\mathbf{M}\mathbf{W})^2} \tag{3.6.1}$$

Setting $dQ/d\mathbf{W}_r^m$ and $dQ/d\mathbf{W}_j^m$ to zero for all m gives the two equations

$$\mathrm{Re}\,(\mathbf{R}_s\mathbf{W}) = \frac{\mathbf{W}^H\mathbf{R}_s\mathbf{W}}{\mathbf{W}^H\mathbf{M}\mathbf{W}}\,\mathrm{Re}\,(\mathbf{M}\mathbf{W})$$

$$\mathrm{Im}\,(\mathbf{R}_s\mathbf{W}) = \frac{\mathbf{W}^H\mathbf{R}_s\mathbf{W}}{\mathbf{W}^H\mathbf{M}\mathbf{W}}\,\mathrm{Im}\,(\mathbf{M}\mathbf{W}) \tag{3.6.2}$$

and adding these two equations, after multiplying the second by j, yields

$$\mathbf{R}_s\mathbf{W} = \mathrm{Re}\,(\mathbf{R}_s\mathbf{W}) + j\,\mathrm{Im}\,(\mathbf{R}_s\mathbf{W}) = \frac{\mathbf{W}^H\mathbf{R}_s\mathbf{W}}{\mathbf{W}^H\mathbf{M}\mathbf{W}}\,\mathbf{M}\mathbf{W} \tag{3.6.3}$$

This equation is recognisable as a joint eigenvector problem and we seek the solution

$$\mathbf{R}_s\mathbf{W} = \lambda\mathbf{M}\mathbf{W} \tag{3.6.4}$$

with the largest λ (if there is a choice) and we also find the output signal-to-noise ratio Q is equal to

$$Q = \frac{\mathbf{W}^H\mathbf{R}_s\mathbf{W}}{\mathbf{W}^H\mathbf{M}\mathbf{W}} = \lambda \tag{3.6.5}$$

Eqn. (3.6.4) can be solved by a standard eigenvalue method if the equation is premultiplied throughout by $\mathbf{M}^{-1}$ (assumed to be of full rank) to get,

$$\mathbf{M}^{-1}\mathbf{R}_s\mathbf{W} = \lambda\mathbf{W} \tag{3.6.6}$$

In the special case that $\mathbf{R}_s$ is a dyad of the form $\pi_s\mathbf{S}^*\mathbf{S}^t$, as in eqn. (2.6.19), a much simpler solution results. Substitution into eqn. (3.6.6) gives

$$\pi_s\mathbf{M}^{-1}\mathbf{S}^*\mathbf{S}^t\mathbf{W} = \lambda\mathbf{W} \tag{3.6.7}$$

in which $\mathbf{S}^t\mathbf{W}$ is, of course, a scalar and eqn. (3.6.7) has the solution

$$\mathbf{W} = \lambda^{-1}\pi_s\mathbf{S}^t\mathbf{W}\mathbf{M}^{-1}\mathbf{S}^* \tag{3.6.8}$$

which is of the same form as eqn. (3.4.24) within a scaling factor (scaling of **W** does not affect the SNR).

3.7 Effect of matrix transformations of the data[22,52,80]

Sometimes the element signals are not directly observable because the vector of data passes through a matrix transformation, an example being a beamformer matrix. Matrix transformations may be deliberately introduced in order to facilitate linear constraint systems (Chapter 6) or to modify the eigenvalues of the covariance matrix (Section 7.12). Let N by N complex invertible matrix $\mathbf{B}$ represent the transformation which outputs a new vector $\mathbf{Y}(t)$ when the original array data vector $\mathbf{X}(t)$ is input:

$$\mathbf{Y}(t) = \mathbf{B}\mathbf{X}(t) \tag{3.7.1}$$

The steering vector appropriate for the $\mathbf{X}(t)$ data is supposed to be $\mathbf{S}^*$. Vector $\mathbf{S}$ is also passed through the transformation to yield a new signal vector $\mathbf{V}$

$$\mathbf{V} = \mathbf{B}\mathbf{S} \tag{3.7.2}$$

Note that in the LMS algorithm this steering vector transformation will occur spontaneously, since the vector of correlations in eqn. (3.3.9) will become

$$\mathbf{V} = E\{\mathbf{Y}(t)d(t)^*\} = \mathbf{B}E\{\mathbf{X}(t)d(t)^*\} = \mathbf{B}\mathbf{S} \tag{3.7.3}$$

In the original array space, the optimal weighting vector is given by, for a linearly constrained system, [eqn. (3.4.24)],

$$\mathbf{W}_x = \lambda \mathbf{R}_x^{-1}\mathbf{S}^*/(\mathbf{S}^t\mathbf{R}_x^{-1}\mathbf{S}^*) \tag{3.7.4}$$

Using the transformed data and steering vector, however, yields

$$\mathbf{W}_y = \lambda \mathbf{R}_y^{-1}\mathbf{V}^*/(\mathbf{V}^t\mathbf{R}_y^{-1}\mathbf{V}^*) \tag{3.7.5}$$

where

$$\mathbf{R}_y = E[\mathbf{Y}(t)^*\mathbf{Y}(t)^t] \tag{3.7.6}$$

We require to show that this solution in $\mathbf{Y}$-space is equivalent to the solution (3.7.4) in $\mathbf{X}$-space. First of all observe that

$$\mathbf{R}_y = E[\mathbf{B}^*\mathbf{X}(t)^*\mathbf{X}(t)^t\mathbf{B}^t] = \mathbf{B}^*\mathbf{R}_x\mathbf{B}^t \tag{3.7.7}$$

and that

$$\mathbf{R}_y^{-1} = (\mathbf{B}^t)^{-1}\mathbf{R}_x^{-1}(\mathbf{B}^*)^{-1} \tag{3.7.8}$$

Substituting eqns. (3.7.2) and (3.7.8) into (3.7.5) yields

$$\mathbf{W}_y = \frac{\lambda(\mathbf{B}^t)^{-1}\mathbf{R}_x^{-1}(\mathbf{B}^*)^{-1}\mathbf{B}^*\mathbf{S}^*}{\mathbf{S}^t\mathbf{B}^t(\mathbf{B}^t)^{-1}\mathbf{R}_x^{-1}(\mathbf{B}^*)^{-1}\mathbf{B}^*\mathbf{S}^*} = (\mathbf{B}^t)^{-1}\mathbf{W}_x \tag{3.7.9}$$

and the output signal of the $\mathbf{Y}$-space system is

$$\begin{aligned} f_y(t) = \mathbf{Y}(t)^t\mathbf{W}_y &= \mathbf{X}(t)^t\mathbf{B}^t(\mathbf{B}^t)^{-1}\mathbf{W}_x \\ &= \mathbf{X}(t)^t\mathbf{W}_x \\ &= f_x(t) \end{aligned} \tag{3.7.10}$$

Thus the output signal waveforms for the optimal **Y**-space and **X**-space systems are identical; consequently output SNRs and SIRs must be equal too. As a check it can also be shown that the directional patterns of the two systems are identical. If a signal with space vector $\mathbf{S}(\theta)$ is applied to the **Y**-space antenna, the amplitude gain is

$$g_y = \mathbf{S}(\theta)^t\mathbf{B}^t\mathbf{W}_y \tag{3.7.11}$$

from eqn. (3.7.1); but substituting for $\mathbf{W}_y$ from eqn. (3.7.9) gives

$$g_y = \mathbf{S}(\theta)^t\mathbf{B}^t(\mathbf{B}^t)^{-1}\mathbf{W}_x = \mathbf{S}(\theta)^t\mathbf{W}_x$$

$$= g_x \tag{3.7.12}$$

Therefore the directional patterns of the optimal systems agree.

Aperture shading: Aperture shading is an elementary data transformation in which the matrix **B** is a diagonal matrix of weighting coefficients:

$$\mathbf{B} = \mathbf{\Lambda} = \text{diag}\,(a_1, a_2, \ldots, a_N) \tag{3.7.13}$$

with the sequence $\{a_i\}$ being Chebychev or similar weights which reduce the sidelobe levels of the nonadaptive array. This form of shading is very important in circular arrays where sidelobe levels are intrinsically higher than for line arrays.

Clearly, from the discussion above, the optimal solution of the antenna, if it is allowed to adapt by the introduction of further variable coefficients following the fixed shading ones, will not be changed; the algorithm will simply negate the effect of the shading coefficients by equalisation and apply an optimal solution as though they did not exist. Thus the shading is of little value.

Applebaum[22] has described a modification of the technique in which the shading coefficients continue to be effective in conjunction with an optimal array weighting algorithm. In his procedure, the element outputs are unweighted, except by the adaptive weights, but the steering vector in a linearly constrained algorithm is shaded as though it were a vector of antenna shadings. The optimal solution is now

$$\mathbf{W} = \mathbf{R}^{-1}\mathbf{\Lambda}S^* \tag{3.7.14}$$

If the adaptive weights are initialised with the steering vector elements then some form of shading is present throughout adaptation.

There are two problems in operation of this technique which may render it unsuitable in certain circumstances. First, there is a degree of mismatch introduced between the desired signal vector and the steering vector which, at high input desired signal SNRs could lead to signal suppression (see Chapter 6). Secondly, if a main-lobe interference is present, the system will steer a retro-lobe as in Fig. 2.5.2 whose pattern will have conventional-type sidelobe levels. These will destroy the previous low sidelobe levels.

3.8 Optimisation of power pattern

The preceding Section indicates that optimisation can be performed in beam space and the optimal beam pattern is a linear combination of the beams formed by a beamformer of arbitrary characteristics. A slightly different approach is to expand the optimal beam pattern as a linear combination of orthogonal patterns, however these will not be intrinsically orthogonal but will be chosen so as to be orthogonal with the interference angular power spectrum as a weighting function. Let

$$G_i(\theta),\ i = 0, 1, \ldots, N \tag{3.8.1}$$

be a set of real functions of $\theta = \sin(\alpha)$ where α is bearing angle relative to broadside for a line array. These functions have the orthogonality relationship

$$\begin{aligned}\int_{-1}^{+1} G_i(\theta)G_j(\theta)P(\theta)\, d\theta &= \phi_i, \qquad i = j\\ &= 0, \qquad i \neq j\end{aligned} \tag{3.8.2}$$

where the weighting function $P(\theta)$ is the angular power density spectrum of the incident interference. If the array pattern is represented as

$$g(\theta) = \sum_{n=0}^{N} c_n G_n(\theta) \tag{3.8.3}$$

where the coefficients c_n are real, then the array output power due to the interference is

$$p = \int_{-1}^{+1} [g(\theta)]^2 P(\theta)\, d\theta = \int_{-1}^{+1} \left[\sum_{n=0}^{N} c_n G_n(\theta)\right]^2 P(\theta)\, d\theta \tag{3.8.4}$$

Let the pattern be constrained to have a fixed amplitude gain of unity at bearing θ_0, $g(\theta_0) = 1$, and add a corresponding Lagrange term $\lambda[\sum_{n=0}^{N} c_n G_n(\theta_0) - 1]$ to p. The derivatives of this augmented power with respect to the coefficients c_n are

$$dp/dc_i = 2\int_{-1}^{+1} \left[\sum_{n=0}^{N} c_n G_n(\theta)\right] G_i(\theta)P(\theta)\, d\theta + \lambda G_i(\theta_0) \tag{3.8.5}$$

and because of the orthogonality relationship of eqn. (3.8.2), this reduces to

$$dp/dc_i = 2c_i\phi_i + \lambda G_i(\theta_0) \tag{3.8.6}$$

and equating to zero yields

$$c_i = -\lambda G_i(\theta_0)/2\phi_i, \qquad i = 0, \ldots, N \tag{3.8.7}$$

Substituting the c_i values into the constraint equation $\sum_{n=0}^{N} c_n G_n(\theta_0) = 1$ produces

$$\lambda = \frac{-2}{\sum_{n=0}^{N} G_n^2(\theta_0)/\phi_n} \tag{3.8.8}$$

and finally

$$c_i = \frac{G_i(\theta_0)/\phi_i}{\sum_{n=0}^{N} G_n^2(\theta_0)/\phi_n} \tag{3.8.9}$$

Thus the optimal minimum output power pattern has the form

$$g(\theta) = \frac{\sum_{n=0}^{N} G_n(\theta) G_n(\theta_0)/\phi_n}{\sum_{n=0}^{N} G_n^2(\theta_0)/\phi_n} \tag{3.8.10}$$

and it is not difficult to show that the minimum output power is given by

$$p_{\min} = \int_{-1}^{+1} |g(\theta)|^2 P(\theta)\, d\theta = \frac{1}{\sum_{n=0}^{N} G_n^2(\theta_0)/\phi_n} = -\tfrac{1}{2}\lambda \tag{3.8.11}$$

In general, these results are of limited use for numerical work as it is easier to proceed via the covariance matrix to find the optimal constrained weighting vector. However, much analytic work can be done for special forms of angular power spectra. For example, if the angular spectrum is 'band limited', i.e. exists only in a small interval around θ_1, then it is often possible to expand the beam pattern as a polynomial in $\theta - \theta_1$ by suitable placing of nulls in the pattern around θ_1 and orthogonal polynomials can then be used as the basis functions. For a rectangular spectral density, the Lagrange polynomials are orthogonal over a finite interval with uniform weighting, for a Gaussian spectrum the Hermite polynomials can be used, and so on.

Chapter 4

Adaptive solutions of optimal antennas

4.1 Meaning of adaptivity

An adaptive antenna is one whose parameters are caused to vary as a function of the interference field so that it always stays in an optimal or fairly optimal condition. The problem with fixed parameter optimal solutions is that interference is seldom constant in space or time and an antenna which is initially optimum in some location will rapidly become very suboptimal as the interference changes either because of the appearance of new interference transmissions, changes in the local scattering environment through movement of vehicles, ships, aircraft etc., variation in the orientation of the antenna if scanning or if it is mounted on a moving vehicle or aircraft, or changes in operating frequency. For example the null steering solutions discussed in the last chapter are obviously extremely sensitive to some of these changes and it is hardly worth while attempting to use them as fixed solutions. The only successful fixed parameter optimised antennas are those with a cautious general purpose optimised function for example arrays with Taylor or Chebychev tapers to reduce sidelobe levels and those operating in a very stable local interference background. Sidelobe levels of better than -30 dB can be achieved in small arrays of 20 elements, with careful design, and better than -40 dB for large arrays but these figures are close to practical limits.

Examples of data adaptive optimal antennas have been given already in Figs. 3.3.3 and 3.4.1, and although these were mainly intended to illustrate mathematical procedures and are basically a block diagram representation of the equations, they are also, as is so often the case, probably the most efficient implementations in terms of adaptation rate and quality of solution. They are also very expensive to implement demanding real time digital data throughput and for practical purposes it is better to adopt cheaper methods, either entirely analogue or hybrid analogue/digital schemes which have a convergence rate penalty but give just as good as solution eventually. In many practical interference scenarios these slower implementations are entirely adequate.

4.2 Gradient methods[56, 83]

Gradient techniques are the most popular method for determining adaptive antenna weighting coefficients. They have been developed from parallel techniques used widely in optimisation theory and practice (References 113, 128, 131). Their use for the latter application is usually because many optimisation problems have a nonlinear criterion or constraint system and gradient techniques offer the only method of solution. For antennas, a closed form solution in the form of eqns. (3.3.7) or (3.4.24) is normally available, if the covariance matrix **R** is estimated on-line, but is often rejected for hardware cost reasons.

To develop a gradient technique we must first specify a cost function which is to be minimised (or maximised in some cases) by the variation of a set of parameters. These variable parameters are element weighting coefficients for the task at hand. Typical cost functions can be mean square error as in eqn. (3.3.2) or interference output power in eqn. (3.4.4). Both of these cost functions are quadratic functions of the weighting coefficients having the general form, in the real N variable case

$$\begin{aligned} J = {} & a_{11} W_1^2 + a_{22} W_2^2 + \cdots + a_{NN} W_N^2 \\ & + 2a_{12} W_1 W_2 + 2a_{23} W_2 W_3 + \cdots + 2a_{N-1,N} W_{N-1} W_N \\ & + 2a_{13} W_1 W_3 + 2a_{24} W_2 W_4 + \cdots + 2a_{N-2,N} W_{N-2} W_N \\ & \cdots\cdots\cdots\cdots \\ & + 2a_{1N} W_1 W_N \\ & + 2b_1 W_1 + 2b_2 W_2 + \cdots + 2b_N W_N \\ & + c \end{aligned} \tag{4.2.1}$$

with a corresponding function for the complex case. Equation (4.2.1) corresponds to the matrix-vector representation

$$J = \mathbf{W}^t\mathbf{A}\mathbf{W} + 2\mathbf{W}^t\mathbf{B} + c \tag{4.2.2}$$

where N by N matrix **A** has elements $A_{ij} = a_{ij}$. Equation (4.2.2) illustrates the economy of effort derived by using vectors and matrices in the analysis and this practice will be adhered to as far as possible.

A gradient method of minimising eqn. (4.2.1) or (4.2.2) consists of finding the partial derivatives of J with respect to each coefficient W_i and making small steps in the value of each coefficient in the direction which reduces the cost function J. Using standard methods the partial derivatives of J in eqn. (4.2.1) are found to be

$$\begin{aligned} \partial J/\partial W_1 &= 2a_{11} W_1 + 2a_{12} W_2 + 2a_{13} W_3 + \cdots + 2a_{1N} W_N + 2b_1 \\ \partial J/\partial W_2 &= 2a_{21} W_1 + 2a_{22} W_2 + \cdots + 2a_{2N} W_N + 2b_2 \\ & \cdots\cdots\cdots\cdots \\ \partial J/\partial W_N &= 2a_{N1} W_1 + 2a_{N2} W_2 + \cdots + 2a_{NN} W_N + 2b_N \end{aligned} \tag{4.2.3}$$

which are seen to be equal to the elements of the vector $\mathbf{G} = 2AW + 2B$ which is the gradient of J, $\nabla_{\mathbf{W}}(J)$. The steps in the coefficient values are

$$W_1(k+1) = W_1(k) - d_1\,\partial J/\partial W_1(k)$$
$$W_2(k+1) = W_2(k) - d_2\,\partial J/\partial W_2(k) \tag{4.2.4}$$
$$W_N(k+1) = W_N(k) - d_N\,\partial J/\partial W_N(k)$$

where the sequence of values taken by the coefficients during the minimisation is $\mathbf{W}(0), \mathbf{W}(1), \mathbf{W}(2), \ldots, \mathbf{W}(k), \mathbf{W}(k+1), \mathbf{W}(k+2), \ldots$. In vector form we have

$$\mathbf{W}(k+1) = \mathbf{W}(k) - \mathbf{DG}(k) = \mathbf{W}(k) - 2\mathbf{D}[\mathbf{AW}(k) + \mathbf{B}] \tag{4.2.5}$$

where $\mathbf{D}$ is a diagonal matrix with element values $d_1, d_2, \ldots, d_N$. The 'feedback' coefficients d_i can take any positive value up to a certain critical point where instability occurs.

Figure 4.2.1 shows typical cost function contours for a problem with two real coefficients. If matrix $\mathbf{A}$ is symmetric and positive definite the contours invariably form a set of ellipses whose eccentricity and orientation is a function of

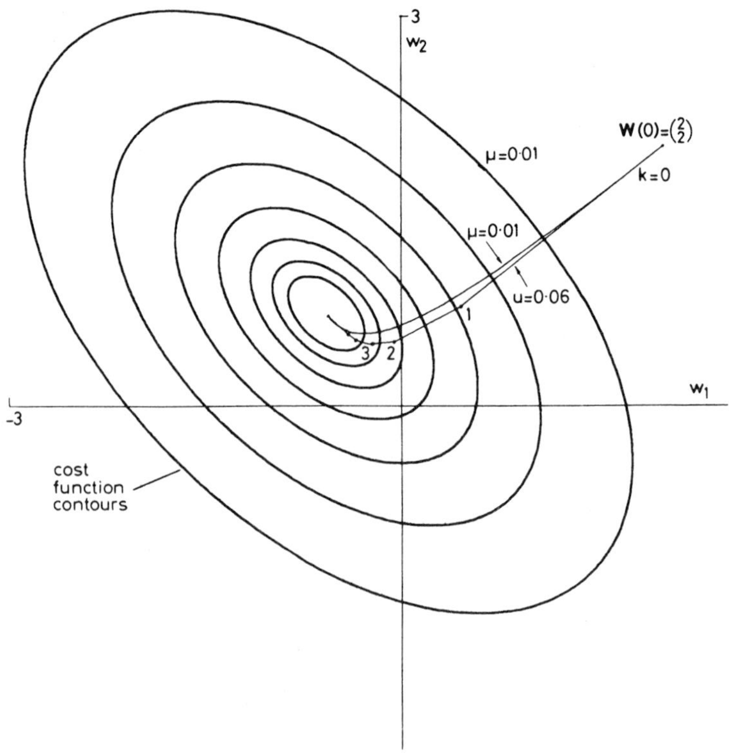

Fig. 4.2.1 *Convergence to optimum point in LMS algorithm*

the eigenstructures of $\mathbf{A}$ (Appendix A1.2). The offset of the centre of the ellipses from the origin might be expected to have co-ordinates b_1, b_2 but this is not the case. The centre of the ellipses is the coefficient vector giving minimum cost function and this problem was solved in eqn. (3.3.7) as

$$\mathbf{W}_{\min} = -\mathbf{A}^{-1}\mathbf{B} \qquad J_{\min} = c - \mathbf{B}^t\mathbf{A}^{-1}\mathbf{B} \tag{4.2.6}$$

Starting at some arbitrary point $\mathbf{W}(0)$ the cost function decreases for any direction pointing into the interior of the associated ellipse and the descent step (4.2.4) moves to such an interior point each time.

The algorithm described above is actually a discrete algorithm in the sense that the coefficients move in discrete jumps. The corresponding implementation in hardware would be digital. It is also possible to define a continuous descent algorithm and this would take the form

$$\begin{aligned} dW_1/dt &= -d_1\,\partial J/\partial W_1 \\ dW_2/dt &= -d_2\,\partial J/\partial W_2 \\ \ldots&\ldots\ldots\ldots \\ dW_N/dt &= -d_N\,\partial J/\partial W_N \end{aligned} \tag{4.2.7}$$

which could be implemented using analogue computer techniques. Both types of algorithm are in common use and are largely equivalent in performance, the preferences depending on hardware constraints.

Steepest descent

In the steepest descent algorithm the direction of movement of the $\mathbf{W}$ vector is that which gives the maximum change of cost function (power) for a given displacement of $\mathbf{W}$. This is an optimum short term strategy. The values of the d_i factors are determined as follows for steepest descent. In the discrete algorithm, consider that the weight coefficients take simultaneous steps of ΔW_1, ΔW_2, $\Delta W_3, \ldots, \Delta W_N$. If these steps are small then the corresponding change in cost function (from Taylor's theorem) is

$$\delta J = \Delta W_1\,\partial J/\partial W_1 + \Delta W_2\,\partial J/\partial W_2 + \cdots + \Delta W_N\,\partial J/\partial W_N \tag{4.2.8}$$

$$= \mathbf{G}^t\boldsymbol{\Delta}_w \tag{4.2.9}$$

where $\boldsymbol{\Delta}_w$ is the vector of steps defined above. From the Cauchy-Schwartz inequality [eqn. (2.4.14)] we have

$$|\delta J| = |\mathbf{G}^t\boldsymbol{\Delta}_w| \leq \|\mathbf{G}\| \cdot \|\boldsymbol{\Delta}_w\| \tag{4.2.10}$$

and $|\delta J|$ attains its maximum value for given $\|\mathbf{G}\|$ and $\|\boldsymbol{\Delta}_w\|$ when $\mathbf{G}$ and $\boldsymbol{\Delta}_w$ are parallel or antiparallel. For most decrease in cost for given $|\boldsymbol{\Delta}_w|$ we set

$$\boldsymbol{\Delta}_w = -|\mu|\mathbf{G} \tag{4.2.11}$$

which corresponds to all the d_i coefficients having the same values and the diagonal **D** matrix in eqn. (4.2.5) being $|\mu|$ times the identity matrix. The discrete steepest descent algorithm is thus

$$\mathbf{W}(k+1) = \mathbf{W}(k) - \mu\mathbf{G}(k) = \mathbf{W}(k) - 2\mu[\mathbf{A}\mathbf{W}(k) + \mathbf{B}] \qquad (4.2.12)$$

and the equivalent continuous algorithm is

$$d\mathbf{W}/dt = -2\mu[\mathbf{A}\mathbf{W} + \mathbf{B}] \qquad (4.2.13)$$

where $d\mathbf{W}/dt$ is a vector whose elements are dW_1/dt, dW_2/dt etc.

Example 4.2.1: The cost function which generates the ellipses of Fig. 4.2.1 is actually

$$J = 4W_1^2 + 4W_2^2 + 4W_1W_2 + \tfrac{12}{7}W_1 - \tfrac{24}{7}W_2 + 24/7 \qquad (4.2.14)$$

which corresponds to

$$\mathbf{A} = \begin{bmatrix} 4 & 2 \\ 2 & 4 \end{bmatrix} \qquad \mathbf{B} = \begin{bmatrix} 6/7 \\ -12/7 \end{bmatrix} \qquad (4.2.15)$$

in eqn. (4.2.2). The gradient vector is thus

$$\begin{aligned} G_1 &= 8W_1 + 4W_2 + 12/7 \\ G_2 &= 4W_1 + 8W_2 - 24/7 \end{aligned} \qquad (4.2.16)$$

Using a feedback coefficient of $\mu = 0{\cdot}06$ in eqn. (4.2.11) yields the sequence of weighting vectors shown in Table 4.2.1. The trajectory of the weighting vector with this feedback is also shown on Fig. 4.2.2. It is seen that the coefficients 'fall' rapidly down the steep gradient of the small diameter of the ellipses then much more slowly down the large diameter finally approaching the minimum cost point $W_{opt} = (-4/7, 5/7)^t$. If the feedback factor is made too great the system will approach instability; this is shown in Fig. 4.2.2 with $\mu = 0{\cdot}14$ and it is apparent that the weighting vector is making wild swings across the smaller diameter of the ellipses while converging quite smoothly along the larger diameter.

If the feedback factor in the discrete algorithm is made very small the convergence trajectory approaches that of the continuous time system eqn. (4.2.13). This trajectory is shown in Fig. 4.2.2 ($\mu = 0{\cdot}01$). The continuous time trajectory remains perpendicular to the contours at all times and the contours and set of trajectories for different starting points form a curvilinear orthogonal coordinate system.

The steepest descent algorithm defined above is not the same as the steepest descent algorithm found in the optimisation literature.[113, 128, 131] There the weight vector is continued along the gradient direction defined at its starting point until it passes through a minimum of power. A new gradient direction is the calculated. This procedure offers faster convergence than the 'damped steepest descent' above but is much more reliant on mathematical computations and is not suitable for simple hardware realisations.

Table 4.2.1 *Convergence of the LMS algorithm in discrete time*

W_1	W_2	G_1	G_2	J
2·000000	2·000000	25·714280	20·571400	47·999
0·457143	0·765716	8·434290	4·525701	6·1681
−0·048914	0·494174	3·299661	0·329134	2·5399
−0·246894	0·474426	1·636831	−0·620769	2·0541
−0·345104	0·511672	1·000137	−0·715639	1·8997
−0·405112	0·554610	0·691825	−0·612165	1·8205
−0·446622	0·591340	0·506668	−0·484364	1·7754
−0·477022	0·620402	0·379715	−0·373470	1·7495
−0·499805	0·642810	0·287085	−0·285335	1·7345
−0·517030	0·659931	0·217765	−0·217275	1·7259
−0·530096	0·672967	0·165383	−0·165247	1·7209
−0·540019	0·682882	0·125658	−0·125620	1·7180
−0·547558	0·690419	0·095491	−0·095480	1·7163
−0·553288	0·696148	0·072571	−0·072567	1·7154
−0·557642	0·700502	0·055153	−0·055152	1·7148
−0·560951	0·703811	0·041917	−0·041916	1·7145
−0·563466	0·706326	0·031857	−0·031856	1·7143
−0·565377	0·708237	0·024211	−0·024210	1·7142
−0·566830	0·709690	0·018400	−0·018400	1·7141
−0·567934	0·710794	0·013984	−0·013984	1·7141
−0·568773	0·711633	0·010628	−0·010628	1·7141
−0·569411	0·712271	0·008077	−0·008078	1·7141
−0·569895	0·712755	0·006139	−0·006138	1·7141
−0·570264	0·713124	0·004666	−0·004665	1·7141
−0·570544	0·713404	0·003546	−0·003545	1·7141

4.3 A real-time least mean square error algorithm[84, 50]

In eqn. (3.3.2) the cost function or mean square error 'p' is

$$p = \overline{[\mathbf{X}(t)^t\mathbf{W} - d(t)]^2} = \overline{e(t)^2} \tag{4.3.1}$$

and the partial derivatives of p with respect to the elements of $\mathbf{W}$ form the vector

$$\mathbf{G} = \nabla_{\mathbf{W}}(p) = \overline{2\mathbf{X}(t)[\mathbf{X}(t)^t\mathbf{W} - d(t)]} = \overline{2\mathbf{X}(t)e(t)} \tag{4.3.2}$$

noting that differentiation inside the averaging is permissable. Thus the gradient of the mean square error is just the vector of correlations between the error signal and the element voltages and the steepest descent direction is the opposite direction or $-\mathbf{G}$. The measurement of this correlation vector is easy to im-

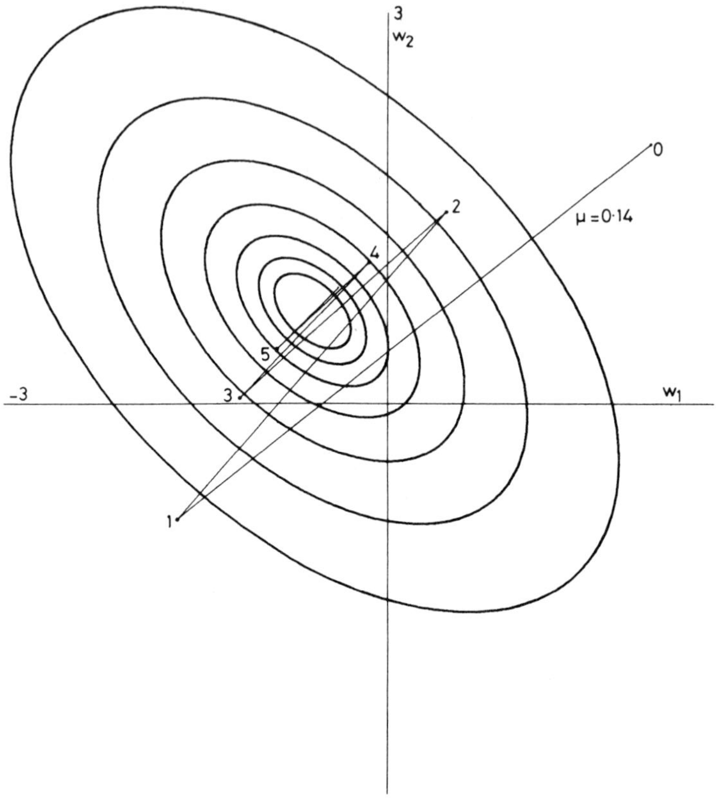

Fig. 4.2.2 *Operation close to instability*

plement, in principle, requiring a set of amplifiers to bring each element voltage up to a level where hardware multipliers can be used in the locations shown in Fig. 4.3.1.[72]

The gradients measured in practice, using limited rather than perfect averaging in eqn. 4.3.2 are actually random time functions whose mean values are the expectation of the gradients. The randomness is caused by short term cross products among the different sources contributing to **G** and the injection of this spurious noise into the noise-free algorithms described above creates jitter of the coefficients, a topic discussed in some depth in Section 4.11 and one peculiar to all adaptive algorithms.

Implementation: Implementation of the steepest descent algorithm eqn. (4.2.13) in the hardware context is done simply by setting

$$\frac{d\mathbf{W}(t)}{dt} = -\mu\mathbf{G}(t) = -2\mu\mathbf{X}(t)e(t) = -2\mu\mathbf{X}(t)\{y(t) - d(t)\} \tag{4.3.3}$$

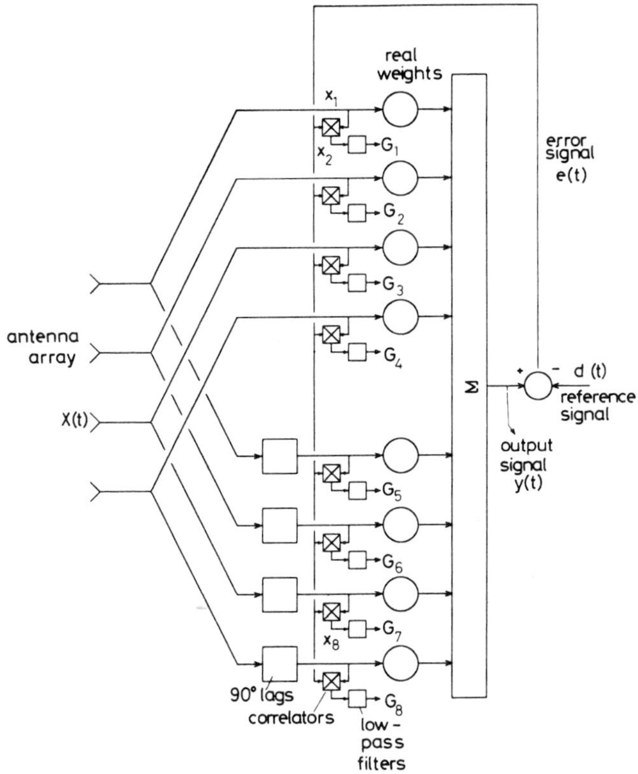

Fig. 4.3.1 *Measurement of gradient vector for real signals*

and solving with a set of integrators:

$$\mathbf{W}(t) = \mathbf{W}(0) - \mu \int_0^t \mathbf{G}(t)\, dt = \mathbf{W}(0) - 2\mu \int_0^t \mathbf{X}(t)e(t)\, dt \tag{4.3.4}$$

One of these is shown in Fig. 4.3.2 with its interconnections, the circuit being very similar to that shown in Fig. 4.6.2. Note that the low-pass filters have been dispensed with, the integrators alone providing sufficient smoothing.[31] The value of μ is chosen to fix the convergence time constant at a suitable value; this is discussed in Section 4.11.

If the expectation or ensemble average of the convergence equation is considered we obtain

$$\begin{aligned} E[\mathbf{X}(t)\, d(t)] &= \mathbf{D} \\ E[\mathbf{X}(t)y(t)] &= E[\mathbf{X}(t)\mathbf{X}(t)'\mathbf{W}(t)] = \mathbf{RW}(t) \end{aligned} \tag{4.3.5}$$

where **D** is the space vector corresponding to the desired signal and **R** is the covariance matrix of the antenna element data. The algorithm in ensemble

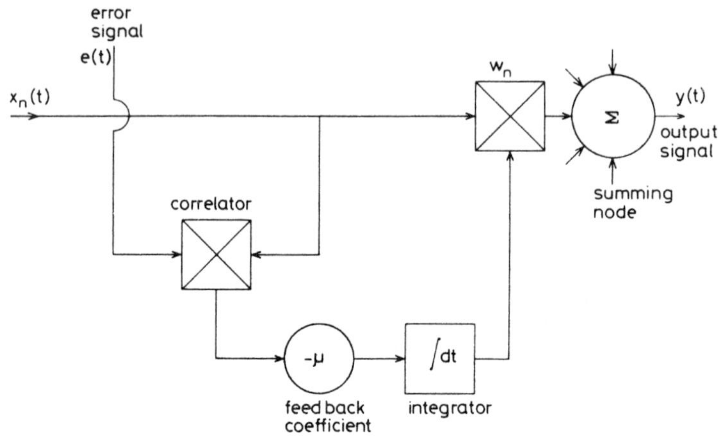

Fig. 4.3.2 *Position of an intergrator in the analogue LMS algorithm*

average form is then

$$\frac{d\mathbf{W}(t)}{dt} = -2\mu\{\mathbf{R}\mathbf{W}(t) - \mathbf{D}\} \tag{4.3.6}$$

An analysis of the convergence properties of this equation is presented in Section 4.9.

Effects of leaky integrators: Integrators with a loss component are often used in the solution of the LMS algorithm. The effect of this is equivalent to injection of local uncorrelated noise into the antenna as far as the optimal solution is concerned but, of course, there is no actual noise produced at the output. A leaky integrator is shown in Fig. 4.3.3, and its time domain transfer function is

$$f(t) = \frac{1}{\beta}\left[g(t) + T\frac{dg(t)}{dt}\right] \tag{4.3.7}$$

where $\beta = R_2/R_1$ and $T = CR_2$. Replacing all integrators by leaky ones modifies the esemble average differential equation to

$$\frac{1}{\beta}\left[\mathbf{W}(t) + T\frac{d\mathbf{W}(t)}{dt}\right] = -2\mu[\mathbf{R}\mathbf{W}(t) - \mathbf{D}] \tag{4.3.8}$$

or

$$\frac{T}{\beta}\frac{d\mathbf{W}(t)}{dt} = 2\mu\left\{\mathbf{D} - \left[\mathbf{R} + \frac{1}{2\beta\mu}\mathbf{I}\right]\mathbf{W}(t)\right\} \tag{4.3.9}$$

which has an equilibrium state, found by setting $d\mathbf{W}(t)/dt = 0$, of

$$\left\{\mathbf{R} + \frac{1}{2\beta\mu}\mathbf{I}\right\}\mathbf{W}_{\text{opt}} = \mathbf{D} \tag{4.3.10}$$

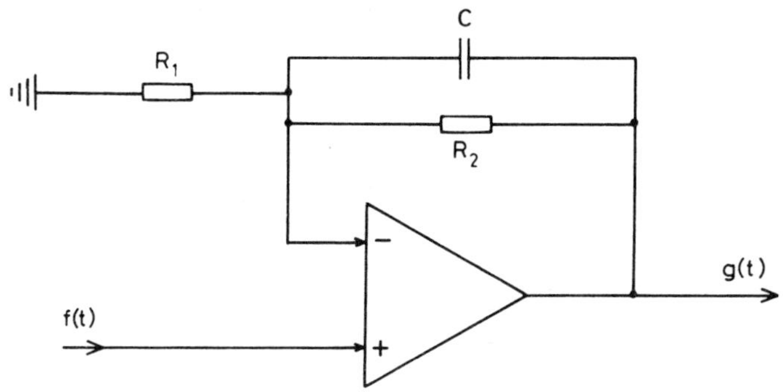

Fig. 4.3.3 *Leaky integrator*

Thus the effect of lossy integrators is equivalent to augmenting the diagonal of the covariance matrix which in turn would be the effect of injecting uncorrelated into the elements of the array.

The effects of this additional term can be predicted from the analysis in Chapter 5: as far as the optimal solution goes, these are quite small. The main effect of the synthetic noise is to speed relaxation of the system when interference sources are switched off, the weights having an in-built bias to sink to zero. In the presence of main lobe jamming the displacement of the weighting vector is reduced though this effect is of most value in the constrained algorithm where main lobe distortion can be reduced. This is discussed in Chapter 6.

Operational features of the LMS algorithm: The LMS algorithm has some interesting properties which may not be anticipated. The first such property is that if the desired signal level falls to zero at the inputs then the weights also fall to zero. This is because the equilibrium state of the system from eqn. (4.3.6) is

$$\mathbf{R}\mathbf{W} = \mathbf{D} \tag{4.3.11}$$

but, from eqn. (4.3.5), $\mathbf{D} = E[\mathbf{X}(t)\, d(t)]$ and is proportional to the amplitude of the desired signal if the amplitude of $d(t)$ is fixed. In the absence of a desired signal $\mathbf{D}$ vanishes and so too, in eqn. (4.3.11), does $\mathbf{W}$ if $\mathbf{R}$ is positive definite. This has important implications if the desired signal is pulsed (as in a radar) or arrives in bursts, for example if the antenna is receiving several signals in a communications net which transmit in turn. In such circumstances, the antenna turns off between signal transmissions and there may be a sluggish turn-on transient when the signal first appears. This property is explored more fully in Section 4.9. The second undesirable property of the antenna is that the weights also turn off in the absence of a reference signal $d(t)$ since the vector $\mathbf{D}$ again goes to zero in eqn. (4.3.7). This means that if the reference signal is obtained from a receiver which takes its input from the adaptive antenna output then

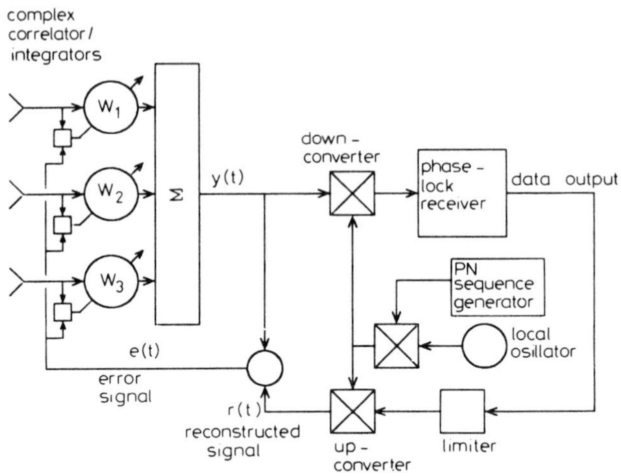

Fig. 4.3.4 *Generation of reference signal by reconstruction of signal*

after a period without a desired signal not only are the weights zero but when the desired signal does appear the antenna blocks its reference component from reaching the pilot receiver (see Fig. 4.3.4). In these circumstances, the system relies upon weight jitter to provide enough signal output to enable pilot receiver lock.

The effect of phase shift in regeneration of the pilot signal from the adaptive antenna output is to set up a steady cycling of the coefficients, a phenomenon discussed by Dicarlo,[44] as the antenna attempts to reduce the phase error in its output. Another problem occurs if the correlators have DC offsets at their outputs, likely in analogue implementations. The effect of this is to bias the measured gradient vector **G** and to shift the optimal solution. If the vector of output offsets is **E**, the equilibrium state of the algorithm is

$$\mathbf{RW} - \mathbf{D} - \mathbf{E} = 0 \tag{4.3.12}$$

instead of eqn. (4.3.11). The effect is related to the steering errors associated with the steered systems discussed in Chapter 6 and similar effects might be observed at high input desired SNRs. As the correlator performance is under the designer's control such problems are avoidable.[36]

Ensemble averaging : Care is needed in defining such terms as 'average' and 'expectation' in connection with adaptive antennas, since many variables have means and variances which change markedly during convergence. For example, the mean square error is likely to decrease by orders of magnitude between initialisation and approach to optimality. Some average measure of these variables is required if any meaningful analysis is to be done and as the time average is unsuitable the ensemble average is a more useful concept. To use this, the existence of an ensemble of identical adaptive arrays is postulated, each one operating on a different realisation of any random variables. It is as though a large number of computer simulations were run under identical conditions with the exception that the seed of the random number generator is set to a different value for each run. A variable which is averaged over all possible random variable realisations at a fixed time after initialisation will be said to take on its ensemble average or expectation. Should the weight vector in the array be fixed then expectation and time averages will have the same value.

Generation of reference signal: Compton[38] has described a technique for reconstruction of a reference signal from the received data itself. A block diagram of the technique is shown in Fig. 4.3.4.

The success of the procedure appears to rely upon the fact that if a binary signal with added variable amplitude Gaussian noise is passed through a hard limiter the mean-aquare difference between the output signal and the input signal (suitably scaled) is minimised if the amplitude of the noise is minimised since a limiter passes a binary signal without distortion. When the spread spectrum signal is bandwidth-compressed all interference is converted to a quasi-Gaussian form, but the desired signal retains its binary character. If this noisy binary signal is hard-limited and used as a reference signal, the LMS criterion corresponds to a maximum ratio of binary signal power to Gaussian noise power and the antenna converges to best SNIR in the usual way. Clearly the bandwidth-compression is an essential feature of the technique since, without this, a binary modulated jamming signal would remain binary at the demodulator output and could capture the limiter. The antenna would then maximise the jammer to signal ratio.

4.4 Steered gradient systems[50, 51]

In Section 3.4 it was noted that provision of a reference waveform was unnecessary if the space vector of the desired signal $\mathbf{S}_w$ were known *a priori*. The creation of an algorithm making use of such a steering vector can conveniently commence with eqn. (4.3.6), which is an ensemble-averaged equation making use of a vector $\mathbf{D}$ which is the average of the correlation between the $\mathbf{X}(t)$ vector at the antenna elements and the reference signal $d(t)$. A half-way stage between this equation and the stochastic eqn. (4.3.4) is the algorithm

$$\mathbf{W}(t) = \mathbf{W}(0) - 2\mu \int_0^t [\mathbf{X}(t)y(t) - \mathbf{D}]\, dt \tag{4.4.1}$$

in which the $\mathbf{D}$ vector is replaced by its ensemble average while the rest goes stochastic. It is fairly obvious that this algorithm will converge to the optimal solution since convergence ceases when

$$\mathbf{D} = \overline{\mathbf{X}(t)y(t)} = \overline{\mathbf{X}(t)\mathbf{X}(t)'\mathbf{W}} = \mathbf{RW} \tag{4.4.2}$$

and the solution is therefore $\mathbf{W} = \mathbf{R}^{-1}\mathbf{D}$. The vector $\mathbf{D}$ is calculated directly from the relative time delays of the desired transmission measured at the receiving elements and this presupposes a knowledge of the direction of the desired source. The correct magnitude of the vector $\mathbf{D}$ is not known but since scaling $\mathbf{D}$ merely scales $\mathbf{W}_{\text{opt}}$ this has no effect on the final SNR and, in fact, offers an opportunity to choose the magnitude of the coefficients. A block diagram to implement the algorithm is shown in Fig. 4.4.1, the elements of the $\mathbf{D}$ vector

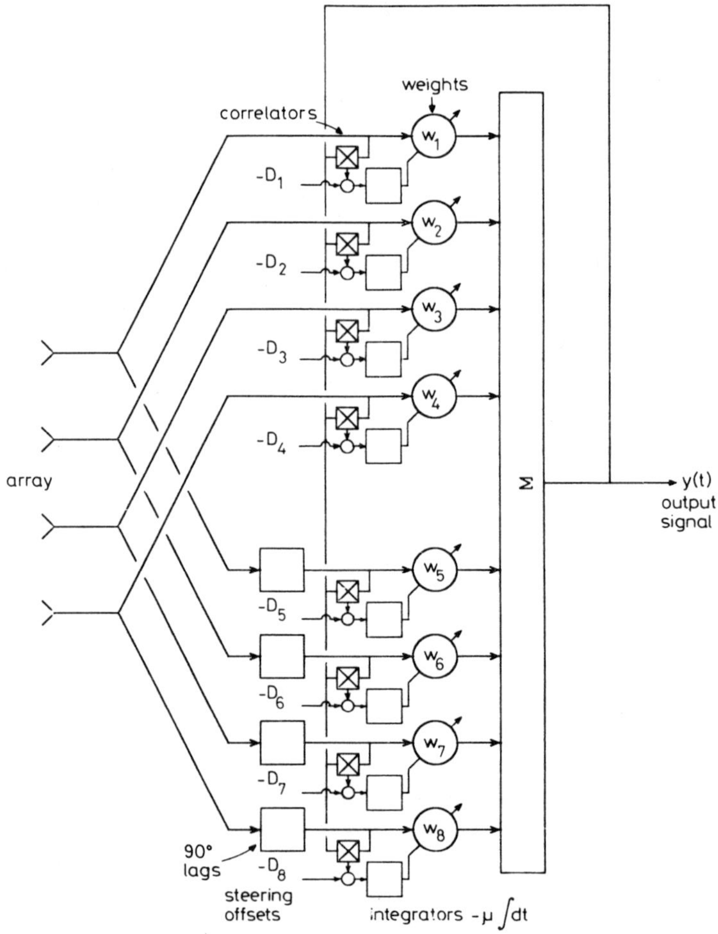

Fig. 4.4.1 *Implementation of steered system*

appear as offsets in the integrator inputs and could be set by potentiometers or via D/A converters from a small computer.

The operational characteristics of this algorithm have some interesting differences from the true reference system.

(*a*) In the absence of a desired signal at the antenna the weights no longer fall to zero but to $\mathbf{W} = \mathbf{M}^{-1}\mathbf{D}$ where $\mathbf{M}$ is the interference-only covariance matrix. Thus the array has finite gain for a signal which suddenly appears and in fact is initially set up with an optimal pattern. This is a very desirable characteristic for a radar receiver where the signals (echo-pulses) are infrequent. In the radar case the direction of the desired signal is known too and **D** is easily found.

(*b*) A sensitivity to errors in the assumed direction of the desired signal becomes very apparent. If such a pointing error exceeds a small fraction of a beamwidth the algorithm will treat the desired signal as an interference and form a main-lobe notch in an attempt to cancel it. This effect is discussed in detail in Section 6.2 where it is concluded that steered systems may be unworkable if the desired-signal to uncorrelated noise ratio exceeds about 10 dB at the element level and this effect tends to restrict the operation of the steered system to application such as radar or communication systems where the signal is almost buried in noise.

There is no significant difference in the convergence properties of this steered system in comparison with the reference signal LMS antenna. The technique, referred to as the 'P-vector' algorithm by Griffiths,[50, 51] is closely related to another steered system called the Howells-Applebaum loop.[48, 56] The transformation from one to the other assumes that the integrators are 'leaky' and have a known DC gain β as in eqn. (4.3.7), and if the **D** vector values are scaled by $-\beta$ in Fig. 4.4.1, the offsets can be injected at the integrator outputs instead of the inputs with no overall change in system functioning. We now have the Howells-Applebaum loop.

The Howells-Applebaum loop has the practical advantage that, in the absence of any supranoise signals, the weights are equal to the **D** vector elements since the integrator outputs vanish. Moreover the gains $(1/T)$ and the leakages $(1/\beta)$ of the integrators can be varied *in situ* without directly affecting the weighting coefficient values thus allowing trimming of adaptation characteristics. In the case of the **P**-vector algorithm the gains could be varied but changing the leakage would scale the quiescent weight coefficient values. However, the Howells-Applebaum loop only approaches the true optimality condition in the limiting case as the leakage vanishes and of course, in this limit, offsets at the integrator outputs are merely constants of integration and have little permanent controlling influence. Therefore, integrator leakage is an inescapable feature of the algorithm and so too is suboptimality though this is not of any practical significance.

If the desired signal is weak and the interferences are all in the sidelobes the weighting coefficients will vary only slightly from their quiescent values and this is of great benefit in hardware oriented systems where dynamic ranges are very limited. The combination of this feature and the flexibility mentioned above explains the popularity of the Howells-Applebaum loop.

4.5 Constrained power minimisation[47]

In Section 3.4, it was found that if a hard linear equality constraint were applied to the weighting coefficients to keep the antenna amplitude gain for the desired signal fixed then an optimal pattern would form if either the interference output power or the interference plus signal output power were minimised. This is a steered system rather like the one in Section 4.4, but there the constraint was a floating one which biased the weight vector but did not attempt to maintain a particular amplitude gain for the desired signal. An algorithm for minimising the output power with a fixed desired-signal amplitude gain is now developed.

The procedure is actually a fairly simple extension of the unconstrained descent mechanism developed in Section 4.2 and by way of illustration the same cost function will be used:

$$p = 4W_1^2 + 4W_2^2 + 4W_1 W_2 \tag{4.5.1}$$

Note that the bias vector **B** is dropped and the centre of the cost function ellipses is now the co-ordinate origin as in Fig. 4.5.1. The linear constraint used

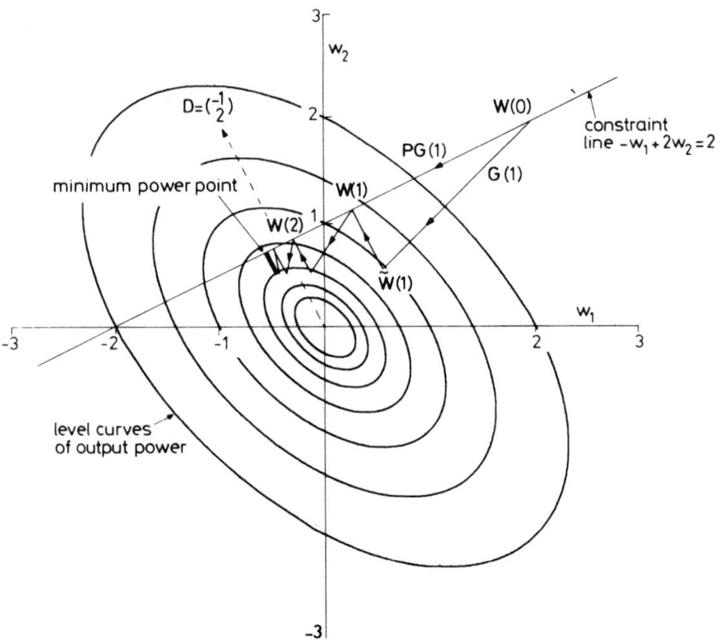

Fig. 4.5.1 *Linearly constrained gradient descent*

is

$$-W_1 + 2W_2 = 2 \tag{4.5.2}$$

which intercepts the W_1 and W_2 axes at -2 and 1. The weight vector is initialised on the constraint line at $\mathbf{W}(0) = (2, 2)^t$.

The first step in the algorithm is an unconstrained steepest descent step. Differentiation of eqn. (4.5.1) gives the gradient vector components as

$$\begin{aligned} \partial J/\partial W_1 &= 8W_1 + 4W_2 \\ \partial J/\partial W_2 &= 4W_1 + 8W_2 \end{aligned} \tag{4.5.3}$$

and using a feedback coefficient μ of 0·06 the descent step is

$$\begin{aligned} \tilde{W}_1(1) &= W_1(0) - 0{\cdot}06\ \partial J/\partial W_1(0) = 2 - 1{\cdot}44 = 0{\cdot}56 \\ \tilde{W}_2(1) &= W_2(0) - 0{\cdot}06\ \partial J/\partial W_2(0) = 2 - 1{\cdot}44 = 0{\cdot}56 \end{aligned} \tag{4.5.4}$$

This point is indicated in the diagram. However, it how lies off the constraint line since $-\tilde{W}_1 + 2\tilde{W}_2 = 0{\cdot}56$. The second phase of the algorithm consists of moving the weight vector back to the constraint line in a perpendicular direction. This perpendicular is parallel to the vector $(-1, 2)^t$ and we correct the coefficients by adding this suitably scaled:

$$\begin{aligned} W_1(1) &= \tilde{W}_1(1) - \alpha \\ W_2(2) &= \tilde{W}_2(2) + 2\alpha \end{aligned} \tag{4.5.5}$$

The value of α is found by substituting eqn. (4.5.5) into (4.5.2)

$$-[\tilde{W}_1(1) - \alpha] + 2[\tilde{W}_2(1) + 2\alpha] = 2 \tag{4.5.6}$$

which has the solution $\alpha = 0{\cdot}29$. Thus the new weight vector $\mathbf{W}(1)$ is, as shown, $(0{\cdot}27,\ 1{\cdot}14)^t$. Repeated application of eqns. (4.5.4) and (4.5.5) causes the weight vector to migrate along the constraint line to the minimum power point, the sequence of values being tabulated in Table 4.5.2.

If vector notation is adopted for multivariable situations, the algorithm may be summarised as

$$\tilde{\mathbf{W}}(k+1) = \mathbf{W}(k) - \mu\mathbf{G}(k) = \mathbf{W}(k) - 2\mu\mathbf{A}\mathbf{W}(k) \tag{4.5.7a}$$

$$\mathbf{W}(k+1) = \tilde{\mathbf{W}}(k+1) + \left[\frac{g - \mathbf{D}^t\tilde{\mathbf{W}}(k+1)}{\mathbf{D}^t\mathbf{D}}\right]\mathbf{D} \tag{4.5.7b}$$

where $\mathbf{D}$ is the vector of coefficients of the constraint surface equation, e.g. $\mathbf{D} = (-1, 2)^t$ above, and g is the constant in the constraint eqn. (4.5.2). In general, the $\mathbf{D}$ vector will be the vector of coefficients required as a weight vector to steer a beam toward the desired signal source and g will be the antenna amplitude gain for the desired signal, i.e. the amplitude in the main lobe when vector $\mathbf{D}$ is used as a weighting vector.

Table 4.5.2 *Convergence of the discrete time linearly constrained gradient algorithm*

W_1	W_2	G_1	G_2	J
2·000000	2·000000	24·000000	24·000000	48·0000
0·272000	1·136000	6·719999	10·175990	6·6938
−0·294784	0·852608	1·052159	5·641727	2·2500
−0·480689	0·759655	−0·806892	4·154487	1·7719
−0·541666	0·729167	−1·416660	3·666671	1·7204
−0·561666	0·719167	−1·616664	3·506668	1·7149
−0·568227	0·715887	−1·682265	3·454187	1·7143
−0·570378	0·714811	−1·703783	3·436973	1·7142
−0·571084	0·714458	−1·710841	3·431327	1·7142
−0·571316	0·714342	−1·713156	3·429475	1·7142
−0·571392	0·714304	−1·713914	3·428868	1·7142
−0·571416	0·714292	−1·714163	3·428669	1·7142
−0·571425	0·714288	−1·714246	3·428602	1·7142
−0·571427	0·714287	−1·714272	3·428582	1·7142
−0·571428	0·714286	−1·714280	3·428575	1·7142
−0·571428	0·714286	−1·714284	3·428573	1·7142
−0·571428	0·714286	−1·714284	3·428572	1·7142
−0·571429	0·714286	−1·714285	3·428571	1·7142
−0·571429	0·714286	−1·714285	3·428571	1·7142
−0·571429	0·714286	−1·714285	3·428571	1·7142

A mathematical simplification of the pair of eqns. (4.5.7) is possible if the first is substituted into the second to yield

$$\mathbf{W}(k+1) = \mathbf{W}(k) - \mu\mathbf{G}(k) + \frac{g - \mathbf{D}^t[\mathbf{W}(k) - \mu\mathbf{G}(k)]}{\mathbf{D}^t\mathbf{D}}\mathbf{D} \tag{4.5.8}$$

then, noting that $\mathbf{D}^t\mathbf{W}(k) = g$ because of the previous iteration, we have

$$\mathbf{W}(k+1) = \mathbf{W}(k) - \mu\mathbf{G}(k) + \frac{\mu\mathbf{D}\mathbf{D}^t\mathbf{G}(k)}{\mathbf{D}^t\mathbf{D}} \tag{4.5.9}$$

$$= \mathbf{W}(k) - \mu\mathbf{P}\mathbf{G}(k) \tag{4.5.10}$$

where

$$\mathbf{P} = \mathbf{I} - \frac{\mathbf{D}\mathbf{D}^t}{\mathbf{D}^t\mathbf{D}} \tag{4.5.11}$$

Equation (4.5.10) is a one step algorithm for updating the weights. The physical significance of the projection is evident in Fig. 4.5.1 where the true gradient is projected into the constraint line (or surface) to eliminate its component in the direction perpendicular to the constraint line. The distinction between the two step algorithms (4.5.7) and the modification (4.5.10) is discussed by Frost[47] who shows that the one step system is subject to accumulation of round off error. A practical point is that eqn. (4.5.7b) can be applied relatively infrequently to reduce computation.

Equation (4.5.10) can be used to find the equilibrium point of the algorithm, which is where $\mu\mathbf{PG} = 0$, and the weight vector remains unchanged over the iteration. Here we have

$$\left(\mathbf{I} - \frac{\mathbf{D}\mathbf{D}^t}{\mathbf{D}^t\mathbf{D}}\right)\mathbf{G} = \underset{\sim}{0} \tag{4.5.12}$$

where $\underset{\sim}{0}$ is a column of zeroes and consequently

$$\mathbf{G} = \mathbf{D}\,\frac{\mathbf{D}^t\mathbf{G}}{\mathbf{D}^t\mathbf{D}} = \lambda\mathbf{D} \tag{4.5.13}$$

Since the gradient $\mathbf{G}$ is equal to $2\mathbf{AW}$ the terminating point is where

$$\mathbf{AW} = \lambda\mathbf{D} \tag{4.5.14}$$

which is an agreement with eqn. (3.4.6) showing that the iterative example converges to the optimal solution.

In the numerical algorithm the theoretical optimum point is from eqn. (3.4.9):

$$\mathbf{W}_{\text{opt}} = \frac{2\mathbf{A}^{-1}\mathbf{D}}{\mathbf{D}^t\mathbf{A}^{-1}\mathbf{D}} = \begin{pmatrix} -4/7 \\ 5/7 \end{pmatrix} \tag{4.5.15}$$

using the numerical values from eqns. (4.5.1) and (4.5.2). The solution is the same as the LMS one and is indicated in the figure.

Unlike the LMS algorithm of Section 4.2, the linearly constrained algorithm does not converge to a point where the gradient is zero but to a point where the gradient is normal to the constraint surface. The residual finite gradient has the effect of causing a limit cycle in the two-step algorithm (4.5.7). This is no drawback because the weight vector would not be used for signal filtering mid-way through the algorithm when it deviates from the constraint.

4.6 Hardware implementations of a linearly constrained algorithm

Fig. 4.6.1 shows a digital processing unit which samples the gradient outputs from the analogue antenna correlators in Fig. 4.3.1, the reference signal being set to zero for this application. The ith correlator output is

$$\hat{G}_i(t) = \frac{2}{T}\int_{t-T}^{t} x_i(t)y(t)\,dt \tag{4.6.1}$$

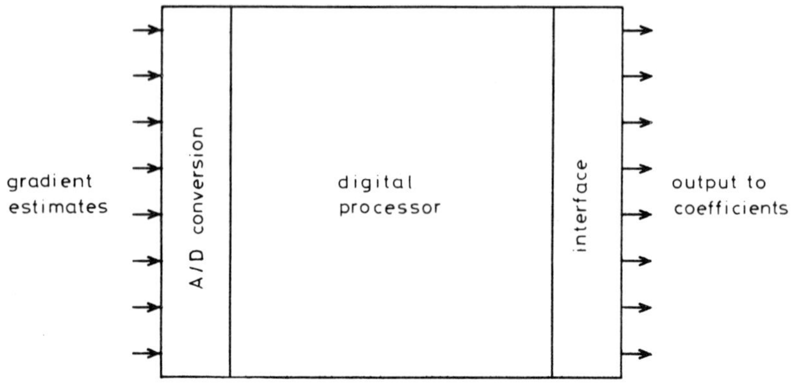

Fig. 4.6.1 *Digital processor for implementation of a gradient descent algorithm*

(treating the LP filter as an integrator) and the vector of sample gradients is

$$\hat{G}(t) = \frac{2}{T}\int_{t-T}^{t} \mathbf{X}(t)y(t)\,dt \tag{4.6.2}$$

which is an approximation to the expectation of the gradient

$$\mathbf{G} = 2\overline{\mathbf{X}(t)y(t)} = 2\overline{\mathbf{X}(t)\mathbf{X}(t)^{t}}\mathbf{W} = 2\mathbf{R}\mathbf{W} \tag{4.6.3}$$

The digital algorithm is just eqn. (4.5.7) or (4.5.10) with an appropriate steering vector **D** derived from the space vector of the desired signal source and with a g value normally close to N the number of antenna elements. The sequence of updated weighting coefficients is transmitted to the variable weighting coefficients either in analogue form after D/A conversion and power amplification for PIN diode control or as straight binary signals if the coefficient control is digital.

The statistical errors due to finite averaging eventually pass to the weighting coefficients where they set up a small amount of coefficient noise, an effect which is discussed later.

This particular hybrid approach combines some good features of the analogue and digital worlds; we have virtually unlimited bandwidth in data throughput and correlation bandwidths and the digital processor can incorporate some logic for fault detection and implementation of complicated constraint systems (for example, Chapter 6). The rate at which the digital processor can update the coefficients is probably much less than 10^6 per second but this may be adequate for many interference situations. Control problems of the weighting coefficients are not too severe since moderate nonlinearity is accommodated by the algorithm without performance degradation so long as the resolution of the coefficients enables any desired set of coefficient weighting values to be set up (this may require 8 or more bits of resolution) and look-up tables can be set up to linearise the control laws of, for example, PIN diodes. The most difficult part of the system is the phase coherent amplification of the element signals to bring them up to a level where analogue correlators can be used.

The convergence properties of the algorithm are discussed in detail in Sections 4.8 and 7.13.

Continuous time version: The continuous time, or analogue version of eqn. 4.5.10 is

$$d\mathbf{W}/dt = -\mu \mathbf{P}\mathbf{G}(t) \tag{4.6.4}$$

which is, however, not easy to implement directly. The best method for handling the projection is to transform the antenna data as discussed in Section 7.11. Fig. 4.6.2 shows such a system in which a 4-port orthogonal beamformer matrix is used to transform the data. The desired signal is assumed to be present in beam 4 only and the look direction constraint reduces to clamping the weighting coefficient in this beam at unity. The quadrature part has been set to zero. The projection $\mathbf{P}$ is transformed into a diagonal projection matrix: $\mathbf{P} \rightarrow \text{diag}(1, 1, 1, 0)$. If it is required to shift the look direction this is most easily done by using the phase shifters shown in Fig. 7.11.2 as the look direction must be specified more

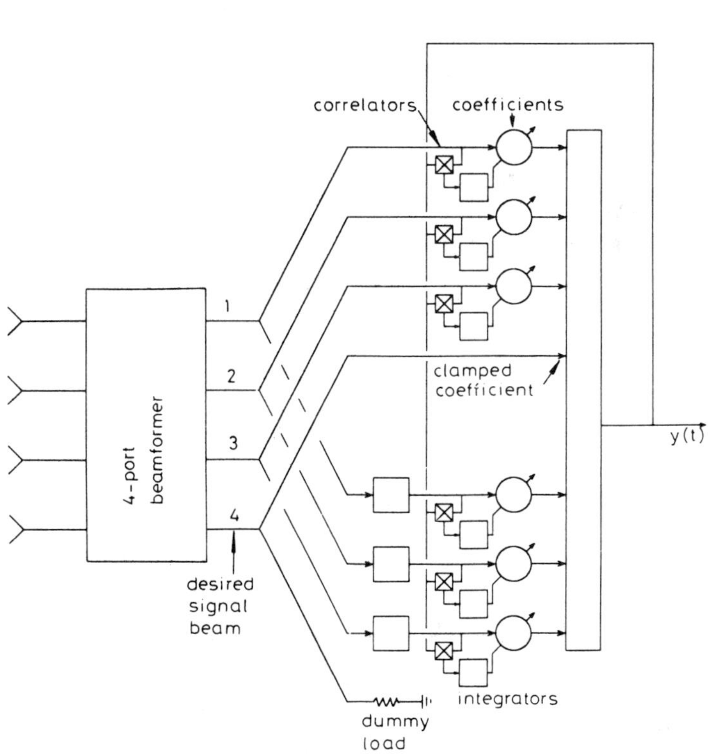

Fig. 4.6.2 *Complete analogue linearly constrained gradient descent algorithm*

accurately than the interbeam separation. Applebaum[22] and Bienvenue[93] have described similar preprocessing.

Simulation of uncorrelated noise : The discrete time algorithm can be operated with 'lossy' discrete integrators just as the analogue system and the end effect is the same. In eqn. (4.5.7), the algorithm is modified to

$$\tilde{\mathbf{W}}(k+1) = \alpha\mathbf{W}(k) - \mu\mathbf{G}(k) \tag{4.6.5}$$

where α is a constant between 0 and 1. To see how this affects the algorithm, it is rewritten in the form

$$\tilde{\mathbf{W}}(k+1) = \mathbf{W}(k) - (1-\alpha)\mathbf{W}(k) - \mu\mathbf{G}(k) \tag{4.6.6}$$

$$= \mathbf{W}(k) - [(1-\alpha)\mathbf{I} + 2\mu\mathbf{R}]\mathbf{W}(k) \tag{4.6.7}$$

and we see that the effective covariance matrix is changed to

$$\tilde{\mathbf{R}} = \frac{(1-\alpha)}{2\mu}\mathbf{I} + \mathbf{R} \tag{4.6.8}$$

The effect of this synthetic noise can be beneficial in the control of main lobe distortion, discussed in Chapter 6, and other effects described in Section 5.2.

Complex signals : If correlation is done at baseband, complex multipliers are used to get the gradients. We have

$$\mathbf{G} = 2\mathbf{R}\mathbf{W} = 2\overline{\mathbf{X}(t)^*\mathbf{X}(t)^t\mathbf{W}} \tag{4.6.9}$$

and the descent algorithm replacing eqn. (4.5.7) is simply

$$\tilde{\mathbf{W}}(k+1) = \mathbf{W}(k) - \mu\mathbf{G}(k) = \mathbf{W}(k) - 2\mu\mathbf{R}\mathbf{W}(k)$$

$$\mathbf{W}(k+1) = \tilde{\mathbf{W}}(k+1) + \frac{g - \mathbf{S}^t\tilde{\mathbf{W}}(k+1)}{\mathbf{S}^t\mathbf{S}^*}\mathbf{S}^* \tag{4.6.10}$$

or

$$\mathbf{W}(k+1) = \mathbf{W}(k) - 2\mu\left(\mathbf{I} - \frac{\mathbf{S}^*\mathbf{S}^t}{\mathbf{S}^t\mathbf{S}^*}\right)\mathbf{R}\mathbf{W}(k) = (\mathbf{I} - 2\mu\mathbf{P}\mathbf{R})\mathbf{W}(k) \tag{4.6.11}$$

where **S** is the steering vector, a complex version of **D**.

Signal limiting : It can be shown that the vector of correlations (4.3.5b)

$$\phi = E[X(t)y(t)] \tag{4.6.12}$$

is unchanged in direction and merely scaled if either $y(t)$ or the elements of $\mathbf{X}(t)$ are hard-limited, provided the signals are Gaussian.[48, 63, 26] If the elements of $\mathbf{X}(t)$ are limited, it is necessary to assume the elements have, initially, uniform variances. In the multiple correlator gradient-measurement system it is common to limit the inputs to the correlators on the element sides corresponding to $\mathbf{X}(t)$.

This has the beneficial effect of reducing the required dynamic range of the correlators; always a difficulty as the dynamic range at their outputs tends to be twice that of the inputs. The scheme is popular in radar sidelobe cancellers where the feedback constant is fixed and the dynamic range of convergence rates is reduced, the effect being similar to using AGC amplifiers in the input channels. The slow convergence associated with disparate eigenvalues in the covariance matrix is not improved however.

In certain gradient systems, the equilibrium correlation between the error signal and the element signal is zero. This is not true of the straightforward linearly constrained algorithm, but is true for the LMS algorithm and for the linearly constrained algorithm if preceded by a preprocessor which filters out the desired signal as discussed in Section 7.11. Under these circumstances it can be shown that the equilibrium state of the antenna is not changed if both sides of the correlator inputs are limited, nor is it necessary to assume Gaussian signals. The convergence behaviour is not much improved by this action which tends to give a constant coefficient displacement norm at each step; however, as the correlators are reduced to exclusive OR gates a very economical design could result.

4.7 Perturbation methods for gradient determination[34, 85, 76, 86, 87]

The gradient of the mean square error in the LMS algorithm or output power in a linearly constrained system is a vector of partial derivatives and can be measured directly by perturbing the weighting coefficients one at a time and measuring the differential changes in power. This is a fundamental modification in technique since the use of correlators to determine gradients has been adopted previously and is motivated by their disadvantages. These include cost, phase and offset errors[36] and dynamic range difficulties.

In the real variable case the elements of the gradient vector **G** are given by

$$G_n = \partial p/\partial W_n \qquad n = 1, \ldots, 2N \tag{4.7.1}$$

and a simple approach to their measurement is to perturb them one at a time by an increment δ_n

$$\hat{G}_n = \frac{p(\mathbf{W} + \delta_n) - p(\mathbf{W})}{\delta_n} \tag{4.7.2}$$

This measurement technique may not be very accurate because the output power is fluctuating and measuring a small difference will produce a noisy result. If the size of the perturbations is made large to overcome this problem the differential coefficient becomes a function of the size itself, i.e. the measurement is biased.

In the most general case [eqn. (4.2.1)] the variation of p with a single coefficient has the form

$$p(W_n) = aW_n^2 + 2bW_n + c \tag{4.7.3}$$

and it can be seen that $\partial p/\partial W_n$ is a function of W_n. The gradient estimate reported by eqn. (4.7.2) is

$$\hat{G}_n = [a(W_n + \delta_n)^2 + 2b(W_n + \delta_n) - aW_n^2 - 2bW_n]/\delta_n \tag{4.7.4}$$

$$= a(2W_n + \delta_n) + 2b \tag{4.7.5}$$

which is a function of δ_n. The δ_n dependence can be cancelled if W_n is perturbed by equal and opposite increments:

$$\hat{G}_n = \frac{p(W_n + \delta_n) - p(W_n - \delta_n)}{2\delta_n} \tag{4.7.6}$$

and substitution of eqn. (4.7.3) into this shows that

$$\hat{G}_n = 2aW_n + 2b \tag{4.7.7}$$

which is equal to the exact differential coefficient in eqn. (4.7.3).

Fig. 4.7.1 shows the parabola corresponding to eqn. (4.7.3). Estimate (4.7.6) above corresponds to finding the gradient of the chord joining the output powers at $W_n - \delta_n$ and $W_n + \delta_n$ which is equal to the gradient of the tangent at the mid-point, W_n.

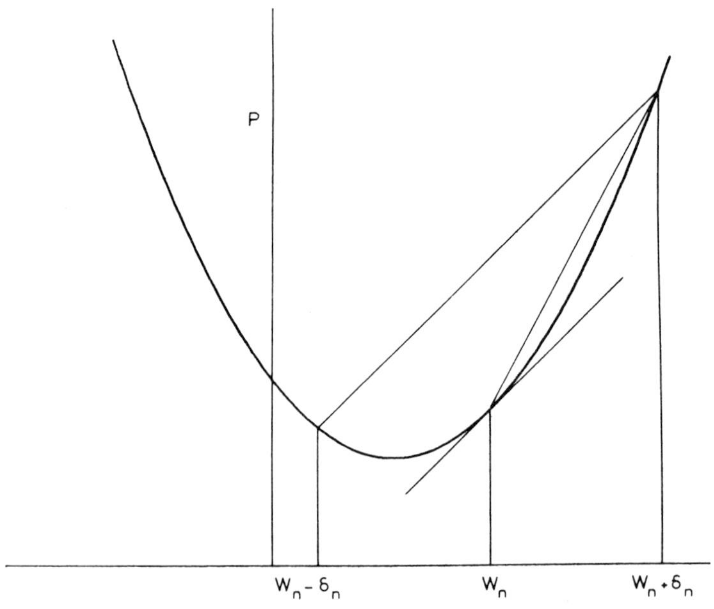

Fig. 4.7.1 *Measurement of gradient by coefficient perturbation*

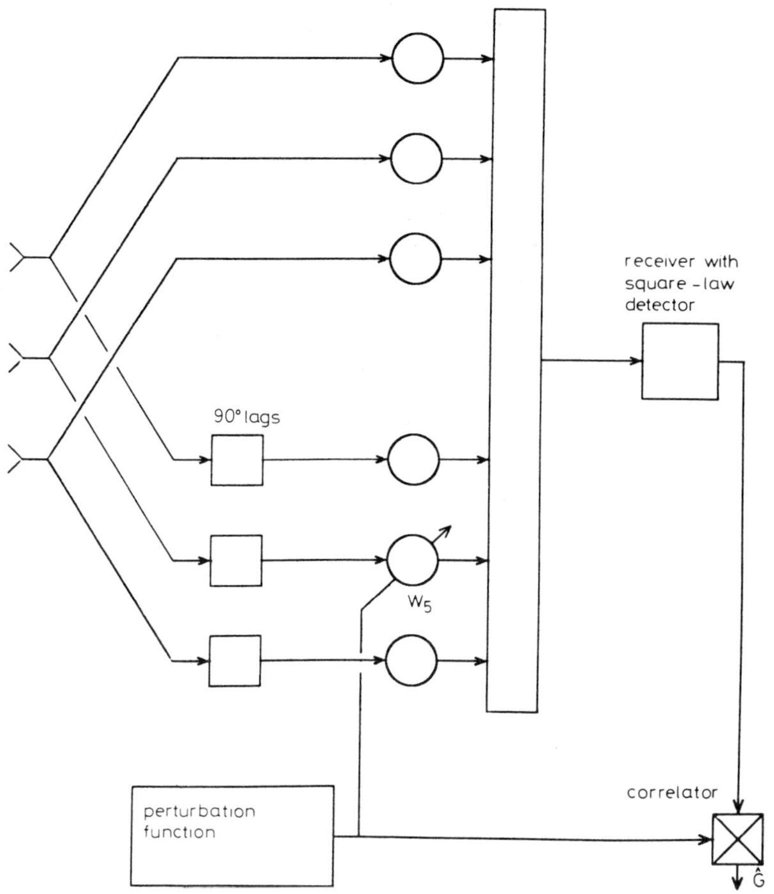

Fig. 4.7.2 *Single receiver perturbation gradient measurement system*

The perturbations can now be made as large as necessary and analysis of the sampling error statistics, eqn. (4.11.24), indicates that the coefficient should be perturbed by a large amount, on the order of $N^{1/2}$ times the value of the coefficient itself, to minimise the effects of measurement noise.

The complete gradient vector is measured by multiplexing the perturbation around the set of $2N$ real coefficients (or $2N$ real and imaginary parts in the complex case). If the rate at which the gradient can be used, for example in a part digital implementation, is limited, it is feasible to perturb the coefficient by a long sequence of steps or a square wave and correlate this with the receiver output power as indicated in Fig. 4.7.2. A low-pass filter will then offer a reduction in the variance of the sampled gradient components. The reduction in variance is given roughly by the ratio of the receiver bandwidth to the filter bandwidth. Another technique is to perturb all of the coefficients simultaneously by orthogonal functions such as the rows of a Walsh function matrix and correlate these functions with the receiver power output in $2N$ correlators.[34]

The estimated gradient can be used in any of the gradient descent algorithms described previously.

Operating disadvantages of single receiver system: There are two main drawbacks to the technique above. The first is that the measurement cycle takes longer than if a set of parallel correlators is available and this reduces the real time speed of convergence of the adaptive algorithm. The second problem is that one of the weights is always perturbed from its nominal setting and the directional pattern is also disturbed allowing interference to get through to the receiver even after a good pattern has developed (it is this interference which allows the detection of the gradient). Thus the output signal to interference ratio is suboptimal during gradient measurement and we must either stop adapting to listen to the desired signal, tolerate a reduced performance, or slave the weights onto a second receiver which is devoted purely to data throughput.

Two-receiver system: The statistical properties of the gradient estimate can be improved by using two receivers to measure the powers $p(W + \delta^n)$ and $p(W - \delta^n)$ simultaneously. Rather than using two complete sets of weighting coefficients one can be fixed at its nominal setting and an RF switch taps into the channel voltages as shown in Fig. 4.7.3, and by suitable combination with the

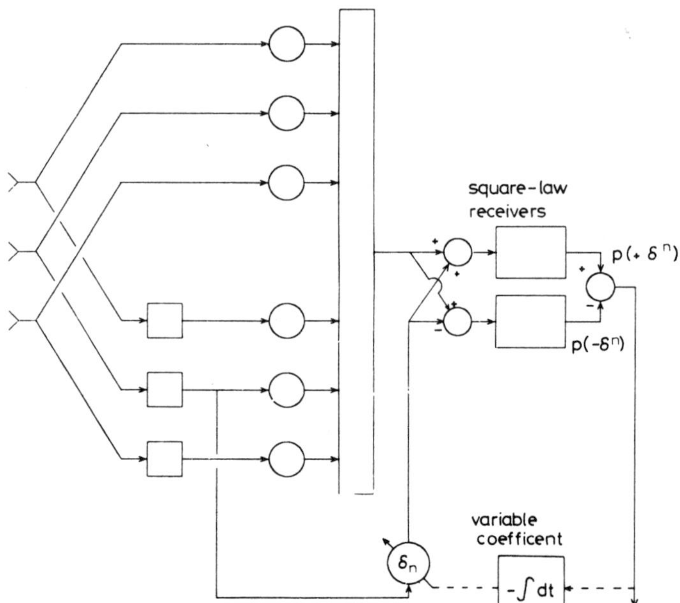

Fig. 4.7.3 *Two receiver gradient measurement system*

main array output the receivers are fed with the same voltages as though they had individual perturbed vectors. The power differencing is done on-line. The overall arrangement will be recognised as a quarter squares multiplier,* once popular in analogue computer circuits and the two receivers are completely analogous to a correlator which takes one input from the array output

$$y(t) = \mathbf{X}(t)^t\mathbf{W} \tag{4.7.8}$$

and the second from one channel. However, this is exactly the correlator connections used in Fig. 4.6.2 and the two-receiver system is identical to a multiplexed correlator if the low-pass filtering following the correlator is the same as the filtering in the receiver power difference.

In the configuration shown in Fig. 4.7.3, it is quite feasible to connect an analogue feedback loop from the receiver power difference output to the single coefficient as indicated in dashed lines. This removes all signal flow from the digital control system and minimises the required speed of the latter.

4.8 Simplified convergence analysis

Whereas the properties of the optimal solution for an adaptive antenna are of great interest, the rate at which the descent algorithms operate is almost as significant since optimal solutions achieved too late may be useless. An exact convergence analysis is given in the next Section, but gives little physical insight into the various factors so in this Section an attempt is made to show how null formation is directly linked to observable parameters of an interference field. The approximations appear to apply best to linearly constrained antennas and, for simplicity, a discrete time algorithm operating with an antenna of this type is considered.

The starting point is eqn. (4.6.11):

$$\mathbf{W}(k+1) = \mathbf{W}(k) - 2\mu\mathbf{PRW}(k) = (\mathbf{I} - 2\mu\mathbf{PR})\mathbf{W}(k) \tag{4.8.1}$$

Expressing the covariance matrix $\mathbf{R}$ as a sum of dyads, as in eqn. 2.6.21, results in

$$\mathbf{W}(k+1) = \mathbf{W}(k) - 2\mu \sum_{i=1}^{M} \pi_i \mathbf{PS}_i^* \mathbf{S}_i^t \mathbf{W}(k) \tag{4.8.2}$$

and the antenna amplitude gain for the nth source is

$$g_n(k+1) = \mathbf{S}_n^t \mathbf{W}(k+1) = \mathbf{S}_n^t \mathbf{W}(k) - 2\mu \sum_{i=1}^{M} \pi_i \mathbf{S}_n^t \mathbf{PS}_i^* \mathbf{S}_i^t \mathbf{W}(k) \tag{4.8.3}$$

The following assumptions are now made. First, that all the sources are in the

* $(x+y)^2 - (x-y)^2 = 4xy$

sidelobes of the conventional antenna pattern and are almost orthogonal to the steering vector **C**. Thus

$$\mathbf{C}^t\mathbf{S}_i^* \simeq 0, \qquad i = 1, \ldots, M \tag{4.8.4}$$

and

$$\mathbf{P}\mathbf{S}_i^* = \left(\mathbf{I} - \frac{\mathbf{C}^*\mathbf{C}^t}{\mathbf{C}^t\mathbf{C}^*}\right)\mathbf{S}_i^* \simeq \mathbf{S}_i^* \tag{4.8.5}$$

Secondly, it is assumed that the sources are almost mutually orthogonal

$$\mathbf{S}_j^t\mathbf{S}_i^* \simeq 0 \qquad i \neq j \qquad \mathbf{S}_i^t\mathbf{S}_i^* = N, \qquad i = 1, \ldots, M \tag{4.8.6}$$

Substituting eqns. (4.8.5) and (4.8.6) into (4.8.3) yields

$$g_n(k+1) = \mathbf{S}_n^t\mathbf{W}(k+1) \simeq \mathbf{S}_n^t\mathbf{W}(k) - 2\mu N\pi_n\mathbf{S}_n^t\mathbf{W}(k) \tag{4.8.7}$$

$$= g_n(k) - 2\mu N\pi_n g_n(k) = (1 - 2\mu N\pi_n)g_n(k) \tag{4.8.8}$$

Thus the amplitude gain for the nth source decreases as a geometric series and after K iterations we have

$$g_n(K) \simeq (1 - 2\mu N\pi_n)^K g_n(0) \tag{4.8.9}$$

The desired signal, lying in the pointing direction, has a constant amplitude gain and the output signal to interference ratio increases monotonically with time.

For stability, the feedback coefficient μ is bounded above by

$$\mu < \frac{1}{N\pi_{\max}} \tag{4.8.10}$$

so that $|1 - 2\mu N\pi_{\max}| < 1$. When the system approaches instability, the weighting vector makes large excursions over the output power contours as shown in Fig. 4.2.2 for the LMS algorithm. For small μs, eqn. (4.8.9) can be approximated as

$$g_n(K) \simeq g_n(0)e^{-2\mu NK\pi_n} \tag{4.8.11}$$

Fig. 4.8.1 shows the beam pattern time history of a linearly constrained antenna having 5 elements at $\lambda/2$ spacing with a 0 dB jammer at $-35°$ and a 6 dB jammer at $+35°$ relative to broadside for $\mu = 0{\cdot}01$, one fifth of the stability limit. The 6 dB jammer causes rapid formation of a null in its direction the theoretical rate being 4·4 dB per iteration from eqn. (4.8.8), whereas the 0 dB jammer is cancelled much more slowly, theoretically at 0·92 dB per iteration.

The effect of finite inner products between the source space vectors, assumed zero in eqn. (4.8.6), is roughly that as a strong jammer is converging and changing the pattern it tends to override the pattern changes due to weaker jammers which are occuring more slowly. Thus eqn. (4.8.9) is an excellent approximation for the interference which at any particular time happens to dominate the output power.

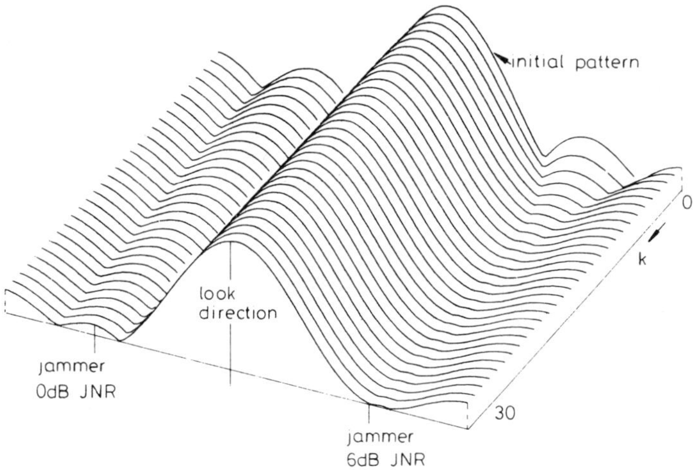

Fig. 4.8.1 *Beam pattern history for a five element array with a look direction constraint* 6 dB JNR interference at 35°, 0 dB JNR at −35° $\mu = 0{\cdot}2/\mathrm{Tr}$

Worst case convergence predictions: Having established that the time constants of the reduction rates of the interferences are inversely proportional to their powers it is possible to calculate some typical order of magnitude results.

Suppose we have a jammer at 50 km range with a directivity of 10 dB, a power of 10^4 W and a bandwidth of 100 MHz. The power density at the receiver is $3{\cdot}2 \times 10^{-6}$ W/m^2 integrated over the bandwidth. The receiver operates at 1 GHz, has a bandwidth of 1 MHz and a noise factor of 10 dB. The capture area of a dipole is about $0{\cdot}12\lambda^2 = 1{\cdot}1 \times 10^{-2}$ m^2 and in the 1 MHz bandwidth the power input to the receiver front end by the jammer is $3{\cdot}4 \times 10^{-10}$ W. The thermal noise available from the dipole is $kTB = 3{\cdot}9 \times 10^{-15}$ and allowing for the noise factor we have an equivalent noise power at the dipole of ten times this.

Thus the input jammer to noise ratio (JNR) is $8{\cdot}7 \times 10^3$ or 39 dB and the ratio of the fastest to the slowest convergence rates, if there are two jammers present, is of the same order. To illustrate this point suppose we set $\mu = 0{\cdot}1/N\pi_{\max}$ for the stronger causing its power gain sequence, eqn. (4.8.11), to become

$$|g(K)|^2 = |g(0)|^2 e^{-0{\cdot}2K} \tag{4.8.12}$$

a decrease of $-1{\cdot}73$ dB per iteration. Its output power will be reduced by 39 dB in 23 iterations by which time it will have ceased to be a threat. For a second jammer of 19 dB input JNR the rate of power reduction is only $e^{-0{\cdot}002K}$ or $-0{\cdot}017$ dB per iteration and to reduce this jammer output power by 19 dB to the equivalent level would require 1117 iterations.

Exact convergence analysis: The algorithm is analysed exactly in terms of the eigenstructure of the covariance matrix in Section 7.13 where it is treated as a special case of a suboptimal processor.

4.9 Continuous time, real data LMS algorithm[31, 48]

The approximations made in the previous Section revolved around the orthogonality of the space vectors of the sources and their orthogonality with the desired signal vector. One way round this approximation is to consider the covariance matrix as being produced by a set of different sources which really are orthogonal and the existence of such a set is guaranteed by the eigenvector expansion

$$\mathbf{R} = \sum_{i=1}^{2N} \lambda_i \mathbf{U}_i \mathbf{U}_i^t \tag{4.9.1}$$

where the λ_i are real and non-negative and can represent source powers and the $\mathbf{U}_i$s can be thought of as space vectors. In general the $\mathbf{U}_i$ vectors will not correspond to plane waves but in conditions where the orthogonality approximations of the previous section are valid they will be very close to this state. If the LMS ensemble averaged equations [eqn. (4.3.6)]

$$\dot{\mathbf{W}} = 2\mu\{\mathbf{D} - \mathbf{RW}\} \tag{4.9.2}$$

are used as a starting point we obtain

$$\dot{\mathbf{W}} = 2\mu\mathbf{D} - 2\mu\left(\sum_{i=1}^{2N} \lambda_i \mathbf{U}_i \mathbf{U}_i^t\right)\mathbf{W} \tag{4.9.3}$$

Let us examine the amplitude gain $\mathbf{w}_n$ of the array for the nth eigenvector source.

$$\dot{\mathbf{w}}_n = \mathbf{U}_n^t \dot{\mathbf{W}} = 2\mu \mathbf{U}_n^t \mathbf{D} - 2\mu\lambda_n \mathbf{w}_n \tag{4.9.4}$$

where the simplification in the latter term arises because of the orthogonality of the eigenvectors. Setting $\mathbf{U}^t\mathbf{D} = \mathbf{d}_n$ we can easily solve this equation to get

$$\mathbf{w}_n = \frac{\mathbf{d}_n}{\lambda_n} - \left\{\frac{\mathbf{d}_n}{\lambda_n} + \mathbf{w}_n(0)\right\} e^{-2\mu\lambda_n t} \tag{4.9.5}$$

Thus the transient behaviour of these eigenvector source amplitude gains is quite elementary.

The situation is not so straightforward when the amplitude gains for the actual sources are examined. If one of the sources has space vector $\mathbf{V}$, represented by

$$\mathbf{V} = \sum_{i=1}^{2N} \mathbf{v}_i \mathbf{U}_i \tag{4.9.6}$$

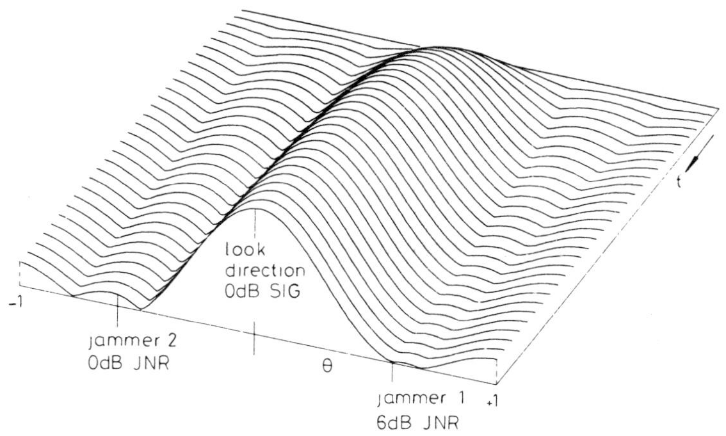

Fig. 4.9.1 *Convergence of beam pattern for 5 element array*
Desired signal 0 dB at 0°, jammers of 0 dB at −35°, 6 dB at +35°. Feedback constant $\mu = 0{\cdot}2/\mathrm{Tr}(\mathbf{R})$

then in eqn. (4.9.5) we have

$$g_v = \mathbf{V}^t\mathbf{W} = \sum_{n=1}^{2N} \mathbf{v}_n \mathbf{w}_n$$

$$= \sum_{n=1}^{2N} \left\{ \frac{\mathbf{v}_n \mathbf{d}_n}{\lambda_n} - \left[\frac{\mathbf{v}_n \mathbf{d}_n}{\lambda_n} + \mathbf{v}_n \mathbf{w}_n(0) \right] e^{-2\mu\lambda_n t} \right\} \tag{4.9.7}$$

which is a sum of $2N$ exponential terms. Fig. 4.9.1 show the history of the beam pattern of a five-element array which is illuminated by a 0 dB desired signal in the broadside direction together with a 0 dB interference at −35° and a +6 dB interference at +35°. The weights are initialised at zero and the amplitude gains for all three sources show an immediate increase corresponding to one of the transient solutions in eqn. (4.9.7). Whereas the amplitude gain in the direction of the desired signal continues to increase the gains for the interferences shows a downturn after a short while the 6 dB interference decaying quite quickly and the 0 dB interference more slowly. These time constants are two further solutions in eqn. (4.9.7).

The approximation used in the previous section that the space vectors approximate to eigenvectors is equivalent, for complex vectors, to

$$\mathbf{U}_n \simeq \mathbf{S}_n^*/\|\mathbf{S}_n\| \qquad \lambda_n \simeq N\pi_n \tag{4.9.8}$$

Maximum feedback factor: The equations above are unconditionally stable but experience shows that the smallest time constant

$$T_{\min} = \frac{1}{2\mu\lambda_{\max}} \tag{4.9.9}$$

should not be less than about 10 times the decorrelation time of the interference signals ($\simeq$ 10/bandwidth). If there is a dominant jammer of power π_{max} then the smallest time constant is associated with it and

$$\lambda_{max} \simeq N\pi_{max} \simeq Tr(\mathbf{R}) \tag{4.9.10}$$

4.10 Eigenvectors of real covariance matrix

The covariance matrix for real data is of the general form shown in eqn. (2.6.4). Suppose that the complex covariance has the form

$$\mathbf{R}_C = \mathbf{R}_r + j\mathbf{R}_j \tag{4.10.1}$$

where $\mathbf{R}_r$ and $\mathbf{R}_j$ are real and symmetric and skew symmetric, respectively. Thus

$$\begin{aligned} \mathbf{R}_r^t &= \mathbf{R}_r \\ \mathbf{R}_j^t &= -\mathbf{R}_j \end{aligned} \tag{4.10.2}$$

The real matrix has the form, for the same data [eqn. (2.6.14)],

$$\mathbf{R}_R = \begin{bmatrix} \mathbf{R}_r & \mathbf{R}_j \\ \mathbf{R}_j^t & \mathbf{R}_r \end{bmatrix} \tag{4.10.3}$$

Suppose $\mathbf{Z} = \mathbf{X} + j\mathbf{Y}$ is an eigenvector of $\mathbf{R}_C$ with real eigenvalue λ:

$$\mathbf{R}_C \mathbf{Z} = \lambda \mathbf{Z} \tag{4.10.4}$$

then

$$\begin{aligned} \mathbf{R}_r \mathbf{X} - \mathbf{R}_j \mathbf{Y} &= \lambda \mathbf{X} \\ \mathbf{R}_r \mathbf{Y} + \mathbf{R}_j \mathbf{X} &= \lambda \mathbf{Y} \end{aligned} \tag{4.10.5}$$

The corresponding eigenvector of $\mathbf{R}_R$ is $\begin{bmatrix} \mathbf{X} \\ -\mathbf{Y} \end{bmatrix}$ since

$$\begin{bmatrix} \mathbf{R}_r & \mathbf{R}_j \\ \mathbf{R}_j^t & \mathbf{R}_r \end{bmatrix} \begin{bmatrix} \mathbf{X} \\ -\mathbf{Y} \end{bmatrix} = \lambda \begin{bmatrix} \mathbf{X} \\ -\mathbf{Y} \end{bmatrix} \tag{4.10.6}$$

as is easily shown using eqn. (4.10.5). However, the vector $\begin{bmatrix} \mathbf{Y} \\ \mathbf{X} \end{bmatrix}$ is also an eigenvector since

$$\begin{bmatrix} \mathbf{R}_r & \mathbf{R}_j \\ \mathbf{R}_j^t & \mathbf{R}_r \end{bmatrix} \begin{bmatrix} \mathbf{Y} \\ \mathbf{X} \end{bmatrix} = \lambda \begin{bmatrix} \mathbf{Y} \\ \mathbf{X} \end{bmatrix} \tag{4.10.7}$$

and is the real equivalent of $j\mathbf{Z}$. Thus although $\mathbf{R}_R$ has twice the dimensions and twice as many eigenvalues as $\mathbf{R}_C$ they occur in equal pairs and with related, although independent, eigenvectors.

4.11 Weight jitter phenomena

Otherwise known as loop noise, weight jitter is a consequence of using noisy gradient estimates. In a real-time application, it is not feasible to smooth the gradient estimates sufficiently to remove the effect because of lack of time for smoothing. The difference between the gradient estimates and the true gradient is a random vector function, uncorrelated with the true gradient, and accumulation of this error in the weighting coefficients, noting that the coefficients act as integrators in all algorithms, produces a vector valued random walk effect. Unless corrected this would tend to make the weight vector deviate from its desired value, the mean-square error, on average, increasing linearly with time. The correcting factor is the descent algorithm itself. The deviation of the weighting vector sets up changes in the directional pattern of the antenna and, if the amplitude gain in the direction of a discrete source increases, then the output signal or error signal due to that source is detected by the correlators and a steady component appears in the gradient which tends to drive the error power to zero. Thus weight jitter effects are an equilibrium condition balanced between the driving function, the gradient error, and the constraining influence, the distribution of sources.

Analysis of the problem is very complicated and only results are presented here. Widrow,[83, 84] Brennan,[31] and Berni[27] have published details of calculations.

Discrete time LMS algorithm:[85] Using the algorithm

$$\mathbf{W}(k+1) = \mathbf{W}(k) - 2\mu\mathbf{X}(k)\{\mathbf{X}(k)^t\mathbf{W}(k) - d(k)\} \tag{4.11.1}$$

where the gradient is derived from eqn. (4.3.2) the final result is

$$\text{cov}\,(\mathbf{W}) = \mu p_{\text{opt}}\,\mathbf{I} \tag{4.11.2}$$

where p_{opt} is the output error power when the weights have their optimal values and $\mathbf{I}$ is a $2N$ by $2N$ identity matrix. This simple result indicates that the errors in the coefficients are uncorrelated with each other and all have the same variance.

Let $\mathbf{V}(k)$ be the difference between the actual and optimal weighting vectors:

$$\mathbf{V}(k) = \mathbf{W}(k) - \mathbf{W}_{\text{opt}} \tag{4.11.3}$$

The excess error power due to the finite variance is given by

$$\Delta_p = E\,|\,\mathbf{X}(k)^t\mathbf{V}(k)\,|^2 = E[\mathbf{X}(k)^t\mathbf{V}(k)\mathbf{V}(k)^t\mathbf{X}(k)] \tag{4.11.4}$$

and since cov $(\mathbf{V})$ = cov $(\mathbf{W})$ substitution of eqn. (4.11.2) into (4.11.4) yields

$$\Delta_p = \mu p_{\text{opt}}\,E[\mathbf{X}(k)^t\mathbf{X}(k)] = \mu p_{\text{opt}}\,Tr(\mathbf{R})$$

where $\mathbf{R}$ is the covariance matrix of the antenna data. Widrow defines a misadjustment M which is the fractional increase in output error power:

$$M = \frac{\Delta_p}{p_{\text{opt}}} = \mu Tr(\mathbf{R}) \tag{4.11.6}$$

For stability reasons, the value of μ is normalised by $Tr(\mathbf{R})$:

$$\mu = \beta / Tr(\mathbf{R}), \qquad 0 < \beta < 1 \tag{4.11.7}$$

thus

$$M = \beta \tag{4.11.8}$$

It is possible to relate β to the time constants in the convergence characteristics of the algorithm and show that small M values arise when the convergence is slow. This indicates that convergence rate must be traded against misadjustment.

Contribution of individual sources to Δp: The output power of some source with array covariance matrix $\pi_i \mathbf{S}_i^* \mathbf{S}_i^t$ (temporarily assumed to be complex) is

$$\Delta_p(i) = \pi_i E[\mathbf{V}^H \mathbf{S}_i^* \mathbf{S}_i^t \mathbf{V}] = \pi_i E[\mathbf{S}_i^H \mathbf{V}^* \mathbf{V}^t \mathbf{S}_i] = \pi_i \mathbf{S}_i^H \mathbf{S}_i \mu p_{\text{opt}} \tag{4.11.9}$$

For $\mathbf{S}_i^H \mathbf{S}_i = N$, the number of elements, π_i is the source power at the elements. Thus the fraction of error due to source 'i' is

$$\frac{\Delta_p(i)}{\Delta_p} = \frac{N\pi_i}{Tr(\mathbf{R})} = \frac{\pi_i}{\sum \pi_j + \sigma^2} \tag{4.11.10}$$

where σ^2 is the uncorrelated noise level and the error power is directly proportional to the power of the source. In any scenario the strongest interference source will produce most of the error.

Estimation of p_{opt}: In the best of worlds, p_{opt} would be close to the level associated with uncorrelated noise in the absence of interference [eqn. (5.1.18)] as discrete interferences can normally be cancelled below the noise level. Factors which might increase p_{opt} are finite bandwidth effects (Section 5.3), coefficient quantisation (Section 5.6) or main-lobe jamming. Actually, in the LMS algorithm, main lobe jamming will reduce p_{opt} below the uncorrelated noise level but will also suppress the desired signal [eqn. (5.2.15)], the net effect being the same as the increase of uncorrelated noise associated with main lobe jamming in a linearly constrained antenna, eqn. (6.2.7).

Continuous time LMS algorithm:[27, 31] The analytic situation is less satisfactory here.[27] In the 'leaky integrator' algorithm, eqn. (4.3.8),

$$(\mathbf{W} + T\dot{\mathbf{W}}) = -2G(\mathbf{X}\mathbf{X}^t\mathbf{W} - d) \tag{4.11.11}$$

or the ideal integrator algorithm

$$\dot{\mathbf{W}} = -2\mu(\mathbf{X}\mathbf{X}'\mathbf{W} - d) \qquad (\mu = G/T) \tag{4.11.12}$$

We have

$$\Delta_p = \mu t_c \, Tr(\mathbf{R}) p_{opt} \tag{4.11.13}$$

where t_c is the decorrelation time of the interferences. From eqns. (4.9.9) and (4.9.10)

$$\mu = \frac{1}{2Tr(\mathbf{R})T_{min}} \tag{4.11.14}$$

where T_{min} is the shortest time constant in eqn. (4.9.5) and we obtain

$$M = \frac{\Delta_p}{p_{opt}} = \frac{t_c \, Tr(\mathbf{R})}{2Tr(\mathbf{R})T_{min}} = \frac{t_c}{2T_{min}} \tag{4.11.15}$$

In general, M would be kept less than 0·1, otherwise coefficient jitter begins to have deleterious effects. Unfortunately, the jitter is associated with the largest eigenvalues of **R** whereas slow convergence is caused by the smaller eigenvalues. The requirements for low jitter and fast convergence are directly opposed.

Discrete time linearly constrained algorithm: Here the algorithm is [eqns. (4.5.10) and (4.6.9)]

$$\mathbf{W}(k+1) = \mathbf{W}(k) - 2\mu\mathbf{P}\mathbf{X}(k)^*\mathbf{X}(k)^t\mathbf{W}(k) \tag{4.11.16}$$

assuming a complex representation and

$$\text{Cov}\,(\mathbf{W}) = \text{Cov}\,(\mathbf{V}) = \mu p_{opt}\,\mathbf{P} \tag{4.11.17}$$

where **P** is the projection defined in (4.6.11). The null eigenvalue of **P** has an eigenvector parallel to the constraint vector, and eqn. (4.11.17) indicates that there is no weight jitter in directions parallel to the constraint vector as we might expect. Apart from this difference the covariance is similar to that for the LMS discrete time algorithm. By using a normalised feedback coefficient

$$\mu = \beta/\text{Tr}(\mathbf{PRP}), \qquad 0 < \beta < 1 \tag{4.11.18}$$

where the limits on the right guarantee stability [eqn. (7.13.11)] we obtain

$$\Delta_p = E|\mathbf{X}^t\mathbf{V}|^2 = E[\mathbf{X}^H\mathbf{V}^*\mathbf{V}^t\mathbf{X}] = \text{Tr}\,[\mathbf{PRP}\,.\,\text{Cov}\,(\mathbf{V})]$$

$$= \mu p_{opt}\,\text{Tr}\,(\mathbf{PRP}) = \beta p_{opt} \tag{4.11.19}$$

As in the LMS algorithm the excess output power is distributed in proportion to the powers of the individual sources, excluding the desired signal.

The major difference between this result and that for the LMS algorithms is that the p_{opt} in (4.11.19) includes the desired signal output power whereas in the LMS version, eqn. (4.11.5), the desired signal power hardly appears at all since the reference signal cancels the signal output from the antenna in the feedback loops. Thus the misadjustment in the linearly constrained algorithm can be much greater than the LMS algorithm if the optimal output SNIR is high and the desired signal dominates the output power. In fact this enables a simple estimate to be made of output SNIR in the presence of weight jitter. If, in eqn. (4.11.19), we get

$$p_{\text{opt}} = \text{desired signal output power} = p_d \tag{4.11.20}$$

then

$$\Delta_p = \beta p_d \tag{4.11.21}$$

and the output SNIR is

$$Q \simeq p_d/\Delta_p = 1/\beta \tag{4.11.22}$$

This result tends to indicate that if high quality output signals are required, with SNIRs of the order of 30 dB, the feedback coefficient β must be kept less than 10^{-3} and convergence will be very slow.

If the desired signal power in the constrained antenna is much weaker than the uncorrelated noise the lower limit of p_{opt} will be fixed by interference rejection and uncorrelated noise as in the LMS algorithms. Ideally, interferences can be suppressed below the uncorrelated noise at the antenna output if there is no limitation due to finite bandwidth or coefficient control errors and p_{opt} will be on the order of $N\sigma^2$. [eqn. (7.2.8) with $K = M = N$].

Figure 4.11.1 shows the fluctuation in the pattern of a nine-element array. Rather extreme conditions have been chosen to give an easily visible result with a μ value of $0{\cdot}75/\text{Tr}\,(\mathbf{R})$, a 6 dB source at the look direction to produce a high p_{opt}, and a 3 dB source at 30° (powers relative to uncorrelated noise). The fluctuations in the pattern are reasonably isotropic as would be expected since

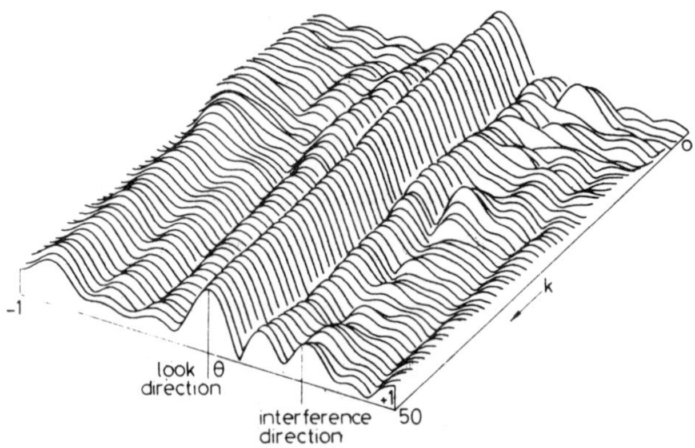

Fig. 4.11.1 *Fluctuations in the beam pattern of a nine element line array due to weight coefficient jitter*
6 dB look direction signal, 3 dB interference at 30°, 0 dB uncorrelated noise

transforming Cov (**W**) into bearing space via an orthogonal matrix will yield a beam covariance of the form $\tilde{\mathbf{P}}$ where $\tilde{\mathbf{P}}$ is orthogonal to the look direction. Thus equal variance fluctuations are expected everywhere except near the look direction. Although the variance of the pattern is the same at the interference bearing ($\theta = 0{\cdot}5$) the rate of fluctuation is increased.

Weight jitter in perturbation algorithm: Attention is restricted to the linearly constrained minimum output power descent procedure. Cantoni[34] and Widrow[85] have discussed gradient estimation errors in the perturbation technique though Widrow has only considered the small perturbation limit.

Single-receiver system: The variances of the gradient component estimates yielded by eqn. 4.7.6 with noisy power measurements are [144, 145]

$$\mathrm{Var}\,(G_n) = \frac{F}{4\delta_n^2}[\mathrm{Var}\,(\mathrm{p}+) + \mathrm{Var}\,(\mathrm{p}-)],\ n = 1,..,2N$$

when statistical independence of the output powers associated with positive (p+) and negative (p−) coefficient perturbations is assumed. F is a smoothing factor which accounts for the variance reducing effect of low-pass filtering of the receiver output power ($F \simeq$ filter bandwidth/receiver bandwidth). If all sources are Gaussian and the receiver measures the squared modulus of the complex output envelope then eqn. A1.6.6 may be used to yield $\mathrm{Var}\,(\mathrm{p}\pm) = E(\mathrm{p}\pm^2) - (E\mathrm{p}\pm)^2 = (E\mathrm{p}\pm)^2$ and using the result $\mathrm{p}\pm = (W \pm \delta_n)^H M (W \pm \delta_n)$ while adding the variances of the real and imaginary gradient components yields complex gradient variance values of

$$F^{-1}\,\mathrm{Var}\,(G_n^C) = \frac{\bar{\mathrm{p}}^2}{\delta_n^2} + |G_n^C|^2 + 2\bar{\mathrm{p}}M_{nn} + \delta_n^2 M_{nn}^2,\ n = 1,..,N \qquad (4.11.23)$$

in which M_{nn} is the nth diagonal element of M and $\bar{\mathrm{p}}$ is the unperturbed output power $W^H M W$. Minimisation of Var (G_n^C) by variation of δ_n yields $\delta_n(opt) = (\bar{\mathrm{p}}/\mathrm{M}_{nn})^{1/2}$ and an associated smallest covariance matrix of

$$\mathrm{Cov}\,(\mathbf{G}^C) = \mathrm{F\ Diag}\,(4M_{nn}\bar{\mathrm{p}} + |G_n^C|^2) \simeq \frac{4}{N}\bar{\mathrm{p}}\ \mathrm{Tr}\,(\mathbf{R})\mathrm{F}\,\mathbf{I} \qquad (4.11.24)$$

where the last approximation applies close to optimality where gradients G_n^C are small and the M_{nn} values, i.e. the channel powers, are all the same.

The optimal value for δ_n may correspond to large coefficient perturbations, especially during a cold start when $\bar{\mathrm{p}}/M_{nn}$ may be large. The use of Walsh function perturbations[34] may be of value here, distributing the steps over all of the coefficients and reducing individual coefficient steps.

Dual-receiver system: If two receivers are used with analogue power differencing, as shown in Fig. 4.7.3, then the description of the statistics is much more straightforward and the covariance of the complex estimate is

$$\mathrm{Cov}\,(\hat{\mathbf{G}}) = \frac{4}{N}\, p_{\mathrm{opt}}\, \mathrm{Tr}\,(\mathbf{R}) F \mathbf{I} \tag{4.11.25}$$

where p_{opt} is the equilibrium output power in the optimal solution without jitter.

Beam pattern effects : Using the constrained descent algorithm

$$\mathbf{W}(k+1) = \mathbf{W}(k) - \mu \mathbf{P}\hat{\mathbf{G}}(k) \tag{4.11.26}$$

with $\mathrm{Cov}\,(\hat{\mathbf{G}}) = \sigma_g^2 \mathbf{I}$ the covariance of the weighting vector can be shown to be

$$\mathrm{Cov}\,(\mathbf{W}) \simeq \tfrac{1}{4}\sigma_g^2\, \mu (\mathbf{PRP})^+ \tag{4.11.27}$$

in which $(\ )^+$ is a pseudoinverse. This result differs from the multiple correlator result of eqn. (4.11.17) significantly and this has important repercussions on the behaviour of the system.

The reason for the matrix inversion appearance in eqn. (4.11.27) runs something like the following. The error in the gradient estimates are uncorrelated from coefficient to coefficient (unlike the multiple correlator system) and application of the descent algorithm, eqn. (4.11.26), causes accumulation of uncorrelated noise samples in the coefficient values; a classical random walk model. If the weighting coefficients are transformed into their beam pattern then, if orthogonal beams are used, the random walk in the coefficients transforms into an uncorrelated random walk of the beam amplitudes, i.e. the array pattern will fluctuate in a fairly isotropic fashion. It will also increase in average power as more noise is added to the coefficients. The mechanism which restrains the pattern from growing to huge magnitudes is the sources in the far field plus the uncorrelated noise set up by the receivers. If the amplitude increases in the direction of an interference source the output power of the antenna increases and the algorithm detects a gradient, a mean value component in $\hat{\mathbf{G}}$, causing the gain to decrease in the source direction. Thus notches in the pattern will be preserved in the directions of sources but there is nothing to prevent the pattern from growing to large amplitudes in source-free directions unless it gets so large that uncorrelated noise begins to dominate the output power. Additionally the dynamic range of the coefficients will tend to limit the amplitude to some finite value. These directions in which the pattern can grow to large amplitudes are associated with the small eigenvalues of $\mathbf{R}$, which, in the inverse $(\mathbf{PRP})^+$, give rise to large $\mathbf{W}$ fluctuations in the corresponding eigenvector directions.

Fig. 4.11.2 shows an example of beam pattern random walk in the presence of an interference. Superficially the pattern looks similar to that for the multiple correlator example in Fig. 4.11.1. However, close inspection reveals that the pattern jitter in the direction of the interference is now relatively small and a good null is preserved for reasons just explained. This does not imply that in absolute terms the jitter is less, merely that the jitter is greater in source free directions than in the multiple correlator algorithms.

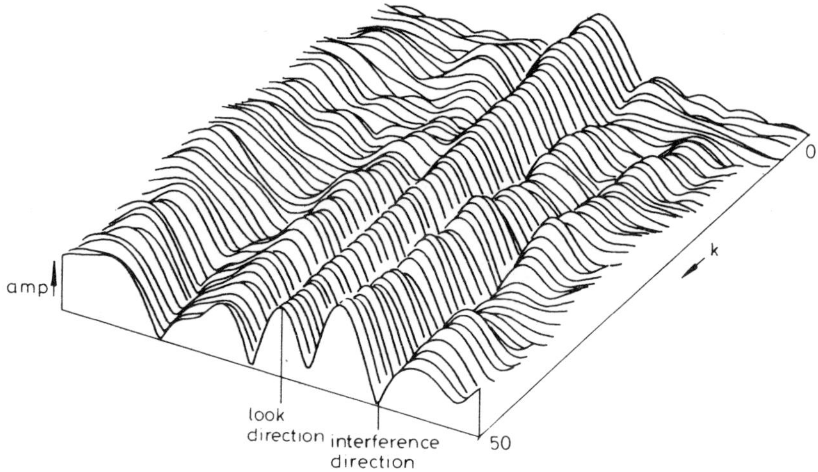

Fig. 4.11.2 *Pattern jitter in perturbation algorithm*
6 dB interference at 30°, uncorrelated noise at −30 dB

Excess output power: This is given by

$$\Delta_p = \tfrac{1}{4}\sigma_g^2\,\mu E[\mathbf{X}^H(\mathbf{PRP})^+\mathbf{X}] = \tfrac{1}{4}\sigma_g^2\,\mu\ \mathrm{Tr}\ [\mathbf{PRP}(\mathbf{PRP})^+]$$

$$= \tfrac{1}{4}\sigma_g^2\,\mu(N-1) = \frac{\tfrac{1}{4}\sigma_g^2\,\beta(N-1)}{\mathrm{Tr}\ (\mathbf{PRP})} \qquad (4.11.28)$$

However, this result assumes, if fact, that the pattern fluctuations have reached an equilibrium state and very large lobes are present in source-free direction. If the coefficients have insufficient dynamic range to bring up the uncorrelated noise to the same order as the interference output then it is more reasonable to assume that only the discrete interference sources contribute to Δ_p and with this assumption we have

$$\Delta_p \simeq \tfrac{1}{4}n_s\,\beta\sigma_g^2/\mathrm{Tr}\ (\mathbf{PRP}) \qquad (4.11.29)$$

where n_s is the number of discrete interference sources.

Using the σ_g^2 values above gives the following excess output powers:

(*a*) *Single-receiver system*

$$\Delta_p \simeq n_s\,\beta N^{-1}\bar{\mathrm{p}}F\ \mathrm{Tr}\ (\mathbf{R})/\mathrm{Tr}\ (\mathbf{PRP}) \qquad (4.11.30)$$

Close to optimality we have $\bar{\mathrm{p}} \simeq p_{\mathrm{opt}}$ and for a weak look direction signal, moreover, Tr (**PRP**) $\simeq$ Tr (**R**); thus

$$\Delta_p \simeq \frac{p_{\mathrm{opt}}F n_s\beta}{N} \qquad (4.11.31)$$

yielding an output SNIR of

$$Q \simeq \frac{p_{\text{opt}}}{\Delta_p} \simeq \frac{N}{Fn_s\beta} \tag{4.11.32}$$

Example 4.11.1: A receiver has a 1 MHz pass band, the low-pass filter for the gradient determination has a 1 kHz pass band, $N = 10$, and $\beta = 0{\cdot}1$. Calculate the output in SNIR if there is one interference present.

Substituting into (4.11.32) yields

$$Q \simeq \frac{10}{10^{-3} \times 1 \times 0{\cdot}1} = 10^5 \text{ or } 50 \text{ dB}$$

(*b*) *Dual-receiver system*

From eqn. (4.11.25) we obtain

$$\Delta_p \simeq 4p_{\text{opt}}\,\pi_{\text{chan}}\frac{Fn_s\beta}{4\,\text{Tr}\,(\mathbf{R})} \simeq \frac{p_{\text{opt}}\,Fn_s\beta}{N} \tag{4.11.33}$$

yielding an output SNIR of

$$Q \simeq \frac{p_{\text{opt}}}{\Delta_p} = \frac{N}{F\beta n_s} \tag{4.11.34}$$

Example 4.11.2: Using the data of example 4.11.1, calculate the output SNIR for a dual-receiver system for $n_s = 3$.

$$Q = \frac{10}{10^{-3} \times 10^{-1} \times 3} = 3 \times 10^4 \text{ or } 45 \text{ dB}$$

Distribution of excess power

For the multiple correlator system it has been shown that the excess output power is distributed among the interference sources in proportion to their powers. A rather different conclusion applies for the present system. If there is a source with representation $m_n(t)\mathbf{S}_n$ at the array it produces an excess power of

$$\Delta_p(n) = \overline{|m_n(t)|^2}\mathbf{S}_n^H\overline{\mathbf{V}^*\mathbf{V}^t}\mathbf{S}_n = \sigma_v^2\,\pi_n\,\mathbf{S}_n^H(\mathbf{PRP})^+\mathbf{S}_n \tag{4.11.35}$$

where $\sigma_v^2 = \frac{1}{4}\sigma_g^2\,\mu$ is the variance of the coefficient fluctuations. We have, however,

$$\mathbf{PRP} = \sum_{i=1}^{N}\pi_i\,\mathbf{PS}_i^*\mathbf{S}_i^t\mathbf{P} \tag{4.11.36}$$

as the covariance matrix where extra sources have introduced to represent the uncorrelated noise, and

$$\sum_{i=1}^{N} \pi_i \mathbf{S}_i^H(\mathbf{PRP})^+\mathbf{S}_i = \mathrm{Tr}\left[(\mathbf{PRP})^+ \sum_{i=1}^{N} \pi_i \mathbf{PS}_i^*\mathbf{S}_i^t\mathbf{P}\right]$$

$$= \mathrm{Tr}\,[(\mathbf{PRP})^+\mathbf{PRP}] = (N-1) \tag{4.11.37}$$

If $\mathbf{S}_i^H(\mathbf{PRP})^+\mathbf{S}_i = f(\pi_i)$ then from eqn. (4.11.37) we have, for any π_i

$$\sum_{i=1}^{N} \pi_i\, f(\pi_i) = (N-1) \tag{4.11.38}$$

which is only satisfied if $f(\pi_i) = N - 1/N\pi_i$. Thus, in eqn. (4.11.35),

$$\Delta_p(n) = \sigma_v^2(N-1)/N \tag{4.11.39}$$

indicating that the excess output power is distributed uniformly among all the sources, including the uncorrelated noise. There is actually some approximation involved in eqn. (4.11.36) and, to be quite accurate, sources only produce uniform excess output power if they exceed the uncorrelated noise level by a reasonable margin. As noted before, it is unlikely that excess output power due to uncorrelated noise will be observed in practice, due to weighting coefficient magnitude limitations.

Modulation of signals by coefficients: An effect allied to coefficient jitter is coherent variation of the weighting coefficients which can manifest itself when signals of narrow and wide bandwidths are mixed. An example of the effect* is when the desired signal is a sine wave at frequency ω_s and there is an interference signal at some bearing α_i which consists of a second sinusoid at frequency $\omega_i = \omega_s + \Delta$. If the feedback coefficient in the analogue linearly constrained algorithm is high, the convergence time constant may be smaller than the beat period between the signal and the interference. If the antenna steers a phase modulated retrobeam in the direction of the interference with amplitude gain

$$g(\alpha_i, t) = g_0 e^{-j\Delta t} \tag{4.11.40}$$

then the interference is heterodyned to the signal frequency and can produce partial cancellation of the signal by suitable selection of g_0. The severity of the cancellation will depend upon the frequency difference Δ. Similar effects are easily set up in simulations with discrete time algorithms if the desired signal and interferences are sinusoidal rather than being modulated by uncorrelated noise and the directional pattern of the adaptive weights will show cyclic fluctuation.

* Unpublished talk by B. Widrow

In a real-time algorithm, there could be undesirable effects if, for example, the desired signal were an amplitude modulated carrier and the carrier component at the antenna output were partially suppressed by the combination of the adaptive algorithm and a CW jammer very close to the carrier frequency. It is unlikely that carrier-free modulation systems such as four-phase binary or SSB will be disturbed by this effect.

4.12 Control of quantised coefficients[57]

The steps in gradient algorithms are likely to be so small that if the coefficients are quantised none of them will change during an update. This effect is avoided by separating the hardware weighting vector from the one in the algorithm and having fine quantisation in the internal vector. Thus in eqn. (4.6.11), consider that **W** is the hardware weighting vector and **V** is an internal vector which is quantised in transmission to **W**, for example by sending the most significant bits only. The algorithm is modified to

$$\begin{aligned} \mathbf{V}(k+1) &= \mathbf{V}(k) - \mu \mathbf{P}\mathbf{G}(k) \\ \mathbf{W}(k+1) &= Q_u[\mathbf{V}(k+1)] \\ \mathbf{G}(k+1) &= 2\mathbf{R}\mathbf{W}(k+1) = 2\mathbf{X}(k+1)^*\mathbf{X}(k+1)^t\mathbf{W}(k+1) \end{aligned} \tag{4.12.1}$$

where Q_u is the quantisation procedure. This technique allows any small gradients components in **G** to be integrated in **V** until they cause a coefficient to change. The algorithm is rather unpredictable when close to convergence, the weight vector showing a tendency to jump among a number of states close to the optimal vector (which it cannot reach). However, these steps do not increase the interference output power very much on average above what the quantisation alone would produce. In fact the antenna output power may be a few decibels less than the uncorrelated quantisation theory (Section 5.6) would predict. This tends to occur when there are excess degrees of freedom and indicates that in such cases there may be points on the hyperplane orthogonal to the interference space vectors and forming a subspace of the constraint surface at which the quantisation error is quite small. The quantisation theory in Section 5.6 simply assumes that the weighting vector moves to the closest quantisation point from the optimal solution and is clearly only a partial solution.

4.13 Sample matrix inversion[29, 54, 55, 69, 73, 95, 96, 104, 115]

An alternative to gradient methods is sample matrix inversion (SMI) and this is the fastest known method for estimating an optimal weighting vector. In the simplest description of the algorithm, we postulate that K samples of the antenna data have been sampled, digitised, and input to a digital processor. The processor computes the maximum likelihood estimate of the covariance matrix[95]

$$\hat{\mathbf{R}}_K = \frac{1}{K} \sum_{k=1}^{K} \mathbf{X}_k^* \mathbf{X}_k^t \tag{4.13.1}$$

then, given information about the steering vector coefficients $\mathbf{S}^*$, it inverts $\hat{\mathbf{R}}_K$ and calculates the optimal weighting vector

$$\mathbf{W}_K = \lambda \hat{\mathbf{R}}_K^{-1} \mathbf{S}^* \tag{4.13.2}$$

This vector is resident in the processor memory at this point and can be used in three ways: (i) it may be used as part of a digital element weighting system in which the weighting vector is transmitted to on-line digital multipliers;[25] (ii) it may be used to control analogue binary attenuators and phase shifters; (iii) it may be converted to an analogue voltage and used to drive PIN diodes in quadrature attenuators.[87]

If an LMS algorithm is required the processor must also estimate a vector of correlations between the antenna data vectors and the reference signal which is then used in eqn. (3.3.15)[55]

$$\hat{\mathbf{C}}_k = \frac{1}{K} \sum_{k=1}^{K} \mathbf{X}_k \, d_k^* \tag{4.13.3}$$

As all the inputs to the processor are in the form of RF cross products and appear at base band it is possible to use analogue correlators to avoid digital multiplication.[45]

The idea of sample matrix inversion is a very old one, originating with least squares methods,[123] but its use in signal processing applications is more recent[95, 96] and its speed advantages over descent methods have been recognised more recently still.[73]

The minimum number of samples necessary to get a good estimate of $\mathbf{W}_{opt}$ is discussed by Capon[96] Reed,[73] and Boronson.[29] Normally, about $2N$ samples are required to get good performance but if there are only $N_s \ll N$ strong sources present, this may be relaxed to $2N_s$ if the estimated covariance matrix $\hat{\mathbf{R}}$ is rendered invertible by augmenting its diagonals by a $\lambda\mathbf{I}$ supplement:

$$\hat{\mathbf{R}}_K = \lambda \mathbf{I} + \frac{1}{K} \sum_{k=1}^{K} \mathbf{X}_k^* \mathbf{X}_k^t \qquad K < N \tag{4.13.4}$$

If λ is small $\mathbf{R}_K^{-1}$ has the action of a projection matrix and makes the computed optimal weight vector orthogonal to all preceding samples $\mathbf{X}_1 - \mathbf{X}_K$ [see eqn. (5.2.28)]. Thus it is only suitable for use on succeeding samples before they are incorporated into $\hat{\mathbf{R}}_K$. A similar procedure has been described by Bühring.[33]

The SMI scheme is of most value in complicated interference environments where the number of interference sources is large. Its performance is almost independent of the eigenvalue spread of $\mathbf{R}$ and similar to a gradient descent

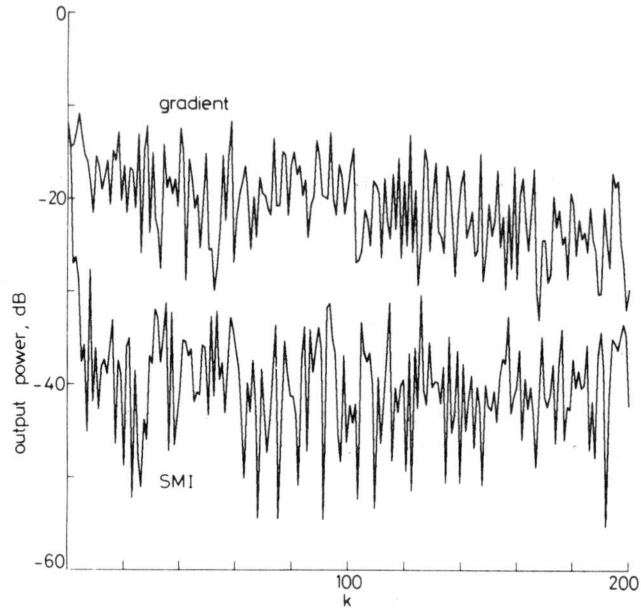

Fig. 4.13.1 *Comparative convergence of gradient and SMI algorithm in an ill-conditioned environment*

algorithm operating with an **R** having equal eigenvalues—a condition which can be achieved by preprocessing to some extent as discussed in Section 7.12. Fig. 4.13.1 shows the convergence of a 9-element array in an ill-conditioned environment: five jammers of powers and bearings of 10 dB/−45°, 10 dB/−43°, 0 dB/−30°, −10 dB/30°, −20 dB/45° with uncorrelated noise of −30 dB. In the gradient algorithm μ is set to 0·2/Tr (**R**). After a few decibels of power reduction the slow convergence modes dominate and output power decays at about 0·05 dB/sample. In the SMI system with the same data, the power output is mostly limited by uncorrelated noise after only 10 iterations.

In Fig. 4.13.2, the covariance matrix is well conditioned, produced by equal power sources at −45°, +45°, and 90° relative to broadside and there is not so much difference between the performances.

Data deweighting : Matrix estimate (4.13.1) is only appropriate for stationary statistics. If the interference parameters are going to vary then a deweighted matrix estimate is more suitable:

$$\hat{\mathbf{R}}_K = (1 - \alpha) \sum_{n=0}^{\infty} \alpha^n \mathbf{X}^*_{K-n} \mathbf{X}^t_{K-n} \qquad 0 < \alpha < 1 \tag{4.13.5}$$

which may be estimated recursively as follows:

$$\hat{\mathbf{R}}_K = \alpha \hat{\mathbf{R}}_{K-1} + (1 - \alpha) \mathbf{X}^*_K \mathbf{X}^t_K \tag{4.13.6}$$

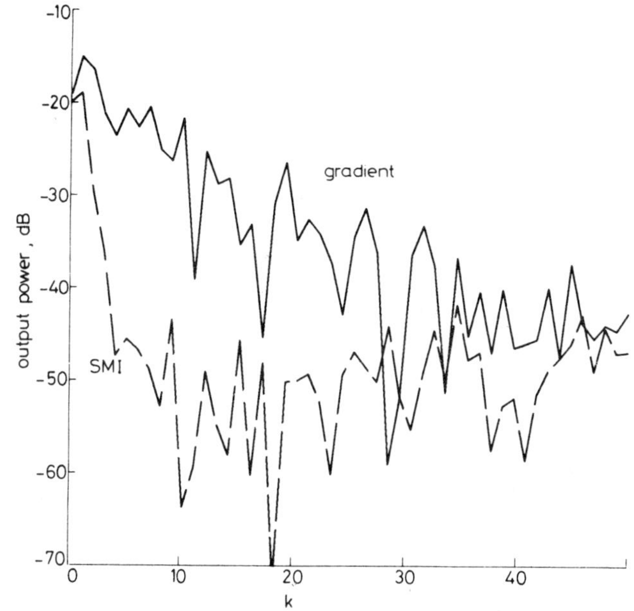

Fig. 4.13.2 *Comparative convergence in a well-conditioned situation*

The matrix inverse may be found recursively by the inversion formula (Appendix A1.2) as

$$\mathbf{R}_K^{-1} = \alpha^{-1}\mathbf{R}_{K-1}^{-1} - \frac{(1-\alpha)\alpha^{-2}\hat{\mathbf{R}}_{K-1}^{-1}\mathbf{X}_K^*\mathbf{X}_K^t\mathbf{R}_{K-1}^{-1}}{1+(1-\alpha)\alpha^{-1}\mathbf{X}_K^t\hat{\mathbf{R}}_{K-1}^{-1}\mathbf{X}_K^*} \tag{4.13.7}$$

and substituted into eqn. (4.13.2) to get a recursive update for $\mathbf{W}_K$. This system was described by Lunde.[104] The numerical load is about N^2 complex multiplications per data sample. If it is not required to update the covariance matrix for every sample of data, a number M of data samples may be stored in a block $\mathbf{Z}$:

$$\mathbf{Z} \triangleq [\mathbf{X}_0 | \mathbf{X}_{-1} | \mathbf{X}_{-2} | \cdots | \mathbf{X}_{-M+1}] \tag{4.13.8}$$

and a formula in Appendix A1.2 may be used to update $\mathbf{R}^{-1}$ every Mth data sample.

$$(\mathbf{R} + \mathbf{Z}^*\mathbf{Z}^t)^{-1} = \mathbf{R}^{-1} - \mathbf{R}^{-1}\mathbf{Z}^*(\mathbf{I}_M + \mathbf{Z}^t\mathbf{R}^{-1}\mathbf{Z}^*)^{-1}\mathbf{Z}^t\mathbf{R}^{-1} \tag{4.13.9}$$

which requires inversion of an M by M matrix only.

Analogue matrix inversion: It is not feasible to invert a matrix completely by analogue methods but systems of linear equations are solveable by analogue computer techniques. Thus given the $2N$ simultaneous linear equations in the general optimal solution, eqn. (3.4.6)

$$\mathbf{MW} = -\tfrac{1}{2}\lambda\mathbf{D} \tag{4.13.10}$$

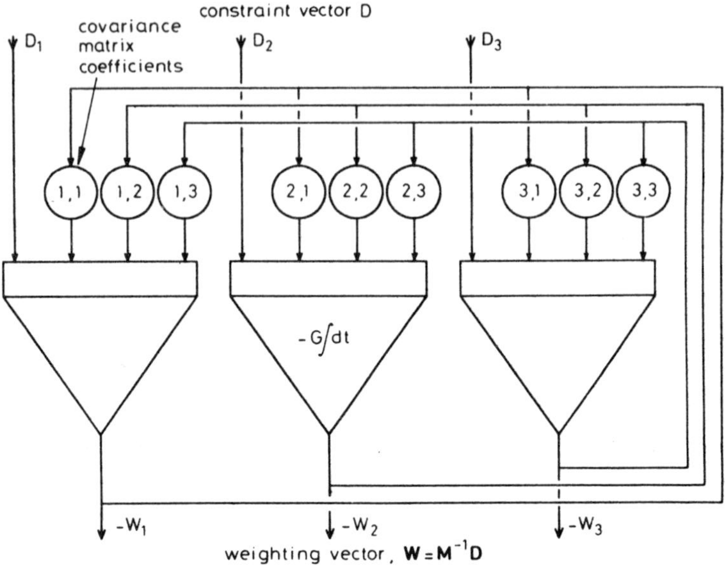

Fig. 4.13.3 *Solution for optimal weighting vector by analogue network*

we can solve them by the linear network shown in Fig. 4.13.3. It is unlikely that the system will handle very ill-conditioned situations due to amplifier overload but is probably suitable for use in preprocessor type systems which precede an adaptive antenna as errors here are not too important (Section 7.12).

Another analogue 'matrix inversion' scheme is described by Compton.[143]

Recursive solutions:[59, 23] If the digitally computed weighting coefficients are transmitted to some external analogue weighting unit it is unlikely that the accuracy of the solution will survive intact. In such a system it is necessary to monitor the state of the coefficients as part of the algorithm in order to correct errors. The 'open-loop' SMI processor is completely unable to do this. Modification of the solution to conform to a Kalman filter algorithm will enable such feedback to be installed.

It is shown in Sections 7.1 and 7.2 that a fully adaptive linearly constrained processor has the solution

$$\mathbf{W}_{\text{opt}} = \mathbf{U} - (\mathbf{PRP})^{+}\mathbf{PRU} = \mathbf{U} + \mathbf{PV} \tag{4.13.11}$$

where $\mathbf{U}$ is the conventional weighting vector, $\mathbf{P}$ is a projection matrix $\mathbf{I} - \mathbf{C}^*\mathbf{C}^t/\mathbf{C}^t\mathbf{C}^*$ with $\mathbf{C}$ being the steering vector, and $(\ )^+$ is a pseudo-inverse. The equivalence of this solution and the usual one [eqn. (3.4.29)] is demonstrated by showing that, in eqn. (4.13.11) $\mathbf{PRW}_{\text{opt}} = 0$ [eqn. (7.13.3)] whence $\mathbf{RW}_{\text{opt}} = \lambda\mathbf{C}$. If the matrix estimator (4.13.6) is substituted into (4.13.11) we can, with a little difficulty, prove that vector $\mathbf{V}$ can be updated recursively by the equations:

$$\mathbf{V}_K = \mathbf{V}_{K-1} - \alpha(\mathbf{PR}_K\mathbf{P})^+\mathbf{PX}_K^*\mathbf{X}_K^t\mathbf{W}_{K-1} \qquad 0 < \alpha < 1$$

with

$$(\mathbf{PR}_K\mathbf{P})^+ = \mathbf{Q} - \frac{\alpha\mathbf{QPX}_k^*\mathbf{X}_k^t\mathbf{PQ}}{1 + \alpha\mathbf{X}_k^t\mathbf{PQPX}_k}$$

$$\mathbf{Q} = \frac{1}{1-\alpha}(\mathbf{PR}_{K-1}\mathbf{P})^+ \tag{4.13.12}$$

In the first of the equations above, the scalar $\mathbf{X}_K^t\mathbf{W}_{K-1} = y_K$ is the antenna output signal using the previous coefficients on the new data vector. An opportunity to sample the hardware antenna coefficient outputs can be taken here, replacing y_K, the digitally calculated output, by $\tilde{y}_K$, the measured output.[59] The system is quite stable if the coefficients are not too nonlinear and, unlike the open loop system, the depth of nulls is quite unaffected by drift and offset of the hardware coefficients. Figure 4.13.4 indicates the necessary feedback connections.

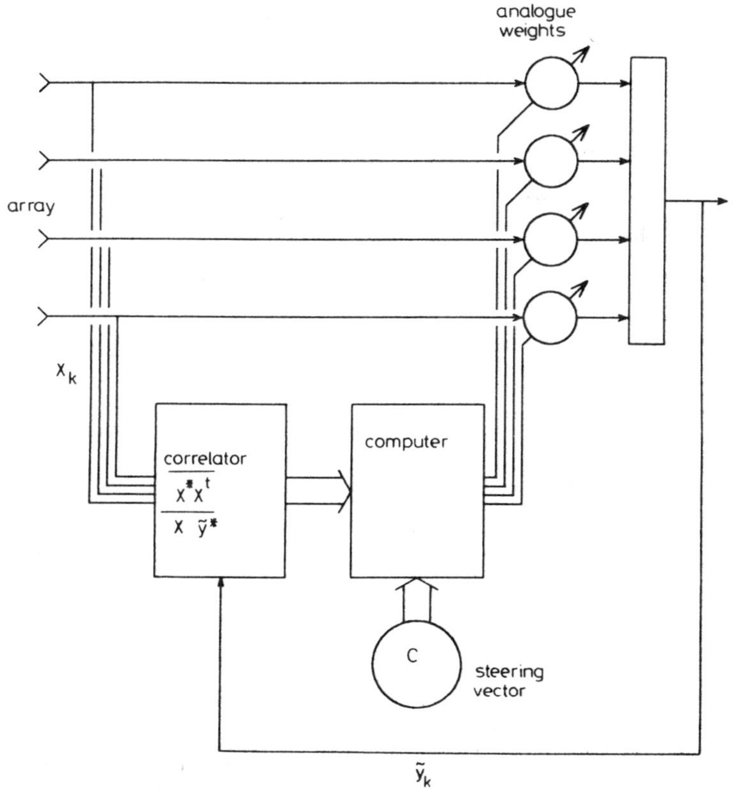

Fig. 4.13.4 *Kalman filter implementation*

Newton Raphson acceleration.[58, 111, 140, 113, 131] It turns out that the inverse matrix update [4.13.12(b)] need not be very accurate or very frequent. In fact any matrix estimate $\hat{\mathbf{R}}$ can be used in [4.13.12(a)] and will probably have a significant acceleration effect. The Kalman filter then has a close resemblence to Newton Raphson acceleration of a gradient descent procedure. A lot of ingenuity can be used to estimate the covariance matrix in a hardware compatible form avoiding the heavy computation load of the full Kalman algorithm.

The application of the Newton Raphson method is discussed by Mueller[140] for a line equalisation algorithm and by Hudson[58] for an antenna. The assessment of performance relative to a simple gradient algorithm is heavily involved with the estimation of the effects of coefficient jitter which are much increased by the acceleration and reduce output SNIR.

4.14 Diagonalisation methods

There is much scope for hardware implementation of mathematical diagonalisation techniques. A technique described by Giraudon[99] and others is very close to Gauss-Jordan matrix inversion and enables parallel operation of

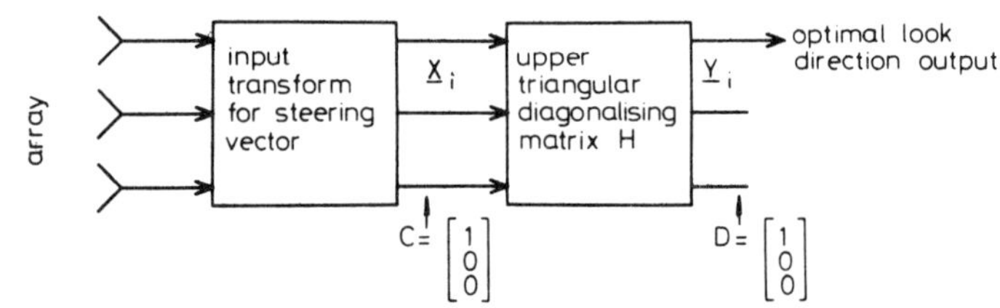

Fig. 4.14.1 *Sequence of transformations in practical prewhitening*

similar hardware units. In Fig. 4.14.1 a matrix fixed preprocessor converts the steering vector into a clamping on a single channel so that all other weights can vary without constraint. Channel 3 is now added into channels one and two with variable weighting such that channels 1 and 2 become decorrelated from 3. This action diagonalises one row and column of the covariance matrix. For an initial covariance matrix of the form

$$\mathbf{R}_1 = \begin{bmatrix} r_{11} \, r_{12} \, r_{13} \\ r_{21} \, r_{22} \, r_{23} \\ r_{31} \, r_{32} \, r_{33} \end{bmatrix} \tag{4.14.1}$$

the weights on channel 3 required to decorrelate channels 1 and 2 are $-r_{13}/r_{33}$ and $-r_{23}/r_{33}$ and the combination is done in a network of the type shown in

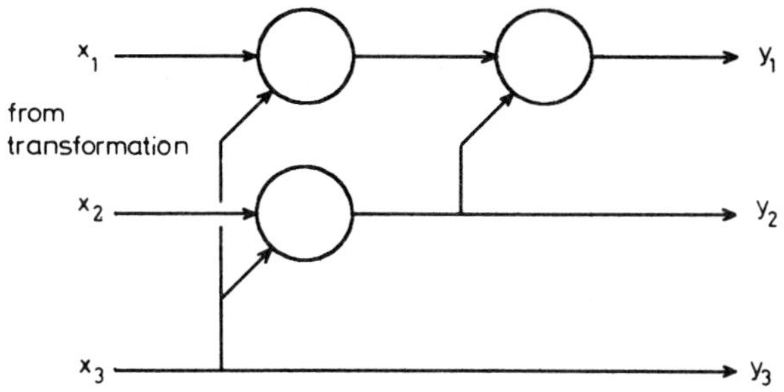

Fig. 4.14.2 *Diagonalising system due to Giraudon*

Figs 4.14.2 and 4.14.3. In matrix terms the effect of this first column of coefficients can be converted into a transformation matrix:

$$\mathbf{K}_1 \triangleq \begin{bmatrix} 1 & 0 & -\dfrac{r_{13}}{r_{33}} \\ 0 & 1 & -\dfrac{r_{23}}{r_{33}} \\ 0 & 0 & 1 \end{bmatrix} \tag{4.14.2}$$

and the covariance matrix $\mathbf{R}_2$ of the signals following the decorrelation is given by

$$\mathbf{R}_2 = \mathbf{K}_1\,\mathbf{R}_1\,\mathbf{K}_1^H = \begin{bmatrix} r'_{11} & r'_{12} & 0 \\ r'_{21} & r'_{22} & 0 \\ 0 & 0 & r_{33} \end{bmatrix} \tag{4.14.3}$$

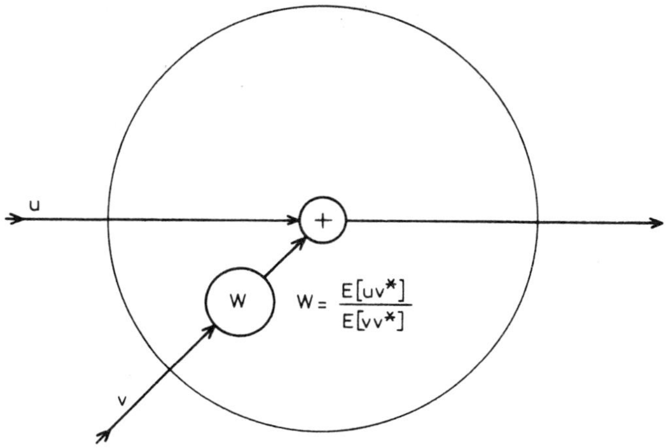

Fig. 4.14.3 *Details of weights in Fig. 4.14.2*

Next, channel 2 is decorrelated from channel 1 with weighting $-r'_{12}/r'_{22}$ equivalent to premultiplication by the matrix $\mathbf{K}_2$:

$$\mathbf{K}_2 = \begin{bmatrix} 1 & -\dfrac{r'_{12}}{r'_{22}} & 0 \\ 0 & 1 & 0 \\ 0 & 0 & 1 \end{bmatrix} \tag{4.13.4}$$

The resulting covariance matrix $\mathbf{R}_3$ is given by

$$\mathbf{R}_3 = \mathbf{K}_2 \mathbf{R}_2 \mathbf{K}_2^H = \begin{bmatrix} r''_{11} & 0 & 0 \\ 0 & r'_{22} & 0 \\ 0 & 0 & r''_{33} \end{bmatrix} \tag{4.13.5}$$

The transformation $\mathbf{K}_n$ do not affect the steering vector, i.e.

$$\mathbf{K}_2 \mathbf{K}_1 \begin{bmatrix} 1 \\ 0 \\ 0 \end{bmatrix} = \begin{bmatrix} 1 \\ 0 \\ 0 \end{bmatrix} \tag{4.13.6}$$

so that it is only necessary to use the output from channel 1 as the optimal output.

The total numerical load involved in finding the optimal solution, which resides in approximately $\frac{1}{2}N(N-1)$ variable coefficients is about the same task as inverting $\mathbf{R}$ directly, in fact the matrices $\mathbf{K}_n$ above are standard forms for inverting matrices. Thus the advantage of this particular system is that it gives a convenient method for paralleling the processing so that $\frac{1}{2}N(N-1)$ simple units replace the serial program of a digital processor.

The system is not very stable if the input covariance matrix $\mathbf{R}$ is ill-conditioned in the sense that some of the coefficients at a late column may attempt to assume large numerical values. Thus some sort of limiting of coefficients moduli might be advisable. It is also slow to respond to changes in an interference environment, depending on how the coefficients are calculated. In this respect it is similar to the SMI algorithms and in fact simulation tests show that its performance, with careful sequencing of the updates is hardly distinguishable.

Chapter 5

Performance of optimal antennas

If it assumed that an adaptive antenna has converged to its optimal condition then the performance, in terms of output SNIR or total noise and interference output power is purely a function of the interference environment and the size and shape of the receiving array. The weighting solutions determined previously have been couched in terms of inverse matrices and steering vectors and, apart from the fact that they give the best signal-to-noise ratio, provide no further insight into the absolute performance, i.e. the values of output SNIR and how it varies with array size and shape and the angular separation of the interference and desired signal. The purpose of this chapter, in contrast, is to get some solutions for optimal antennas which hopefully will provide such insights. Naturally this demands that the interference situations be fairly simple and while the general two interference problem has been solved by Kwok et *al.*[61, 62] and Morgan[68] and the general form of the associated inverse matrix is indicated in eqn. (2.7.6), we will restrict the discussion mostly to a single interference in a background of uncorrelated noise. The performance of the LMS and steered antennas is addressed here, equivalent results for constrained systems are easily derived by analogy and are also special cases of the suboptimal arrays treated in Section 7.2 taking $K = M = N$ there.

Subsequent sections deal with the effects of singular covariance matrices, finite bandwidths, jammer polarisation, coherent multipath and quantisation and the aim is that an intuitive feel for the operational characteristics of optimal antennas should be obtained. On the first reading the more difficult parts of Sections 5.2 and 5.3 could well be omitted.

5.1 LMS algorithm–single signal and interference

Initially we look at the problem of the LMS algorithm with a unit amplitude reference signal when the antenna receives the sum of the desired signal, an interference, and uncorrelated noise at power levels of π_s, π_v, and σ^2, respectively. The covariance matrix of this combination is

$$\mathbf{R} = \sigma^2\mathbf{I} + \pi_s \mathbf{S}^*\mathbf{S}^t + \pi_v \mathbf{V}^*\mathbf{V}^t \tag{5.1.1}$$

It is assumed that the elements of $\mathbf{S}$ and $\mathbf{V}$ have unity modulus so that $\mathbf{S}^t\mathbf{S}^* = \mathbf{V}^t\mathbf{V}^* = N$ = number of elements. The optimal solution for the LMS algorithm, with unit amplitude reference, is

$$\mathbf{RW} = \pi_s^{1/2}\mathbf{S}^* \tag{5.1.2}$$

i.e.

$$\sigma^2\mathbf{W} + \pi_s \mathbf{S}^*\mathbf{S}^t\mathbf{W} + \pi_v \mathbf{V}^*\mathbf{V}^t\mathbf{W} = \pi_s^{1/2}\mathbf{S}^* \tag{5.1.3}$$

Putting all vectors parallel to $\mathbf{S}$ to the right yields

$$\sigma^2\mathbf{W} + \pi_v \mathbf{V}^*\mathbf{V}^t\mathbf{W} = \mathbf{S}^*(\pi_s^{1/2} - \pi_s \mathbf{S}^t\mathbf{W}) \tag{5.1.4}$$

or

$$\mathbf{MW} = \lambda\mathbf{S}^* \tag{5.1.5}$$

Equation (5.1.4) can be solved for the desired signal and interference amplitude gains as follows. For g_s ($= \mathbf{S}^t\mathbf{W}$) we premultiply throughout by $\mathbf{S}^t$ getting

$$\sigma^2\mathbf{S}^t\mathbf{W} + \pi_v \mathbf{S}^t\mathbf{V}^*\mathbf{V}^t\mathbf{W} = \mathbf{S}^t\mathbf{S}^*(\pi_s^{1/2} - \pi_s \mathbf{S}^t\mathbf{W}) \tag{5.1.6}$$

and setting $\mathbf{S}^t\mathbf{V}^* = N\rho_{sv} = N\rho_{vs}^*$ we eventually obtain

$$g_s(\sigma^2 + N\pi_s) = N\pi_s^{1/2} - N\pi_v \rho_{sv} g_v \tag{5.1.7}$$

In the case of the interference gain we have

$$\sigma^2\mathbf{V}^t\mathbf{W} + \pi_v \mathbf{V}^t\mathbf{V}^*\mathbf{V}^t\mathbf{W} = \mathbf{V}^t\mathbf{S}^*(\pi_s^{1/2} - \pi_s g_s) \tag{5.1.8}$$

which yields

$$g_v(\sigma^2 + N\pi_v) = N\rho_{vs}(\pi_s^{1/2} - g_s \pi_s) \tag{5.1.9}$$

These simultaneous equations in g_v and g_s may be solved as required; the case for g_s is considered in eqns. (5.2.13) and (5.2.15). It is fairly clear from eqn. (5.1.9) that for $\pi_v \to \infty$ the product $g_v \pi_v$ tends to a constant limit and the interference output power, proportional to $|g_v|^2\pi_v$, tends to zero.

Output SNIR: If we only require to calculate the output SNIR (signal to noise plus interference ratio) simplification is possible since scaling of the weighting vector does not effect SNIR. Thus we may use eqn. (5.1.5) with $\lambda = 1$ for determination of $\mathbf{W}_{opt}$. Applying the matrix inversion formula (Appendix A1.2) to $\mathbf{M}$ and setting σ^2 to unity for convenience yields

$$\mathbf{M}^{-1} = \mathbf{I} - \frac{\pi_v \mathbf{V}^*\mathbf{V}^t}{1 + \pi_v \mathbf{V}^t\mathbf{V}^*} \tag{5.1.10}$$

which, in eqn. (5.1.5), results in

$$\mathbf{W} = \mathbf{M}^{-1}\mathbf{S}^* = \left\{\mathbf{I} - \frac{\pi_v \mathbf{V}^*\mathbf{V}^t}{1 + \pi_v \mathbf{V}^t\mathbf{V}^*}\right\}\mathbf{S}^* \tag{5.1.11}$$

Note if $\pi_v \rightarrow \infty$ then

$$\mathbf{W} \rightarrow \left\{\mathbf{I} - \frac{\mathbf{V}^*\mathbf{V}^t}{\mathbf{V}^t\mathbf{V}^*}\right\}\mathbf{S}^* = \mathbf{PS}^* \tag{5.1.12}$$

which will be recognised as a projection of the desired signal vector onto a subspace orthogonal to **V**, a procedure discussed in Section 2.5. This shows that the projection operation is optimal, for a single interference, in the output SNIR sense when the interference power is high. This conclusion is generalised to multiple interferences in eqn. (5.2.28).

The output SNR and SNIR of the antenna are now examined as the interference power level is varied. The amplitude gain for the desired signal is found from eqn. (5.1.11) as

$$g_s = \mathbf{S}^t\mathbf{W} = \mathbf{S}^t\mathbf{S}^* - \frac{\pi_v \mathbf{S}^t\mathbf{V}^*\mathbf{V}^t\mathbf{S}^*}{1 + \pi_v \mathbf{V}^t\mathbf{V}^*} = N\left\{1 - \frac{N\pi_v|\rho|^2}{1 + N\pi_v}\right\} \tag{5.1.13}$$

where $\rho = \rho_{sv}$ as defined above. The desired signal output power is thus

$$p_s = |g_s|^2\pi_s = \frac{N^2\pi_s[1 + N\pi_v(1 - |\rho|^2)]^2}{(1 + N\pi_v)^2} \tag{5.1.14}$$

The interference output power is found from g_v and π_v. From eqn. (5.1.11) we obtain

$$g_v = \mathbf{V}^t\mathbf{W} = \mathbf{V}^t\mathbf{S}^* - \frac{\pi_v \mathbf{V}^t\mathbf{V}^*\mathbf{V}^t\mathbf{S}^*}{1 + \pi_v \mathbf{V}^t\mathbf{V}^*} = \frac{N\rho_{vs}}{1 + N\pi_v} \tag{5.1.15}$$

and the interference output power is

$$p_v = \pi_v|g_v|^2 = \frac{N^2\pi_v|\rho|^2}{(1 + N\pi_v)^2} \tag{5.1.16}$$

Evaluating $dp_v/d\pi_v$ in this equation and setting to zero determines the input interference power where the output interference power is maximised. It is easily shown that $\pi_v = N^{-1}$ for this and the associated p_v is $\frac{1}{4}N|\rho|^2$. For sidelobe interference $|\rho|^2 \ll 1$ and, as will be seen below, the uncorrelated noise output power is on the order of N; therefore the interference output power will always be well below the uncorrelated noise output power. If the algorithm uses lossy integrators as in eqns. (4.3.10) or (4.6.8) the input interference power level where the output is maximised can be varied at the will of the designer.

Finally, the uncorrelated noise output power is determined. For unity uncorrelated noise variance we have, from eqn. (5.1.11),

$$p_n = \mathbf{W}^H\mathbf{W} = \left\| \frac{\mathbf{S}^*(1 + N\pi_v) - \mathbf{V}^* N\rho_{vs}\pi_v}{1 + N\pi_v} \right\|^2$$

$$= \frac{\mathbf{S}'\mathbf{S}^*(1 + N\pi_v)^2 + \mathbf{V}'\mathbf{V}^* N^2|\rho|^2\pi_v^2 - 2N^2|\rho|^2\pi_v(1 + N\pi_v)}{(1 + N\pi_v)^2}$$

$$= \frac{N[1 + (1 - |\rho|^2)(2N\pi_v + N^2\pi_v^2)]}{(1 + N\pi_v)^2} \qquad (5.1.17)$$

The combined interference and noise output power is the sum of eqns. (5.1.16) and (5.1.17):

$$p_{n+i} = \pi_v|g_v|^2 + \mathbf{W}^H\mathbf{W} = \frac{N(1 + N\pi_v)[1 + N\pi_v(1 - |\rho|^2)]}{(1 + N\pi_v)^2} \qquad (5.1.18)$$

It is interesting that if the equivalent result for a linearly constrained antenna is derived [e.g. eqn. (7.1.33)] it is found that p_{n+i} always increases monotonically for increasing interference power, an effect produced by the constraints. The same is not true for the equation above. For example if $|\rho|$ is very close to unity it is clear that p_{n+i} becomes very small if π_v is large and this can only happen if all the coefficients become small.

The output signal to noise ratio is the ratio of eqn. (5.1.14) to eqn. (5.1.18):

$$Q = \frac{p_s}{p_{n+i}} = \frac{N\pi_s[1 + N\pi_v(1 - |\rho|^2)]}{1 + N\pi_v} = N\pi_s\left(1 - \frac{N\pi_v|\rho|^2}{1 + N\pi_v}\right) \qquad (5.1.19)$$

This is a monotonically decreasing function of π_v and for the large interference power limit we have

$$Q \simeq N\pi_s(1 - |\rho|^2) \qquad (5.1.20)$$

which is the output SNR for a conventionally weighted antenna in the absence of interference ($N\pi_s$) reduced by the factor $(1 - |\rho|^2)$ the loss being due to the pattern distortion when the null is formed in the interference direction which reduces the uncorrelated noise gain of the antenna.

Sidelobe jamming: In this case $|\rho|^2$, the normalised interference amplitude gain, is small. A mean square estimate of $|\rho|^2$ for an array of N elements is obtained by putting the antenna into an isotropic noise field and calculating the proportion of output power which is due to noise arriving from the sidelobe directions. For $\lambda/2$ interelement spacing a spherically isotropic noise field produces uncorrelated signals at all element spacings except zero (Fig. 2.8.1) and p_{out} is just $N\sigma^2$ where the noise variance is σ^2 in each channel. Line arrays whose patterns approximate to sin $(x)/x$ have about 5/6 of their total power response in the main lobe (i.e. the beam efficiency is 5/6) and 1/6 in the sidelobes, therefore of the total power of $N\sigma^2$ in an isotropic noise field only $N\sigma^2/6$ is due to the sidelobes. Thus

$$\overline{|\rho|^2} \simeq \frac{N\sigma^2/6}{N^2\sigma^2} = \frac{1}{6N} \tag{5.1.21}$$

in which the denominator of the second expression represents the antenna output power were all the noise to originate in the centre of the main lobe.

Substituting eqn. (5.1.21) into (5.1.20) yields

$$\bar{Q} \simeq N\pi_s(1 - 1/6N) \tag{5.1.22}$$

which will only be a fraction of a decibel inferior to the conventionally weighted antenna, in the absence of interference, if N is greater than 3 or so. Thus the sidelobe jamming cancellation performance of the adaptive antenna is extremely good.

Main-lobe jamming: For main-lobe jamming the interference gets so close to the pointing direction that $|\rho| \simeq 1$. This condition is most easily examined for the case that the antenna is a uniformly spaced line array having a directional pattern of the form shown in eqn. (2.4.12):

$$\rho = \frac{\sin \frac{1}{2}Nkd\theta}{N \sin \frac{1}{2}kd\theta} \simeq 1 - \tfrac{1}{24}(N^2 - 1)k^2d^2\theta^2 \tag{5.1.23}$$

in which θ is the sine of source bearing relative to broadside, d is the interelement spacing, and $k = 2\pi/\lambda = \omega/c$ is the wavenumber.

Substituting approximation (5.1.23) into eqn. (5.1.20) yields

$$Q \simeq \tfrac{1}{12}N(N^2 - 1)\pi_s k^2d^2\theta^2 \simeq \tfrac{1}{12}N^3\pi_s k^2d^2\theta^2 \tag{5.1.24}$$

which shows that the output SNIR increases as the square of the bearing difference between the interference and the desired signal. If the ratio of the output SNIR to the conventional SNR in the absence of interference is formed we obtain

$$Q/Q_{conv} \simeq \tfrac{1}{12}N^2k^2d^2\theta^2 \simeq \frac{\pi^2}{3}\,\theta^2/\theta_0^2, \qquad \theta^2 \ll \theta_0^2 \tag{5.1.25}$$

where θ_0 is the 3 dB beamwidth of the conventional antenna and $\pi = 3{\cdot}1415\ldots$. Thus there is an intimate relationship between conventional antenna beamwidth and adaptive antenna mainlobe cancellation performance. Figure 7.3.3 shows the output SNIR for a five element array as a strong jammer is swept through the direction of the desired signal.

Grating lobes: If the separation of the antenna elements is much larger than $\lambda/2$ secondary main lobes may appear in the conventional pattern. For each such lobe $|\rho| \rightarrow 1$ and the antenna cannot steer nulls at interferences which might be there without, at the same time, steering a null at the desired signal in the primary main lobe which is just what the LMS antenna would do. These secondary lobes may be broken up by irregular element spacings to a certain extent. Even if the elements of the antenna are directional and the magnitude of the secondary main lobes is much less than the primary main lobe the effect will appear because it is the direction of the interference space vector which is important, not its magnitude.

Two-dimensional apertures: The fundamental results in eqns. (5.1.19) and (5.1.20) apply equally to 2-D apertures but the relationship between ρ and N becomes more complex. For the main lobe jamming case suppose the aperture is N by M elements and the jammer moves in the 'N' direction away from the look direction. Then in eqn. (5.1.23) we have the same expression for ρ but in eqn. (5.1.20) we replace N by NM. Thus the main-lobe jamming approximations in eqns. (5.1.24) and (5.1.25) become

$$Q \simeq \tfrac{1}{12} NM^2 \pi_s k^2 d^2 \theta^2 \quad \text{and} \quad Q/Q_{\text{conv}} \simeq \frac{\pi^2}{3} \theta^2/\theta_0^2 \tag{5.1.26}$$

5.2 Solutions for ill-conditioned covariance matrices

The covariance matrix may, more often than not, be ill-conditioned for either or both of the following reasons.

(i) In circumstances where the interference is due to a small number of discrete source, less than the number of array elements, and the uncorrelated noise component is relatively weak. In principle, covariance matrices of analogue data are never singular due the inevitable presence of thermal noise but when the interference to thermal noise (or other uncorrelated noise) ratio exceeds 30–40 dB, the matrix may have to be treated as singular for practical purposes.

(ii) In sample matrix inversion algorithms [eqn. (4.13.1)] the number of data samples may be less than the number of elements. For K samples we have Rank $(\hat{\mathbf{R}}) \leq K$ and if $K < N$ then $\hat{\mathbf{R}}$ is inevitably singular. Given that $\mathbf{R}$ is singular two questions arise. First, what is the optimum weighting vector if $\mathbf{R}^{-1}$ does not exist; second, if a gradient algorithm is used, will it converge to this optimal condition. Inspection of the optimality criteria discussed in Chapter 3 shows that some of them, eqns. (3.1.13) and (3.5.5), appear to require invertibility while others like the LMS algorithm eqn. (3.3.14), the linearly constrained algorithm eqns. (3.4.6) and (3.4.20), and the best SNR solution eqn. (4.6.4) merely use inversion as a solution technique the basic optimality criteria being, from eqn. (3.3.14)

$$\mathbf{R}\mathbf{W}_{\text{opt}} = \mathbf{S}^* \tag{5.2.1}$$

for the LMS result, and from eqn. (3.4.20)

$$\mathbf{R}\mathbf{W}_{\text{opt}} = \lambda\mathbf{S}^* \tag{5.2.2}$$

for the linearly constrained system. Techniques for the solution of these equations which do not use matrix inversion will be discussed.

LMS criterion: The condition where the covariance matrix is intrinsically singular, as in case (1) above, is first addressed. A vanishingly small bandwidth must be assumed too or the dispersion of the space vectors of the sources, discussed in Section 5.3 will increase the rank of the covariance matrix. For M interference sources and a single desired source (indexed 0) we may write

$$\mathbf{R} \triangleq \sum_{i=0}^{M} \pi_i \mathbf{S}_i^* \mathbf{S}_i^t \tag{5.2.3}$$

and, if the vectors $\mathbf{S}_i$ are linearly independent, $\mathbf{R}$ has rank $M + 1$. Substitution of this representation into eqn. (5.2.1) yields, for unit reference amplitude,

$$\sum_{i=0}^{M} \pi_i \mathbf{S}_i^* g_i = \pi_0^{1/2} \mathbf{S}_0^* \tag{5.2.4}$$

where

$$g_i = \mathbf{S}_i^t \mathbf{W}_{\text{opt}}, \qquad i = 0, 1, \ldots, M \tag{5.2.5}$$

are the amplitude gains of the antenna for the $M + 1$ sources. For the assumed linear independence of the space vectors we have, by definition,

$$\sum_{i=1}^{M} \alpha_i \mathbf{S}_i^* \neq \mathbf{S}_0^* \tag{5.2.6}$$

for any combination of αs and, in order to satisfy eqn. (5.2.4), we must select

$$\pi_i g_i = 0, \qquad i = 1, \ldots, M \tag{5.2.7}$$

and

$$\pi_0^{1/2} g_0 = 1 \qquad \text{i.e. } g_0 = \pi_0^{-1/2} \tag{5.2.8}$$

Since the π_is are assumed non-zero we further have

$$g_i = 0, \qquad i = 1, \ldots, M \tag{5.2.9}$$

Equation (5.2.9) implies that perfect nulls are steered toward the interference sources and eqn. (5.2.8) that the desired signal is output without amplitude error; in fact, the output signal error is zero and the SNIR is infinite.

If the level of the desired signal varies then the weighting vector is inversely scaled to compensate and if π_0 becomes very small, g_0 is very large, and if π_0 is zero there is no solution to eqn. (5.2.8) and hence neither to eqn. (5.2.4). The consequences of the lack of solution in a gradient algorithm are pursued later. It is not strictly necessary for the signal to become zero, if the weighting coefficients reach some limit on their values for sufficiently small signal amplitude, for deleterious effects to be observed; nor need $\mathbf{R}$ be exactly singular.

The behaviour of the weighting coefficients as the desired signal and uncorrelated noise powers vary is conveniently examined in detail for the case that there is a single interference since the theory of Section 5.1 is adequate for this limited objective. In eqn. (5.1.4) we have, after dividing throughout by σ^2 to fix the noise power,

$$\left[I + \frac{\pi_v}{\sigma^2} \mathbf{V}^*\mathbf{V}^t\right]\mathbf{W}_{\text{opt}} = \mathbf{S}^*\left[\frac{\pi_s^{1/2}}{\sigma^2} - \frac{\pi_s}{\sigma^2} g_s\right] \tag{5.2.10}$$

which is of the form

$$\mathbf{M}\mathbf{W}_{\text{opt}} = \alpha\mathbf{S}^* \tag{5.2.11}$$

having the solution, analogous to eqn. (5.1.11), of

$$\mathbf{W}_{\text{opt}} = \alpha\mathbf{M}^{-1}\mathbf{S}^* = \left[\frac{\pi_s^{1/2}}{\sigma^2} - \frac{\pi_s}{\sigma^2} g_s\right] \times \left[\mathbf{S}^* - \frac{\pi_v \sigma^{-2}\mathbf{V}^*\mathbf{V}^t\mathbf{S}^*}{1 + \pi_v \sigma^{-2}\mathbf{V}^t\mathbf{V}^*}\right] \tag{5.2.12}$$

and the desired signal amplitude gain is

$$g_s = \mathbf{S}^t\mathbf{W}_{\text{opt}} = \left[\frac{\pi_s^{1/2}}{\sigma^2} - \frac{\pi_s g_s}{\sigma^2}\right] \times \left[\mathbf{S}^t\mathbf{S}^* - \frac{\pi_v \sigma^{-2}\mathbf{S}^t\mathbf{V}^*\mathbf{V}^t\mathbf{S}^*}{1 + \pi_v \sigma^{-2}\mathbf{V}^t\mathbf{V}^*}\right] \tag{5.2.13}$$

$$\underset{\pi_v \to \infty}{\simeq} \left[\frac{\pi_s^{1/2}}{\sigma^2} - \frac{\pi_s g_s}{\sigma^2}\right]\left[\mathbf{S}^t\mathbf{S}^* - \frac{\mathbf{S}^t\mathbf{V}^*\mathbf{V}^t\mathbf{S}^*}{\mathbf{V}^t\mathbf{V}^*}\right] \tag{5.2.14}$$

and using the normed properties $\|\mathbf{S}\| = \|\mathbf{V}\| = N^{1/2}$ and $\mathbf{S}^t\mathbf{V}^* = N\rho$ we obtain

$$g_s = \frac{N(1 - |\rho|^2)\pi_s^{1/2}\sigma^{-2}}{1 + N\pi_s(1 - |\rho|^2)\sigma^{-2}} \tag{5.2.15}$$

It can be seen from the result that g_s, which may be taken as a measure of the magnitude of the coefficients, is proportional to $\pi_s^{1/2}$ for small π_s but, after peaking as π_s increases, eventually falls as $\pi_s^{-1/2}$. The latter result is expected since the array is in a signal-matching regime if the input and output SNRs are high. Calculating $dg_s/d\pi_s$ in eqn. (5.2.15) and equating to zero determines the input signal power at which g_s is maximised, the result being

$$\hat{\pi}_s = \frac{\sigma^2}{N(1 - |\rho|^2)} \tag{5.2.16}$$

and, at this point, the desired signal amplitude gain is

$$\hat{g}_s = \tfrac{1}{2}\pi_s^{-1/2} = \frac{1}{2\sigma}\{N(1 - |\rho|^2)\}^{1/2} \tag{5.2.17}$$

This result indicates that if the uncorrelated noise variance σ^2 is small (to be exact, small in comparison with the reference signal power) then $\hat{g}_s$ can assume large values, if the desired signal is weak, and as this demands large coefficient moduli there is a strong likelihood that a hardware limitation will occur. The signal power which produces maximum g_s is, from eqn. (5.2.16), the value which would produce unity output SNR if the weights had conventional values and there were no interference (assuming $|\rho|^2$ is small). As an LMS algorithm is likely to use spread-spectrum techniques to derive the reference (Fig. 4.3.4) whereas the adaptive algorithm operates at the full interference bandwidth, the particular operating conditions are not unreasonable, i.e. a subzero input SNR to the antenna and a high SNR after bandwidth compression. In order to prevent the development of large coefficients either the antenna output must be scaled by AGC action after the summing node in the antenna or the reference amplitude must be scaled down.

LMS descent algorithm convergence: Having discussed some of the properties of the 'ideal' solution in the case of a singular covariance matrix we now direct attention to how the LMS descent algorithm, for example eqn. (4.3.6), will operate in these circumstances. Convergence ceases in this equation when

$$\dot{\mathbf{W}} = -2\mu(\mathbf{RW} - \mathbf{D}) = 0$$

i.e.

$$\mathbf{RW} = \mathbf{D} \tag{5.2.18}$$

which agrees with eqn. (5.2.1) (after conversion to complex form) but this result is easily shown to be non-unique since if any vector $\mathbf{Z}$ external to the range space of $\mathbf{R}$ is taken:

$$\mathbf{RZ} = \mathbf{0} \tag{5.2.19}$$

it may be added to $\mathbf{W}$ without violating eqn. (5.2.18) since immediately

$$\mathbf{R}(\mathbf{W} + \mathbf{Z}) = \mathbf{D} \tag{5.2.20}$$

This superimposed $\mathbf{Z}$ component does not affect the amplitude gains for any sources and its directional pattern, if used as a weighting vector alone, would have nulls directed toward every source. Should the covariance matrix have a finite component due to uncorrelated noise then the $\mathbf{Z}$ vector will increase the error power and, in principle, it will gradually decay to zero. However, the rate of decay may be slow as to be negligible in practice. The same modification of the solution in eqn. (5.2.4) is permissible; that is not unique either.

Thus, for practical purposes, it appears that the LMS algorithm will converge to the optimum solution when the data covariance matrix is ill-conditioned or singular. Equation (5.2.8) might seem to indicate that an infinite look direction amplitude gain should appear if the desired signal vanishes but in practice either a finite amount of uncorrelated noise is allowed, as in eqn. (5.2.15), and the desired signal amplitude gain vanishes or, failing this, since the reference is uncorrelated with the element signals there is no mechanism for varying the desired signal amplitude gain which stays constant.

If the LMS algorithm is modified to the steered circuit shown in Fig. 4.4.1, the equilibrium solution is now eqn. (5.2.1) with a constant **S*** vector, one independent of desired signal amplitude, and the solution for g_s is derived from eqn. (5.2.10) if $\pi_s^{1/2}$ is replaced by 1. This leads to the result, for $\pi_v \to \infty$,

$$g_s = \frac{N(1 - |\rho|^2)\sigma^{-2}}{1 + N\pi_s(1 - |\rho|^2)\sigma^{-2}} \tag{5.2.21}$$

and we now find that g_s tends to a large value as π_s vanishes. What the antenna is doing is increasing its gain sufficiently that the I/0 thermal noise correlation vector offsets the **D** vector of biasing at the integrator inputs in Fig. 4.4.1. If the desired signal is appreciably mismatched with the steering vector or there are correlator offsets[36] rather similar effects occur though the increase in g_s is less marked as much of the required correlator outputs can be obtained by the I/0 correlation of the desired signal. The optimal weighting vector using this algorithm is a scaled version of that obtained by a linear look direction gain constraint and power minimisation and the desired signal mismatch phenomena discussed in detail in Chapter 6 apply to it.

The problems of the steered algorithm are reduced by supplementing the naturally occuring uncorrelated noise with a synthetic effect obtained by leaky integrators as discussed in Section 5.1 and even injection of true uncorrelated noise.[81] If the integrators are leaky the steering vector **D** may be applied at the integrator outputs after multiplying by the DC gain of the integrators and in this configuration the circuit is often called the 'Howells-Applebaum loop'.[56]

Linearly constrained algorithm: The solutions in the case of the algorithm with a hard look direction gain equality constraint are rather less complicated than the LMS ones because the constraint provides a fixed point of reference. If eqn. (5.2.4) is examined, the solutions (5.2.9) still apply but eqn. (5.2.8) is changed by virtue of the λ coefficient in eqn. (5.2.2) which is varied to fix the desired signal gain. We finish up with the same pattern, i.e. one with perfect nulls directed at interferences, but scaled for a fixed amplitude gain in the desired signal direction rather than one which varies with the desired signal power. Again any vector **Z** orthogonal to **R** may be added to **W** without destroying the nulls or changing the desired signal gain.

When implemented as a descent algorithm, adaptation ceases when, from eqn. (4.5.10),

$$\mathbf{PG}(k) = 2\mathbf{PRW}(k) = 0 \tag{5.2.22}$$

which corresponds either to $\mathbf{RW}(k) = \lambda \mathbf{S}_0^*$ as in eqn. (4.5.14) (with a change in variables) or

$$\mathbf{RW}(k) = 0 \tag{5.2.23}$$

and one or other of these equations always has a solution. In the presence of a non-zero desired signal eqn. (4.5.14) can be satisfied just as for the LMS algorithm above but in its absence eqn. (5.2.23) is found and the weight vector is orthogonal to every interference space vector though still having a certain gain in the assumed direction of the desired signal.

General weak uncorrelated noise solutions: If uncorrelated noise of power σ^2 is added to the data, when this has an intrinsically singular covariance matrix, and σ^2 is allowed to become very small then direct inversion solutions apply. Let **M** be the singular data covariance matrix and

$$\mathbf{R} = \mathbf{M} + \sigma^2\mathbf{I} \tag{5.2.24}$$

the augmented one. Only the case where the desired signal is absent is considered since the nature of the approximations to be used will not permit its inclusion. From eqn. (5.2.3), excluding $\mathbf{S}_0$, we have

$$\mathbf{R} = \sigma^2\mathbf{I} + \sum_{i=1}^{M} \pi_i \mathbf{S}_i^*\mathbf{S}_i^t = \sigma^2\mathbf{I} + \mathbf{Q}^*\mathbf{\Lambda}\mathbf{Q}^t \tag{5.2.25}$$

where N by M rectangular matrix **Q** is defined as

$$\mathbf{Q} \triangleq [\mathbf{S}_1 | \mathbf{S}_2 | \cdots | \mathbf{S}_M] \tag{5.2.26}$$

and $\mathbf{\Lambda} \triangleq \text{diag}(\pi_1, \pi_2, \ldots, \pi_M)$ replaces matrix **P** of eqn. (2.7.5). As in eqn. (2.7.6) we have

$$\mathbf{R}^{-1} = \sigma^{-2}\mathbf{I} - \sigma^{-4}\mathbf{Q}^*\{\mathbf{\Lambda}^{-1} + \sigma^{-2}\mathbf{Q}^t\mathbf{Q}^*\}^{-1}\mathbf{Q}^t \tag{5.2.27}$$

and, as σ^2 tends to zero,

$$\mathbf{R}^{-1} \to \sigma^{-2}\{\mathbf{I} - \mathbf{Q}^*(\mathbf{Q}^t\mathbf{Q}^*)^{-1}\mathbf{Q}^t\} \tag{5.2.28}$$

which will be recognised as a projection matrix **P**, eqn. (2.5.20), which projects vectors onto a subspace orthogonal to the vectors $\mathbf{S}_1, \ldots, \mathbf{S}_M$. Substitution into eqn. (5.2.2) yields

$$\mathbf{W}_{\text{opt}} = \lambda\mathbf{M}^{-1}\mathbf{S}^* = \lambda\sigma^{-2}\mathbf{P}\mathbf{S}^* \tag{5.2.29}$$

in which, for a linearly constrained solution, as $\sigma^2 \to 0$, λ will also decrease to compensate. Thus in the absence of a desired signal the weighting vector converges to a projection solution and this is known to be a minimum norm weighting vector for steering a certain number of nulls.

Undersampled covariance matrix—pseudoinverse: In the case that an SMI algorithm is implemented, Section 4.13, the covariance matrix is singular until, in eqn. (4.13.1), $K \geq N$ yet intuitively we may feel that sometimes an optimised solution must be definable for $K < N$. Consider the case of a three element array for which a good pattern can be formed with SMI using six data samples if there are two discrete interferences and the desired signal is relatively weak. If the size of the array is now increased to 7 elements but six data samples only are taken then the estimated covariance matrix is only of rank six and not invertible, as it is of order seven. However, using the same data, an 'optimised' solution is available of a sort even though it involves ignoring four of the elements in the worst case. It is tempting to use a pseudoinverse in this situation, taking

$$\mathbf{W}_{\text{opt}} = \hat{\mathbf{R}}_K^+ \mathbf{S}^* \tag{5.2.30}$$

as the optimal vector and the pseudoinverse can easily be calculated by forming the N by K matrix $\mathbf{Q}$:

$$\mathbf{Q} \triangleq [\mathbf{X}_1 | \mathbf{X}_2 | \cdots | \mathbf{X}_K] \tag{5.2.31}$$

consisting of the K data samples and using the formula[126]

$$\hat{\mathbf{R}}_K^+ = (\mathbf{Q}^* \mathbf{Q}^t)^+ = \mathbf{Q}^* (\mathbf{Q}^t \mathbf{Q}^*)^{-2} \mathbf{Q}^t \tag{5.2.32}$$

which assumes that $\mathbf{Q}$ is of rank K and $K \leq N$. Using eqn. (5.2.30) is equivalent to reducing the dimensionality of the problem since $\hat{\mathbf{R}}_K$ may be diagonalised as

$$\mathbf{\Lambda}_K = \mathbf{Z}^H \hat{\mathbf{R}}_K \mathbf{Z} = \text{diag}\,(\lambda_1, \lambda_2, \ldots, \lambda_K, 0, 0, 0, \ldots, 0) \tag{5.2.33}$$

where $\mathbf{Z}$ is the eigenvector matrix and the pseudoinverse is just

$$\hat{\mathbf{R}}_K^+ = \mathbf{Z} \,.\, \text{diag}\,(\lambda_1^{-1}, \lambda_2^{-1}, \ldots, \lambda_K^{-1}, 0, 0, 0, \ldots, 0) \,.\, \mathbf{Z}^H \tag{5.2.34}$$

If the transformation $\mathbf{Z}$ were used as a preprocessor, as in Fig. 3.2.1, the number of channels in $\hat{\mathbf{X}}(t)$ which are non-zero over the K input samples would be just K and the preprocessing would be equivalent to dimensionality reduction—but in eigenspace rather than physical array space. The solution for the eigenspace weighting vector $\mathbf{U}$ is, from eqn. (5.2.30),

$$\begin{aligned} \mathbf{U} = \mathbf{Z}^H \mathbf{W}_{\text{opt}} &= \mathbf{\Lambda}_K^+ \mathbf{Z}^H \mathbf{S}^* = \mathbf{\Lambda}_K^+ \mathbf{s}^* \\ &= \sum_{i=1}^{K} \frac{\mathbf{s}_i^*}{\lambda_i} \end{aligned} \tag{5.2.35}$$

where vector $\mathbf{s}^* = \mathbf{Z}^H \mathbf{S}^*$ contains the amplitudes of the projections of $\mathbf{S}^*$ onto the eigenvectors while $\mathbf{U}$ contains the same elements but inversely weighted by the eigenvalues. Thus the weight vector is drawn into a direction which lies among the eigenvectors of $\hat{\mathbf{R}}_K$ with small eigenvalues and tends to be almost orthogonal to those with large eigenvalues. In general the latter are caused by the strongest interferences—hence the null steering effect. The weighting vector is precisely orthogonal to the eigenvectors with null eigenvalues though, in general, $\mathbf{S}^*$ will have components in such directions which are thus ignored. Substitution of $\mathbf{W}_{\text{opt}}$ from eqn. (5.2.30) into eqn. (5.2.1) yields

$$\hat{\mathbf{R}}_K \mathbf{W}_{\text{opt}} = \hat{\mathbf{R}}_K \hat{\mathbf{R}}_K^+ \mathbf{S}^* = \mathbf{P}\mathbf{S}^* = \tilde{\mathbf{S}}^* \tag{5.2.36}$$

where $\mathbf{P}$ is a projection matrix which sets to zero any vector components external to the range space of $\hat{\mathbf{R}}_K$ ($\mathbf{P}$ is an Hermitian idempotent).[132] Thus we may interpret $\tilde{\mathbf{S}}^*$ as a closest approximation to $\mathbf{S}^*$ in the range space of $\hat{\mathbf{R}}_K$ and eqn. (5.2.36) is a least squares approximation to the expectation result, eqn. (5.2.1), in some sense.

If $K > N$ and $\mathbf{R}$ is intrinsically singular, the pseudoinverse becomes difficult to compute; eqn. (5.2.32) is inapplicable and it is almost impossible to use eqn. (5.2.34) since computation and quantisation errors will perturb the null eigenvalues.

Undersampled matrix—projection solution: A technique which works, without modification, for both undersampled and intrinsically singular covariance matrices is one which minimises the antenna output power over the sampling interval.[59] Indeed, because a singular covariance matrix has one or more eigenvectors with null eigenvalues, a weighting vector which is a linear combination of such eigenvectors will be orthogonal to all the data samples and the antenna will output zero energy over the sampling interval.

The operation yielding this weighting vector is a projection onto a subspace orthogonal to $\hat{\mathbf{R}}_K$ (or every data sample) and the projection matrix may be found from eqn. (2.5.20) or, probably more conveniently, from eqn. (5.2.24) by direct inversion since this will yield, as in eqn. (5.2.28), the same result apart from a scaling factor. The latter result enables the technique to be related to the pseudoinverse above since if the eigenvalues of $\hat{\mathbf{R}}_K$ are augmented by $\sigma^2\mathbf{I}$ in eqn. (5.2.24) then, in eqn. (5.2.33), all eigenvalues are non-zero and the pseudoinverse reverts to a normal inverse with eigenvalues of either $(\lambda_i + \sigma^2)^{-1}$ for $i = 1, \ldots, K$ or σ^{-2} for $i = k + 1, \ldots, N$. Using these eigenvalues in eqn. (5.2.35) causes $\mathbf{W}_{\text{opt}}$ to be drawn into the subspace spanned by the eigenvectors of $\hat{\mathbf{R}}_K$ which had, initially, null eigenvalues. Thus the effect of augmenting $\hat{\mathbf{R}}_K$ by $\sigma^2\mathbf{I}$ is to activate the subspace which is dormant in the pseudoinverse solution.

The weighting vector derived by this procedure, being orthogonal to all previous data samples, cannot be used to filter old data, i.e. that on the interval $1, \ldots, K$, and must be used on succeeding data samples before they are incorporated into the covariance matrix estimate. If there is an appreciable desired signal component in the data the attempts of the algorithm to form weight vectors orthogonal to its space vector may produce difficulties and this projection approach is best suited to zero-signal adaptation conditions.

This projection technique was used in the early stages of sampling for Figs. 4.13.2, with σ^2 approximately 40 dB below the strongest interference in eqn. (4.13.4), and it is clear that significant cancellation of the interference is taking place after only three or four samples have been used to estimate $\hat{\mathbf{R}}_K$ in spite of the antenna having nine elements.

Nonlinear approach: For rapid convergence in the presence of a desired signal a feasible algorithm consists of minimising the antenna output power over the sampling interval combined with the application of a norm bound to the weighting vector. If the numerical algorithm described in Appendix 4, which uses an eigenstructure analysis, is employed there is no difficulty with singular matrices and the same procedure can be used irrespective of number of data samples or intrinsic conditioning of the data. The action of the norm bound will prevent desired signal suppression but is unlikely to improve the output SNIR above the usual SMI result.[29]

5.3 Finite bandwidth effects[66, 77]

Free-field effects: Even without multipath arrivals the bandwidth affects the covariance matrix of the signals and the effect becomes deleterious when the bandwidth exceeds a few percent. The general effect is that each source splits into a spectrum of sources, a number of which may be strong enough to require cancellation. To examine the problem we consider a single signal source which has a frequency dependent space vector $\mathbf{S}(\omega)$. In a small band about the mid-frequency ω_0 a truncated Taylor series representation of $\mathbf{S}(\omega_0 + \Delta)$ is sufficiently accurate:

$$\mathbf{S}(\omega_0 + \Delta) = \mathbf{S}_0 + \Delta\mathbf{S}_1 + \tfrac{1}{2}\Delta^2\mathbf{S}_2 + \cdots \tag{5.3.1}$$

Where $\mathbf{S}_i$ is the vector of ith derivatives of $\mathbf{S}(\omega_0)$ with respect to ω.

Consider the case of a rectangular source spectral density of total power unity extending from $\omega_0 - \Delta_0$ to $\omega_0 + \Delta_0$. The covariance mátrix of this single source is obtained by averaging over frequency with the assumption that all frequency spectrum components are uncorrelated.

$$\mathbf{R} \simeq \frac{1}{2\Delta_0}\int_{\omega_0-\Delta_0}^{\omega_0+\Delta_0} [\mathbf{S}_0 + \Delta\mathbf{S}_1 + \tfrac{1}{2}\Delta^2\mathbf{S}_2]^*[\mathbf{S}_0 + \Delta\mathbf{S}_1 + \tfrac{1}{2}\Delta^2\mathbf{S}_2]^t \, d\omega \tag{5.3.2}$$

$$= \mathbf{S}_0^*\mathbf{S}_0^t + \tfrac{1}{3}\Delta_0^2\mathbf{S}_1^*\mathbf{S}_1^t + \tfrac{1}{20}\Delta_0^4\mathbf{S}_2^*\mathbf{S}_2^t + \tfrac{1}{6}\Delta_0^2(\mathbf{S}_0^*\mathbf{S}_2^t + \mathbf{S}_2^*\mathbf{S}_0^t) \tag{5.3.2}$$

This equation can be rewritten as

$$\mathbf{R} = \left(\mathbf{S}_0 + \frac{\Delta_0^2}{6}\,\mathbf{S}_2\right)^*\left(\mathbf{S}_0 + \frac{\Delta_0^2}{6}\,\mathbf{S}_2\right)^t + \frac{\Delta_0^2}{3}\,\mathbf{S}_1^*\mathbf{S}_1^t + \frac{\Delta_0^4}{45}\,\mathbf{S}_2^*\mathbf{S}_2^t \tag{5.3.3}$$

For suitably chosen phase centre the vector $\mathbf{S}_1$ can be made orthogonal to $\mathbf{S}_0$ and $\mathbf{S}_2$ (examples are given later). If an eigenvector representation of $\mathbf{R}$ is considered:

$$\mathbf{R} = \lambda_0\,\mathbf{U}_0^*\mathbf{U}_0^t + \lambda_1\,\mathbf{U}_1^*\mathbf{U}_1^t + \lambda_2\,\mathbf{U}_2^*\mathbf{U}_2^t + \cdots \tag{5.3.4}$$

then we may equate it term by term with expansion (5.3.3) so long as the terms in (5.3.3) have orthogonal vectors. This fails at the third term as $\mathbf{S}_2$ is not orthogonal to $\mathbf{S}_0$, but as the third term is very small relatively speaking when the bandwidth is small it may be neglected. Thus we have a two term spectral approximation to $\mathbf{R}$ with

$$\lambda_0 \simeq \left\|\mathbf{S}_0 + \frac{\Delta_0^2}{6}\,\mathbf{S}_2\right\|^2$$

$$\lambda_1 \simeq \frac{\Delta_0^2}{3}\,\|\mathbf{S}_1\|^2 \tag{5.3.5}$$

which will be adequate to explain the first order of effects of finite bandwidths.

Line array: For a uniformly spaced line array with element co-ordinates x_n and a source at α,

$$\mathbf{S}_0 = \left\{\exp\left[-j\frac{\omega}{c}x_n\sin(\alpha)\right]\right\}$$

$$\mathbf{S}_1 = \left\{-j\frac{x_n}{c}\sin(\alpha)\exp\left[-j\frac{\omega}{c}x_n\sin(\alpha)\right]\right\}$$

$$\mathbf{S}_2 = \left\{-\frac{x_n^2}{c^2}\sin^2(\alpha)\exp\left[-j\frac{\omega}{c}x_n\sin(\alpha)\right]\right\} \qquad n = 1, \ldots, N \tag{5.3.6}$$

Orthogonality of $\mathbf{S}_1$ and $\mathbf{S}_0$ is obtained if

$$\mathbf{S}_0^H\mathbf{S}_1 = -j\frac{1}{c}\sin(\alpha)\sum_{n=1}^{N}x_n = 0 \tag{5.3.7}$$

which requires that the centre of the array be taken as the phase centre. If the array is uniformly spaced, as assumed, then $\mathbf{S}_2^H\mathbf{S}_1 = 0$ too and all the necessary conditions are satisfied.

From eqns. (5.3.6) and (5.3.5) we have, for small Δ_0^2,

$$\lambda_0 \simeq N \qquad \lambda_1 \simeq \frac{1}{3}\frac{\Delta_0^2}{c^2}\sin^2(\alpha)\sum_{n=1}^{N}x_n^2 \tag{5.3.8}$$

Let F_b be the fractional bandwidth, $2\Delta_0/\omega_0$. Noting that $c = \lambda\omega_0/2\pi$ we have

$$\lambda_1 \simeq \frac{1}{3}F_b^2\pi^2\sin^2(\alpha)\sum_{n=1}^{N}\frac{x_n^2}{\lambda^2} \tag{5.3.9}$$

Circular array: For the array shown in Fig. 5.3.1,

$$\mathbf{S}_0 = \left\{\exp\left[-j\frac{\omega R}{c}\cos(\alpha_i - \alpha_s)\right]\right\}$$

$$\mathbf{S}_1 = \left\{-j\frac{R}{c}\cos(\alpha_i - \alpha_s)\exp\left[-j\frac{\omega R}{c}\cos(\alpha_i - \alpha_s)\right]\right\}$$

$$\mathbf{S}_2 = \left\{-\frac{R^2}{c^2}\cos^2(\alpha_i - \alpha_s)\exp\left[-j\frac{\omega R}{c}\cos(\alpha_i - \alpha_s)\right]\right\} \tag{5.3.10}$$

Orthogonality of $\mathbf{S}_0$ and $\mathbf{S}_1$ requires that

$$\frac{R}{c}\sum_{i=1}^{N}\cos(\alpha_i - \alpha_s) = 0 \tag{5.3.11}$$

which places the phase centre at the centre of gravity of the array. Orthogonality of $\mathbf{S}_1$ and $\mathbf{S}_2$ demands that

$$\sum_{i=1}^{N}\cos^3(\alpha_i - \alpha_s) = 0 \tag{5.3.12}$$

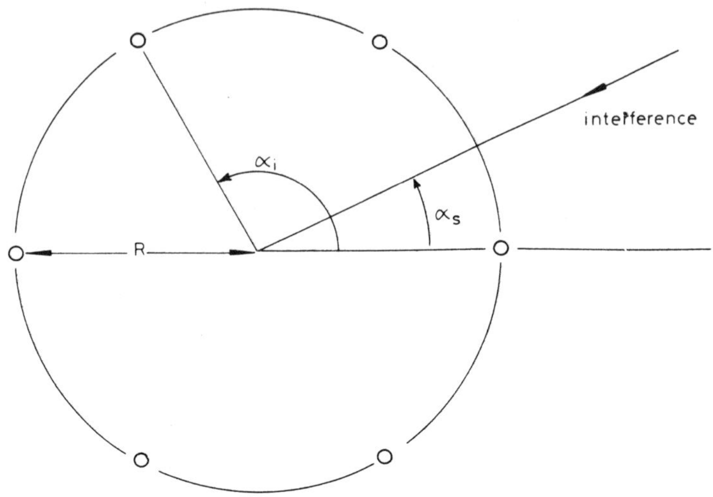

Fig. 5.3.1 *Circular array geometry*

which requires additionally a uniformly spaced array. Substituting (5.3.10) into (5.3.5) yields

$$\lambda_0 \simeq N$$
$$\lambda_1 \simeq \frac{\Delta_0^2}{3} \frac{R^2}{c^2} \sum_{i=1}^{N} \cos^2(\alpha_i - \alpha_s) \simeq \frac{I\pi^2 F_b^2}{3\lambda^2} \tag{5.3.13}$$

where I is the moment of inertia of the array with the line joining the phase centre to the source as axis. For a uniformly spaced circle of N elements

$$\lambda_1 \simeq \frac{\pi^2 N R^2 F_b^2}{6\lambda^2} \tag{5.3.14}$$

For both line and circular arrays the third eigenvalue is expected to vary as F_b^4 and within a few dB is given by

$$\lambda_2 \simeq \frac{\Delta_0^4}{45} \|\mathbf{S}_2\|^2 \tag{5.3.15}$$

Figs. 5.3.2 and 5.3.3 show exact numerical results for the eigenvalues of the covariance matrix for a 7-element line array with $\lambda/2$ mid-band spacing and a 15-element circular array with $R = 1{\cdot}072\lambda$ ($0{\cdot}45\lambda$ element spacing). The approximate theoretical result agrees perfectly with the exact one for bandwidths less than 0·1.

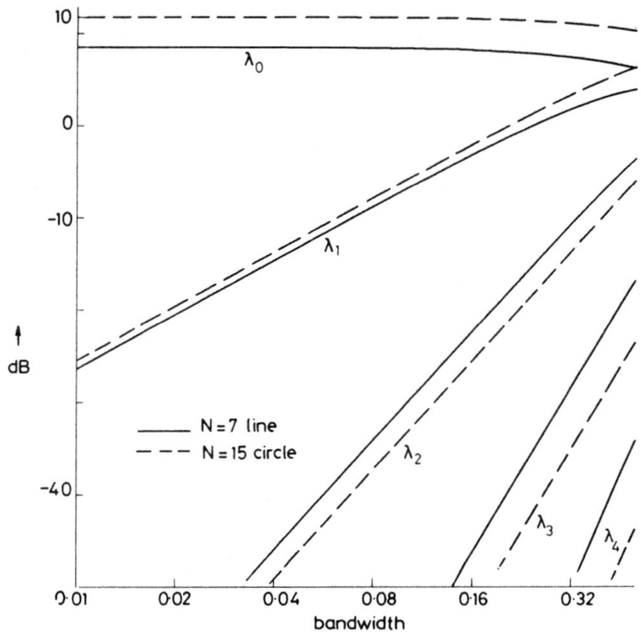

Fig. 5.3.2 *Eigenvalues of R for a single source in a 7-element line and a 15-element circular array*

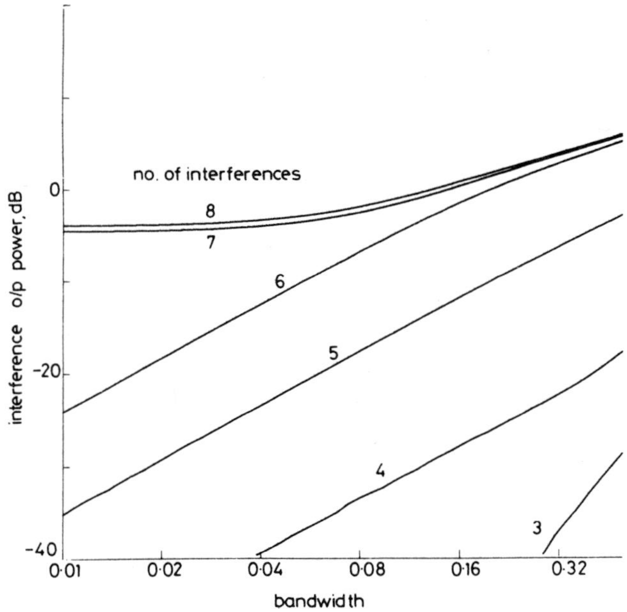

Fig. 5.3.3 *Output power against receiver bandwidth in a 7-element array with up to 8 broad-band sources*

Effect of eigenvalues on cancellation of interference : For very small fractional bandwidths F_b the value of all eigenvalues, except λ_0 will fall below the uncorrelated noise level and as the array treats each source as a mid-band sine wave excellent results will be obtained. As F_b increases λ_1 will eventually emerge from the noise background and if good cancellation is to be maintained the array must steer a null toward it, consuming an extra degree of freedom, and producing a broadened null in the frequency domain. In the case of a line array the frequency and bearing domains are interchangeable to a certain extent and the array will have a broad null in its directional pattern too.[70, 77] This will not happen with the circular array.

If the fractional bandwidth is so wide that λ_2 appears above the uncorrelated noise level then a third degree of freedom must be used to maintain perfect cancellation and the null will get even wider.

If there are several sources present which are well separated in bearing then they will tend to generate independent sets of eigenvalues and the array must cancel them all. If it has limited degrees of freedom it must ignore the weakest eigenvalues and cancel the dominant ones. Thus the residual interference output power will grow either as F_b^4 or as F_b^2. Figure 5.3.3 shows the residual interference output power as a function of fractional bandwidth when 0 dB sources are cumulatively positioned at 90°, 45°, −45°, −30°, 20°, −15°, −20°, and −75°. Interference output power is proportional to F_b^4 for 3 sources present but to F_b^2 if there are 4, 5, or 6. With 3 sources the array can cancel the eigenvectors corresponding to all the λ_0 and λ_1 eigenvalues and the residue is of the λ_2 eigenvalues. With 4 sources there is an uncancelled λ_1 eigenvector source and residue rises as F_b^2. With 7 or more sources the array is overrun even for zero bandwidth and F_b is a secondary factor. However, there is some increase in output power for large F_b.

Apart from their effect on the optimal solution the appearance of small eigenvalues in the covariance matrix is likely to lead to slow convergence in gradient algorithms.

Finite bandwidth propagation effects : The effects discussed above refer to a free-field condition. In the presence of multipath effects from scatters close to the array finite bandwidth effects may have more serious consequences. Simple examples of these effects are discussed in Section 5.5. If an antenna location is being investigated for suitability with adaptive algorithms the simplest experimental procedure would be to measure the covariance matrix of the field of a single far-field source with a pair of movable elements and a two channel correlator. In the absence of such a device the outputs of the elements could be combined after phase shifting and the summed power function measured as the phase is varied. For equal element powers this will provide both the phase and modulus of the correlation (Appendix A1.5).

Other effects can occur within the hardware itself as a consequence of transmission line mismatches, mutual couplings in the array etc.

5.4 Polarisation, near-field scattering, and E/H disparities

The effects of polarisation on adaptive antennas are surprisingly complicated. The simplest case to consider is that the interference source, whether with plane, elliptical, or circular polarisation has only one modulation function, which is narrow-band, and the receiving array is in free space. Under these conditions the signals arriving at the array elements, whatever their orientation, are completely coherent. The same conclusion holds if there is scattering from objects or bodies into the array aperture but these are so close that the modulation functions of the direct and secondary arrivals are completely coherent (Section 2.7). Under these conditions, only one degree of freedom is required to cancel each source.

Consider next that the array remains in free space but the transmitter radiates equal powers on two orthogonal polarisations, either plane or circular, and can be considered as unpolarised statistically speaking. The antenna array now receives two space vectors with different modulations:

$$\mathbf{X}(t) = m_1(t)\mathbf{S}_1 + m_2(t)\mathbf{S}_2 \tag{5.4.1}$$

If the aperture is planar and the array elements are parallel it is probable that the vectors $\mathbf{S}_1$ and $\mathbf{S}_2$ are very similar and steering a null at one modulation signal arrival will simultaneously steer a null at the other. To a first approximation one degree of freedom is enough to cancel the transmitter completely. If the aperture is nonplanar or has diversely oriented elements this is no longer true and the vectors $\mathbf{S}_1$ and $\mathbf{S}_2$ will be quite different. The antenna must now form two nulls, one for $\mathbf{S}_1$ and the other for $\mathbf{S}_2$ and two degrees of freedom are consumed for each transmitter.

If the aperture is planar but there is a near-field scattering point then it is unlikely that the scattering strength will be the same for the two planes of polarisation. Thus we now receive the signal

$$\mathbf{X}(t) = m_1(t)[\mathbf{S}_1 + \mathbf{F}_1] + m_2(t)[\mathbf{S}_1 + \mathbf{F}_2] \tag{5.4.2}$$

where $\mathbf{F}_1$ and $\mathbf{F}_2$ are scattered signal space vectors and the space vectors for the two modulation signals will diverge. Again, two degrees of freedom are required to cancel the one transmitter. If the elements have a joint sensitivity to E and H field components, as is the case for example with resonant loops, then the near-field scattering point will produce a space vector which is different for the E and H fields and moreover which also varies with the plane of polarisation of the incident field. This might tend to accentuate the divergence of the space vectors in eqn. (5.4.2).

If the radiated signal is very broad band it is possible that a near field scattering point will produce a signal in the array aperture which is significantly decorrelated from the direct arrival. In this case the effects discussed above for orthogonal modulation functions will appear even if there is only one modulation function at the transmitter. This indicates the importance of minimising the bandwidth of the receiver in adaptive applications.

When there is a scattering point, for example the sea surface, which is so far from the direct path that the modulation function is uncorrelated with the direct arrival then for all practical purposes it must be treated as a second interference source. The effects discussed above then apply equally to the direct and the secondary arrivals.

5.5 Effects of coherent multipath[109]

The signal from a source may arrive via several paths of different time delays and attenuations. Typical examples might be wing scattering in aircraft[37, 137] and mast and deck scattering for a ship's radar. In the underwater case the sea surface and bottom are the most significant scattering surfaces. For the bandwidths typical of radio systems it is unlikely that motion of the scatterers will be effective in causing additional decorrelation over that due to path length differences. The same might not be true for a passive sonar environment where the receiver bandwidth may be only a few Hz.

Effect on the desired signal: The LMS algorithm, by definition, finds the best weighting vector in the output SNIR sense for an arbitrary input signal and multipath is not expected to be a serious problem here unless the various desired signal arrivals are fluctuating so rapidly that the algorithm cannot follow the changes fast enough. This is particularly likely to be the case if the desired signal is very weak for its convergence is then rather slow as seen in the last chapter.

The linearly constrained minimum output power algorithm is in a very different category since its success in converging to a good weighting vector depends critically upon *a priori* specification of the desired signal space vector. Some indication of the likely magnitude of the problems can be obtained by analysing the gain of the antenna when the desired signal with multipath is present in the absence of interference. Thus let the antenna data vector assuming one secondary arrival be

$$\mathbf{X}(t) = m_1(t)\mathbf{S}_1 + m_2(t)\mathbf{S}_2 \tag{5.5.1}$$

where in general the modulation function m_2 will be a delayed version of m_1

$$m_2(t) = am_1(t - T) \tag{5.5.2}$$

a being the attenuation of the secondary path. The steering vector is assumed to be aligned correctly with $\mathbf{S}_1$:

$$\mathbf{C} = \mathbf{S}_1 \tag{5.5.3}$$

such that the constraint $\mathbf{C}'\mathbf{W} = g$ fixes the amplitude gain of the antenna at g for the primary desired signal arrival.

If the arrival producing vector $\mathbf{S}_2$ comes from a direction significantly different from the primary arrival then the array can produce a lobe of arbitrary magnitude in its direction while keeping the primary amplitude gain at g. Assume that

$$\mathbf{S}_2^t \mathbf{W} = b \tag{5.5.4}$$

defines the amplitude gain in the secondary direction, In general, b will be complex, Since the desired signal and its secondary are the only signals present the antenna output signal is a function of the single parameter b:

$$y(t) = m_1(t)\mathbf{S}_1^t \mathbf{W} + m_2(t)\mathbf{S}_2^t \mathbf{W} = m_1(t) \, . \, g + m_2(t) \, . \, b \tag{5.5.5}$$

and as it minimises its output power we find the minimum of $|y(t)|^2$ by variation of b. We have

$$\begin{aligned} p = E[|y(t)|^2] &= E[|m_1|^2]|g|^2 + |b|^2 E[|m_2|^2] \\ &\quad + 2 \text{ Re } (E[m_1 m_2^* g b^*]) \\ &= \pi_1 |g|^2 + |b|^2 \pi_2 + 2(\pi_1 \pi_2)^{1/2} \text{ Re } (g b^* \rho) \end{aligned} \tag{5.5.6}$$

where ρ is the complex correlation coefficient between m_1 and m_2. Differentiation with respect to b yields

$$\begin{aligned} \partial p/\partial \text{ Re } (b) &= 2\pi_2 \text{ Re } (b) + 2(\pi_1 \pi_2)^{1/2} \text{ Re } (\rho g) \\ \partial p/\partial \text{ Im } (b) &= 2\pi_2 \text{ Im } (b) + 2(\pi_1 \pi_2)^{1/2} \text{ Im } (\rho g) \end{aligned} \tag{5.5.7}$$

Equating to zero gives

$$\text{Re } (b) = -(\pi_1/\pi_2)^{1/2} \text{ Re } (\rho g), \qquad \text{Im } (b) = -(\pi_1/\pi_2)^{1/2} \text{ Im } (\rho g)$$

and finally

$$b = -(\pi_1/\pi_2)^{1/2} \rho g \tag{5.5.8}$$

Substitution back into (5.5.6) gives the minimum output power

$$\begin{aligned} p_{\min} &= \pi_1 |g|^2 + \pi_1 |\rho g|^2 - 2\pi_1 |\rho g|^2 \\ &= \pi_1 |g|^2 (1 - |\rho|^2) \end{aligned} \tag{5.5.9}$$

As the output power of a conventionally weighted antenna would be $\pi_1 |g|^2$ the presence of correlation between the primary and secondary paths signals invariably leads to a reduction in output desired signal power which becomes progressively more serious as $|\rho|$ tends to unity. This will reduce the output signal to interference ratio if interference is also present.

Effect of multipath on interference cancellation : The problems associated with multipath of the interference signals is not quite so serious as with the desired signal, the main effect being that the antenna must steer nulls at several sources rather than one for each interference transmitter. It was shown in Section 2.7

how the covariance matrix of a two source situation could be resolved into its eigenstructure and if the modulation correlation is high one of the eigenvalues may be quite weak.

Estimation of coherence between delayed arrivals: The correlation coefficient ρ is proportional to the complex autocovariance function of the transmitted waveform.[129] If the transmitted waveform, after filtering by the receiver has power density spectrum $P(f)$ the autocovariance ψ is

$$\psi(t) = \int_{-\infty}^{\infty} P(f)e^{j2\pi ft}\, df \tag{5.5.10}$$

and

$$\rho(t) = \psi(t)/\psi(0) \tag{5.5.11}$$

5.6 Coefficient quantisation[57, 25]

Provision of continuously variable coefficients may be inconvenient for certain purposes, such as when an array is used both for transmission and reception and must carry appreciable power. If the coefficients are quantised in uniform steps of quadrature amplitude gain the following simple theory is quite accurate. If phase and amplitude weighting is used the magnitude of the errors tends to be function of the coefficient value and some approximation is involved.

Let $\mathbf{W}_{\text{opt}}$ be the optimal weighting value and suppose that the effective coefficients are given by

$$\mathbf{W} = \mathbf{W}_{\text{opt}} + \sigma\mathbf{U} \tag{5.6.1}$$

where σ is the standard deviation of the coefficient errors and $\mathbf{U}$ is a vector of uncorrelated elements whose variances are unity.

$$\text{Cov}\,(\mathbf{U}) = \mathbf{I} \tag{5.6.2}$$

If the size of the quadrature coefficient steps is Δ then the mean square error in each coefficient is

$$\text{MSE} = \Delta^2/12 \tag{5.6.3}$$

a result obtained from standard quantisation theory. For uncorrelated errors on the quadrature coefficients the total MSE is $\Delta^2/6$. Thus

$$\sigma^2 = \Delta^2/6 \tag{5.6.4}$$

The probability distribution of the elements of $\mathbf{U}$ is rectangular in the real and imaginary parts, extending from $-\sqrt{1{\cdot}5}$ to $+\sqrt{1{\cdot}5}$ on each axis.

The output power of the antenna is given by

$$\begin{aligned} E(p) &= E[\mathbf{W}_{\text{opt}} + \sigma\mathbf{U})^H\mathbf{R}(\mathbf{W}_{\text{opt}} + \sigma\mathbf{U})] \\ &= \mathbf{W}_{\text{opt}}^H\,\mathbf{R}\mathbf{W}_{\text{opt}} + \sigma^2 E(\mathbf{U}^H\mathbf{R}\mathbf{U}) \end{aligned} \tag{5.6.5}$$

Now

$$E(\mathbf{U}^H\mathbf{R}\mathbf{U}) = E[\mathrm{Tr}\,(\mathbf{R}\mathbf{U}\mathbf{U}^H)] = \mathrm{Tr}\,[\mathbf{R}E(\mathbf{U}\mathbf{U}^H)] = \mathrm{Tr}\,(\mathbf{R}) \tag{5.6.6}$$

Thus

$$E(p) = p_{\mathrm{opt}} + \sigma^2\,\mathrm{Tr}\,(\mathbf{R}) = p_{\mathrm{opt}}\left(1 + \frac{\sigma^2\,\mathrm{Tr}\,(\mathbf{R})}{p_{\mathrm{opt}}}\right) \tag{5.6.7}$$

and the output power is increased in the ratio indicated. We now find where this extra output power is coming from. For interference no. n with space vector $\mathbf{S}_n$ and power π_n we have an output power of

$$p_n = \pi_n E[|\mathbf{W}^t\mathbf{S}_n|^2] = \pi_n E[|(\mathbf{W}_{\mathrm{opt}} + \sigma\mathbf{U})^t\mathbf{S}_n|^2] \tag{5.6.8}$$

If we assume that the optimal weighting vector has a deep null directed toward the interference then $\mathbf{W}^t_{\mathrm{opt}}\,\mathbf{S}_n \simeq 0$ and

$$p_n \simeq \pi_n E|\sigma\mathbf{U}^t\mathbf{S}_n|^2 = \sigma^2\pi_n E(\mathbf{S}_n^H\mathbf{U}^*\mathbf{U}^t\mathbf{S}_n) = \sigma^2\pi_n N \tag{5.6.9}$$

if $\mathbf{S}_n$ is normed such that $\mathbf{S}_n^H\mathbf{S}_n = N$. Thus the output power caused by each source is proportional to its input power.

As the interference input power will normally be much greater than the desired signal input power if the adaptive antenna is offering any useful function it is reasonable to assume that all the excess output power is due to interference. If there is a dominant jammer of power $\pi_{\max}$ then $\mathrm{Tr}\,(\mathbf{R}) \simeq N\pi_{\max}$ and the jammer output power due to quantisation is about

$$p_n(\max) = N\sigma^2\pi_{\max} \tag{5.6.10}$$

If we require that the excess output power should equal the thermal noise output power of $N\sigma_n^2$ in the absence of interference, for unit modulus coefficients, then

$$\sigma^2 = \sigma_n^2/\pi_{\max} = (\mathrm{JNR})^{-1} \tag{5.6.11}$$

and substitution into eqn. (5.6.4) results in

$$\Delta = 6^{1/2}(\mathrm{JNR})^{-1/2} \tag{5.6.12}$$

For a dominant interference of 40 dB JNR we have $\Delta = 0{\cdot}0245$ and to cover the amplitude range -1 to $+1$ would require some 82 steps, possibly implemented by 7-bit control lines.

The control of quantised coefficients in a gradient algorithm is difficult as the update steps tend to be so small that they fail to move any of the coefficients. A suitable algorithm is discussed in Section 4.12.

It is desirable to keep the coefficient moduli as close as possible to unity in order that the relative size of the quantisation steps is minimum. This is not difficult to achieve in constrained algorithms where the coefficients do not vary very much from their conventional settings unless there is main-lobe interference. However in the LMS algorithm the weights may tend to vary with the level of the desired signal. In this case a feedback into the receiver AGC should be established in order that the coefficients maintain a reasonably constant amplitude.

The use of the fixed preprocessor described by Griffiths[52] has the potential for optimisation of the dynamic ranges of the variable coefficients (see Fig. 7.11.1).

5.7 Signal quantisation[25, 57]

The possibility of quantisation of the array signals is present in radar and sonar digital processing. We can represent the quantised data vector $\tilde{\mathbf{X}}(k)$ as

$$\tilde{\mathbf{X}}(k) = \mathbf{X}(k) + \mathbf{U}(k) \tag{5.7.1}$$

where $\mathbf{U}(k)$ is a random vector of complex quantisation errors. For quantisation steps of Δ in the A/D converters (referred to the input voltage) we have

$$\text{Cov}\,(\mathbf{U}) = \sigma^2\mathbf{I} = \frac{\Delta^2}{6}\,\mathbf{I} \tag{5.7.2}$$

having twice the variance as for single real channel quantisation. Since $\mathbf{U}$ may be assumed to be uncorrelated with $\mathbf{X}$ if the number of bits of quantisation exceeds 5 or 6 we have

$$\text{Cov}\,(\mathbf{R}) = E(\tilde{\mathbf{X}}^*\tilde{\mathbf{X}}^t) = \mathbf{R} + \sigma^2\mathbf{I} \tag{5.7.3}$$

and the effect of quantisation is equivalent to the input of uncorrelated noise into the elements.

This uncorrelated noise is little problem for the algorithm and in fact is likely to be beneficial. However, it will reduce the output SNR and affect performance. The best strategy is to allow the quantisation noise to relate to the thermal noise. Thus suppose the dominant interference is presented to the A/D converter, whose input range is -1 to $+1$, at a level of -10 dB so that the overload level is $3{\cdot}162\sigma_{\text{int}}$ for Gaussian modulation. Such a high overhead is essential since the effect of transient overload is likely to feed a lot of erroneous updates into the algorithm. For a 2 M step digitiser the quantisation noise is given by eqn. (5.7.2) as $1/6\ \text{M}^2$ while if the input JNR is Q_{in} the thermal noise level is $0{\cdot}1/Q_{\text{in}}$. Setting the quantisation noise power equal to $\frac{1}{4}$ that of the thermal noise yields

$$M \simeq (1{\cdot}5Q_{\text{in}})^{1/2} \tag{5.7.4}$$

Thus if Q_{in} is 10^5 for a 50 dB JNR, then $M = 387$ and 10 bits of accuracy would be required in the A/D converter.

Chapter 6

Main-lobe constraints

6.1 Introduction

Main-lobe constraints are very important in linearly constrained minimum output power arrays of the type discussed in Section 3.4 because the look direction constraint is only exactly effective at a single point in the directional pattern and its influence decreases rapidly as distance from the look direction increases.[97] Lack of control over the beam pattern allows the formation of nulls in the main lobe quite close to the look direction.[89] Although this is beneficial if there is a jammer close to the look direction and the desired signal is exactly at the look direction, improving output signal to interference ratio, it is disasterous if the desired signal strays from the look direction because the array will then cancel it just as effectively as an interference. The 'steered' LMS algorithm discussed in the first part of Section 3.4 is equally prone to main-lobe distortion since it outputs the same SNR as a system with a hard look direction constraint. However, the true LMS algorithm, with a pilot signal sent from the source, is immune to loss of SNR since it creates its own steering vector which tracks the space vector of the desired signal.

In this chapter, techniques for extending the control on the beam pattern to finite angular sectors are developed and this is done by a combination of linear and nonlinear constraints in addition to the simple look direction constraint. However, application of these is not without drawbacks. The ability of the antenna to cancel main-lobe interference is greatly reduced and performance of the antenna in the presence of main lobe jamming is substantially inferior to the LMS algorithm. Consequently one can only aim to improve performance in scenarios where there is sidelobe jamming only.

The constraints are useful in controlling the shape of the main lobe if it used for direction finding purposes, e.g. in a tracking radar since a main-lobe interference, even if it is relatively weak and does not affect detection, will distort the lobe shape of an adaptive antenna and shift the direction of maximum response.

6.2 Formation of main-lobe notches

The conditions under which an antenna will form a notch in its main lobe are now derived. The term notch is used to describe an imperfect null. Much of the theory has already been worked out in Section 5.1 but is repeated with a hard constraint on the look direction amplitude gain as this will turn out to be necessary in the additional constraints used later.

Assume the presence of a source of power π_v in uncorrelated noise of power σ^2 giving rise to a covariance matrix of the form

$$\mathbf{M} = \sigma^2\mathbf{I} + \pi_v \mathbf{V}^*\mathbf{V}^t \tag{6.2.1}$$

with an inverse of

$$\sigma^2\mathbf{M}^{-1} = \mathbf{I} - \frac{\dfrac{\pi_v}{\sigma^2}\mathbf{V}^*\mathbf{V}^t}{1 + \dfrac{\pi_v}{\sigma^2}\mathbf{V}^t\mathbf{V}^*} \tag{6.2.2}$$

From eqn. (3.4.24) we have

$$\mathbf{W}_{\text{opt}} = \frac{N\mathbf{M}^{-1}\mathbf{C}^*}{\mathbf{C}^t\mathbf{M}^{-1}\mathbf{C}^*} \tag{6.2.3}$$

in which $\mathbf{C}$ is the steering vector and the arbitary gain factor g has been set to N. Substituting eqn. (6.2.2) into (6.2.3) yields

$$\begin{aligned}\frac{\mathbf{W}_{\text{opt}}}{N} &= \frac{\mathbf{C}^* - \dfrac{(\pi_v/\sigma^2)\mathbf{V}^*\mathbf{V}^t\mathbf{C}^*}{1 + \pi_v\mathbf{V}^t\mathbf{V}^*/\sigma^2}}{\mathbf{C}^t\mathbf{C}^* - \dfrac{\pi_v\mathbf{C}^t\mathbf{V}^*\mathbf{V}^t\mathbf{C}^*/\sigma^2}{1 + \pi_v\mathbf{V}^t\mathbf{V}^*/\sigma^2}} \\ &= \frac{\mathbf{C}^*(1+\gamma) - \mathbf{V}^*\rho\gamma}{N(1+\gamma) - \gamma N|\rho|^2} \end{aligned} \tag{6.2.4}$$

where $\gamma = \pi_v N/\sigma^2$ and $\rho = \mathbf{V}^t\mathbf{C}^*/N$. It is assumed that $\mathbf{V}$ and $\mathbf{C}$ have unit modulus elements and consequently ρ is a normalised conventional amplitude gain (unity maximum) for the source. γ is the output SNR of the source if it were present in the look direction of the conventionally weighted array.

The amplitude gain of the optimally weighted array for the source is

$$\begin{aligned} g_s = \mathbf{V}^t\mathbf{W}_{\text{opt}} &= N\frac{N\rho(1+\gamma) - N\rho\gamma}{N(1+\gamma) - \gamma N|\rho|^2} \\ &= \frac{N\rho}{1 + \gamma(1 - |\rho|^2)} \end{aligned} \tag{6.2.5}$$

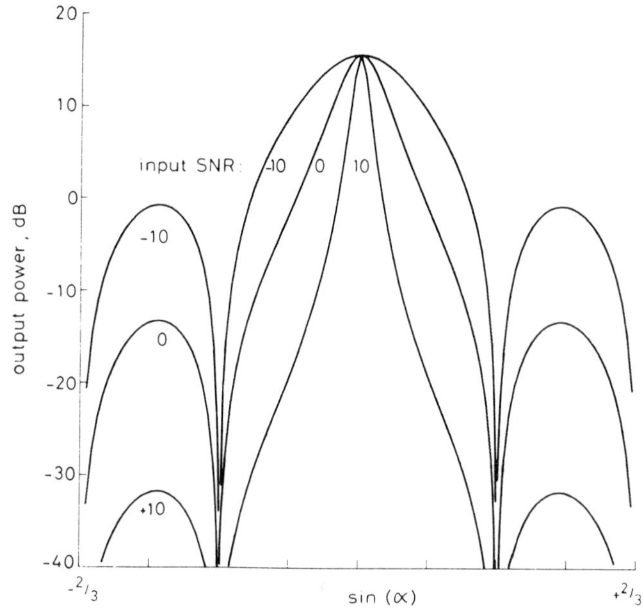

Fig. 6.2.1 *Source output power against source bearing and input SNR in a six-element array*

For a look direction source we have $|\rho| = 1$ and g_s has a value of N. However, if γ is large and $|\rho|$ is slightly less than one the amplitude gain can fall to very low values; in other words, a deep notch may be formed in the main lobe. The output power due to the source is

$$p = \pi_v |\mathbf{V}'\mathbf{W}_{\text{opt}}|^2 = \frac{\pi_v N^2 |\rho|^2}{[1 + \gamma(1 - |\rho|^2)]^2} \tag{6.2.6}$$

The acceptance angle of the optimal array, over which it outputs useful signal power, is obviously small from this result if γ is large, a point noted by Zahm.[89] Thus we have a quantitative description of main lobe distortion and have shown that at high input SNRs the main lobe is rather 'floppy' in the sense that it allows a null to be formed very close to the look direction.

Fig. 6.2.1 shows the way in which the signal output power varies as a 0 dB source is scanned from $-42°$ to $+42°$ relative to broadside in a six element array with uncorrelated noise powers of -10, 0, and $+10$ dB, γ taking the values 60, 6, and 0·6 in eqn. (6.2.6). The pointing accuracy clearly becomes quite critical for high input SNR's.

In fact, the situation is worsened by the formation of very large weighting coefficients when a main-lobe null is created which causes enhanced output of uncorrelated noise. This is investigated by calculating the weight vector norm squared $\mathbf{W}^H\mathbf{W}$ since noise output power is proportional to it, from (2.8.3). From eqn. (6.2.4) we have

$$\|\mathbf{W}_{\text{opt}}\|^2 = N\left\{\frac{(1+\gamma)^2 + |\rho|^2\gamma^2 - 2|\rho|^2\gamma(1+\gamma)}{[1+\gamma(1-|\rho|^2)]^2}\right\} \tag{6.2.7}$$

and using this and eqn. (6.2.6) gives the output SNR as

$$Q = \frac{p|\mathbf{V}^t\mathbf{W}|^2}{\sigma^2\mathbf{W}^H\mathbf{W}} = \frac{\gamma|\rho|^2}{1+\gamma(2+\gamma)(1-|\rho|^2)} \tag{6.2.8}$$

Example 6.2.1: A five-element array receives a desired signal at the −3 dB point in its conventional pattern. Calculate the output SNR if the weights adapt to the minimum output power condition given that the input SNR is −10, 0, or +10 dB and the noise is uncorrelated.

The γs for the three input SNRs are 0·5, 5, and 50. In each case, $|\rho|^2$ is 0·5 and substitution in eqn. (6.2.8) yields output SNRs of 0·216 (−6·6 dB), 0·135 (−8·7 dB), 0·019 (−17 dB). Note that the output SNR decreases as the input SNR increases.

Fig. 6.2.2 shows the complete beam patterns for these three input SNRs and it is clear that the amplitude gain for the desired signal is depressed rapidly as the input SNR increases.

Random channel errors

Pointing errors are not the only contribution to deterioration of adaptive SNR. Even if the array can be lined up exactly with the direction of the desired source

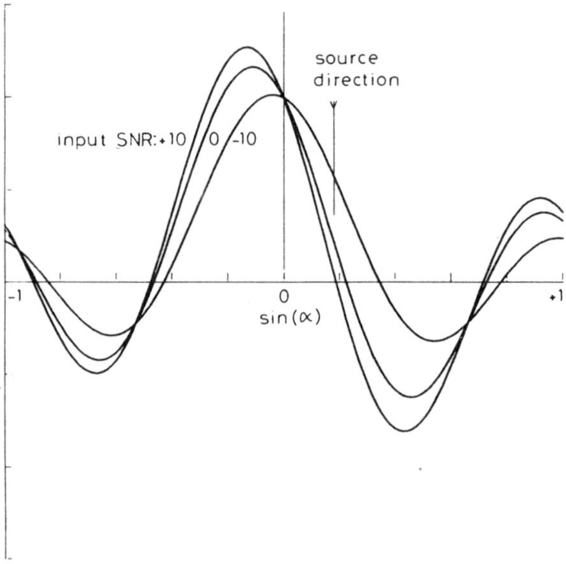

Fig. 6.2.2 *Adapted beam patterns for a source at the −3 dB point of the conventional pattern*

the vector **C**, which the processor takes as the space vector of the desired signal, may not correspond precisely with the observed signal space vector **S** due to random errors which arise from the following sources:

(i) channel-gain errors caused by pre-amp differences, feeder delay mismatches,
(ii) mutual coupling effects in the array which distort **S**,
(iii) scattering from local obstacles such as ship's masts, aircraft wings, the ground which add a coherent multipath component to **S** (see Section 2.7).

Even if **S** and **C** are numerically equal and there are no errors of the above type there may still be a problem if the control laws of the coefficients are slightly random and are not individually calibrated. Although the processor will now output a weighting vector **W** to the control lines of the weighting coefficients which satisfies the constraint $\mathbf{C}^t\mathbf{W} = 1$ accurately, the same is not true for the analogue attenuations which the hardware coefficient controllers set up. Since the output power minimisation algorithms sense the true analogue coefficients this function is not too greatly affected by mild control error; however the setting of the constraint vector, for example eqn. (4.5.7(b)) relies upon accurate control and has no error detection. The effect is that the adaptive algorithm will null out a look direction signal even while it ensures internally that such an event could not take place.

The distorted space vectors under consideration are time invariant. Rather different results are obtained if the channel errors are random time functions.[109]

For small deviations of **S** from the constraint vector we can write it as the constant-norm form

$$\mathbf{S} = (1 - a^2)^{1/2}\mathbf{C} + a\mathbf{PE} \tag{6.2.9}$$

in which a is a real scalar and **E** is a random vector, i.e. one whose elements are drawn from a random distribution. The projection $\mathbf{P} = \mathbf{I} - \mathbf{CC}^H/\mathbf{C}^H\mathbf{C}$ ensures that **PE** is orthogonal to **C** and if $\|\mathbf{PE}\|^2 = \|\mathbf{C}\|^2$ then

$$\|\mathbf{S}\|^2 = N \tag{6.2.10}$$

from eqn. (6.2.9). This result is important and explains the rather cumbersome model for **S** in eqn. (6.2.9) for, if eqn. (6.2.10) is applicable, we may define $\rho = \mathbf{S}^t\mathbf{C}^*/N$ as in eqn. (6.2.4) and all the previous analysis is applicable. In other words, random errors in **S** can be treated as a random pointing error, one which does not necessarily correspond to an azimuth or elevation error—the algorithm is sensitive to deviations of **S** from **C** but not to the precise cause of the difference.

If $\|\mathbf{PE}\|^2 = N$ then a is the RMS value of the elements of the vector $a\mathbf{PE}$ and can be related to physical parameters. For example, if the elements of **S**, as received, have a 10° RMS phase error relative to **C** then we may interpret this as due to an addition to **C** of a random vector, one constrained to change phases only, whose RMS element values are about 1/6, i.e. producing 1/6 radian RMS error, and the value of a would be about 1/6 too. In the presence of combined amplitude errors of 16% added to the existing phase errors the variance would be doubled and a would have a value of about 0·24.

The value of ρ, the normalised conventional antenna response is

$$\rho = \mathbf{S}^t\mathbf{C}^*/N = (1 - a^2)^{1/2} \qquad (6.2.11)$$

and if this value is substituted in, for example, eqn. (6.2.8) the adaptive antenna output SNR in the presence of channel errors can be determined.

Example 6.2.1: An adaptive antenna receives a desired signal at 20 dB input SNR in each channel and it is required that the adaptive output SNR be greater than 0 dB in the presence of error between **S** and **C**. Determine the maximum tolerable standard deviation of the errors.

In eqn. (6.2.8) we set $Q = 1$ and solve for $|\rho|^2$:

$$|\rho|^2 = \frac{1 + 2\gamma + \gamma^2}{3\gamma + \gamma^2} = 0{\cdot}99038 \qquad (6.2.12)$$

Finally, we substitute $|\rho|^2$ into eqn. (6.2.21) and find a:

$$a = (1 - |\rho|^2)^{1/2} = 0{\cdot}098 \qquad (6.2.13)$$

and conclude that about 10% channel errors are permissable. These errors might arise as a combination of 7% amplitude gain errors and 4° phase errors.

Contemplation of examples such as this one emphasises the difficulty of operating steered system under conditions of high SNR such as might occur in communication systems. The problem might be alleviated by reducing desired signal SNR in a spread spectrum configuration, or injection of local noise,[81] or synthesising its effect in the adaptive algorithm [eqns. (4.3.10), (4.6.8)].

It might be thought that radar systems in which the signal is, on average, very weak would not suffer from signal suppression problems. Although it is true that the problem is not so acute, what is likely to happen here is that a strong continuous jammer appears in the main lobe as the radar beam sweeps through it and severely distorts the main lobe pattern. While the jammer cancellation enables an echo to be detected the distortion of the lobe may cause the echo to appear to originate from an erroneous direction which, in certain circumstances, could be just as serious as total loss of the echo.

6.3 Polynomial expansions of antenna responses in the main lobe

Taylor series: As a prelude to the discussion of main-lobe constraint systems, a representation of the directional patterns of arrays in the vicinity of the look direction is required and is conveniently done by a Taylor series expansion in this region. Let $\mathbf{S}(\theta)$ be the space vector corresponding to a source whose bearing sine relative to the broadside direction of a line array is θ. From eqn. (2.2.7) we write

$$\mathbf{S}(\theta) = \{\exp(jkx_i\theta)\} \qquad i = 0, \ldots, N-1 \tag{6.3.1}$$

where $k = 2\pi/\lambda$, and the x_i are the locations of the elements in the array measured along its length. In eqn. (6.3.1), the representation $\{.\}$ implies a column vector whose elements take on the different values of i.

Differentiation of eqn. (6.3.1) with respect to θ gives a sequence of derivative vectors of increasing orders:

$$\begin{aligned}
\mathbf{S}^{\text{i}}(\theta) &= \{jkx_i \exp(jkx_i\theta)\} \\
\mathbf{S}^{\text{ii}}(\theta) &= \{-k^2x_i^2 \exp(jkx_i\theta)\} \\
\mathbf{S}^{\text{iii}}(\theta) &= \{-jk^3x_i^3 \exp(jkx_i\theta)\} \\
\mathbf{S}^{\text{iv}}(\theta) &= \{k^4x_i^4 \exp(jkx_i\theta)\}
\end{aligned} \tag{6.3.2}$$

and we can write $\mathbf{S}(\theta)$ as a Taylor series in the derivatives (6.3.2) as follows:

$$\begin{aligned}
\mathbf{S}(\theta) &= \mathbf{S}(0) + \theta\mathbf{S}^{\text{i}}(0) + \frac{\theta^2}{2!}\mathbf{S}^{\text{ii}}(0) + \frac{\theta^3}{3!}\mathbf{S}^{\text{iii}}(0) \\
&\quad + \frac{\theta^4}{4!}\mathbf{S}^{\text{iv}}(0) + \cdots \\
&= \sum_{m=0}^{\infty} \frac{\theta^m}{m!}\mathbf{S}^m(0)
\end{aligned} \tag{6.3.3}$$

The directional response of the conventionally weighted array for the source, assuming that the pointing direction is broadside, is

$$c(\theta) = \mathbf{S}(0)^H\mathbf{S}(\theta) = \mathbf{S}(0)^H \sum_{m=0}^{\infty} \frac{\theta^m}{m!}\mathbf{S}^m(\theta) \tag{6.3.4}$$

which is equivalent to the scalar Taylor expansion of the beam pattern:

$$c(\theta) = c(0) + \theta c^{\text{i}}(0) + \frac{\theta^2}{2!}c^{\text{ii}}(0) + \frac{\theta^3}{3!}c^{\text{iii}}(0) + \frac{\theta^4}{4!}c^{\text{iv}}(0) + \cdots \tag{6.3.5}$$

in which

$$\begin{aligned}
c(0) &= \mathbf{S}(0)^H\mathbf{S}(0) = N \\
c^{\text{i}}(0) &= \mathbf{S}(0)^H\mathbf{S}^{\text{i}}(0) = \sum_{i=1}^{N} jkx_i \\
c^{\text{ii}}(0) &= \mathbf{S}(0)^H\mathbf{S}^{\text{ii}}(0) = \sum_{i=1}^{N} -k^2x_i^2 \\
c^{\text{iii}}(0) &= \mathbf{S}(0)^H\mathbf{S}^{\text{iii}}(0) = \sum_{i=1}^{N} -jk^3x_i^3 \\
c^{\text{iv}}(0) &= \mathbf{S}(0)^H\mathbf{S}^{\text{iv}}(0) = \sum_{i=1}^{N} k^4x_i^4
\end{aligned} \tag{6.3.6}$$

By suitable selection of the phase centre of the array, from which point x_i is measured, we can make $c^{i}(0) = 0$, i.e. the complex amplitude gain first derivative is zero for a source at the look direction or in other words the derivatives of the modulus and phase of the response are zero. This means that $\mathbf{S}(0)$ and $\mathbf{S}^{i}(0)$ are orthogonal vectors.

Orthogonal basis vector expansion: Although much use will be made of the Taylor expansions above they suffer from the disadvantage that the basis vectors $\mathbf{S}^{m}(0)$ are nonorthogonal, with the exception that for the colinear array under consideration all the odd derivative vectors are orthogonal to the even derivative vectors. The desired orthogonal vector expansion is of the form

$$\mathbf{S}(\theta) = s_0 \mathbf{S}_0 + s_1 \mathbf{S}_1 + s_2 \mathbf{S}_2 + \cdots \tag{6.3.7}$$

and whereas $\mathbf{S}(0)$ and $\mathbf{S}^{i}(0)$ are suitable candidates for $\mathbf{S}_0$ and $\mathbf{S}_1$, $\mathbf{S}^{ii}(0)$ is unsuitable for $\mathbf{S}_2$ since it is not orthogonal to $\mathbf{S}_0$. To remedy this the projection of $\mathbf{S}^{ii}(0)$ into the subspace orthogonal to $\mathbf{S}_0$ is taken as the third vector, and using the Gram–Schmidt method the projection is found by subtracting from $\mathbf{S}^{ii}(0)$ the vector $\alpha\mathbf{S}_0^*$ with α adjusted to produce the required orthogonality. From eqn. (6.3.2), the projected vector is given by

$$\mathbf{S}_2 = \{-k^2x_i^2 - \overline{(-k^2x_i^2)}\} \tag{6.3.8}$$

where

$$\overline{(-k^2x_i^2)} = \frac{1}{N}\sum_{i=1}^{N} -k^2x_i^2 = \frac{1}{N}c^{ii}(0) \tag{6.3.9}$$

The coefficients s_n in eqn. (6.3.7) are found as follows. For s_0 we have

$$\mathbf{S}(0)^H\mathbf{S}(\theta) = s_0\,\mathbf{S}(0)^H\mathbf{S}(0) = s_0 N = c(\theta) \tag{6.3.10}$$

thus

$$s_0 = N^{-1}c(\theta) \tag{6.3.11}$$

For s_1, by comparison with eqn. (6.3.3), we have just

$$s_1 = \theta \tag{6.3.12}$$

and similarly for s_2 we get

$$s_2 = \tfrac{1}{2}\theta^2 \tag{6.3.13}$$

The norms of the basis vectors will be required; these are given by

$$\phi_0 = \|\mathbf{S}_0\|^2 = \|\mathbf{S}(0)\|^2 = N \tag{6.3.14}$$

$$\phi_1 = \|\mathbf{S}_1\|^2 = \|\mathbf{S}^{i}(0)\|^2 = \sum_{i=1}^{N} k^2x_i^2 = -c^{ii}(0) \tag{6.3.15}$$

from eqns. (6.3.2) and (6.3.9) and

$$\phi_2 = \|\mathbf{S}_2\|^2 = \sum_{i=1}^{N} [k^2x_i^2 - \overline{(-k^2x_i^2)}]^2$$

$$= \sum_{i=1}^{N} k^4x_i^4 - \sum_{i=1}^{N} \overline{(-k^2x_i^2)^2}$$

$$= \sum_{i=1}^{N} k^4x_i^4 - \frac{1}{N}\left[\sum_{i=1}^{N} k^2x_i^2\right]^2$$

$$= c^{\text{iv}}(0) - \frac{1}{N}[c^{\text{ii}}(0)]^2 \tag{6.3.16}$$

from eqns. (6.3.8) and (6.3.6).

Fig. 6.3.1 shows the directional patterns associated with the vectors $\mathbf{S}_0$, $\mathbf{S}_1$, and $\mathbf{S}_2$ for a nine-element array with $\lambda/2$ spacing. These patterns are just the conventional directional patterns together with its first and modified second derivatives. The three patterns are orthogonal on the interval -1 to $+1$ in θ space.

An alternative and exact 3-term orthogonal vector basis expansion based on $\mathbf{S}_0$, $\mathbf{S}_1$, and the projection of $\mathbf{S}(\theta)$ orthogonal to $\mathbf{S}_0$ is easily derived but the associated coefficients s_i are difficult to determine except for the elementary line array.

Optimal weighting vectors: It is possible to find expressions for the optimal weighting vectors, in the SNR sense of Chapter 3, when the vector is expressed as a linear combination of the orthogonal basis vectors and this is a powerful technique within the main lobe. Thus we express $\mathbf{W}_{\text{opt}}$ as

$$\mathbf{W} = w_0\mathbf{S}_0^* + w_1\mathbf{S}_1^* + w_2\mathbf{S}_2^* + \cdots \tag{6.3.17}$$

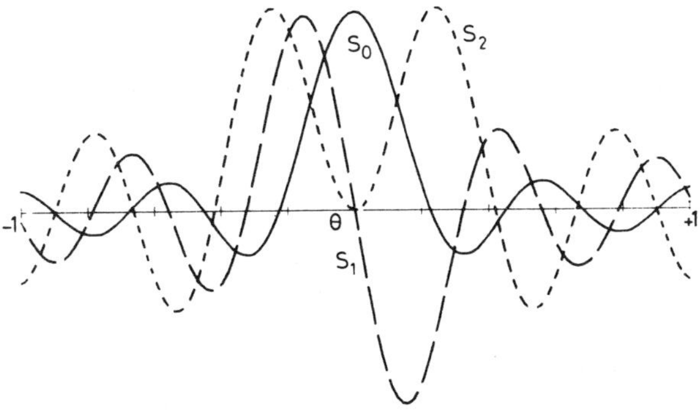

Fig. 6.3.1 *Directional patterns associated with the derivative vectors* $\mathbf{S}_0$, $\mathbf{S}_1$, $\mathbf{S}_2$ *as a function of* θ
Not to scale

where the coefficients w_i are, in general, complex. The directional pattern of this vector is a combination of the directional patterns of the derivative vectors as in Fig. 6.3.1. The output power of the antenna when it is excited by a unit power source in the main lobe is considered:

$$p = \mathbf{W}^H\mathbf{R}\mathbf{W} = \mathbf{W}^H(\mathbf{S}^*\mathbf{S}^t + \sigma^2\mathbf{I})\mathbf{W} = \mathbf{W}^H\mathbf{S}^*\mathbf{S}^t\mathbf{W} + \sigma^2\mathbf{W}^H\mathbf{W} \tag{6.3.18}$$

and substituting eqns. (6.3.7) and (6.3.17) into this expression yields

$$p \simeq \left\| \left[\sum_{i=0}^{2} w_i \mathbf{S}_i^*\right]^t \left[\sum_{i=0}^{2} s_i \mathbf{S}_i\right] \right\|^2 + \sigma^2 \sum \|w_i \mathbf{S}_i^*\|^2 \tag{6.3.19}$$

$$= \left[\sum_{i=0}^{2} w_i s_i \phi_i\right]\left[\sum_{i=0}^{2} w_i^* s_i^* \phi_i\right] + \sigma^2 \sum_{i=0}^{2} w_i w_i^* \phi_i \tag{6.3.20}$$

where

$$\phi_i = \mathbf{S}_i^H \mathbf{S}_i = \|\mathbf{S}_i\|^2$$

We now minimise p by variation of the w_i coefficients (some of these coefficients may be later constrained but this does not affect this step) and since the w_i are complex the partial derivatives with respect to the real and imaginary parts are found separately and set to zero.

$$dp/dw_m^r = \left(\sum_{i=0}^{2} w_i s_i \phi_i\right) s_m^* \phi_m + \left(\sum_{i=0}^{2} w_i^* s_i^* \phi_i\right) s_m \phi_m$$

$$+ \sigma^2(w_m^* \phi_m + w_m \phi_m) \tag{6.3.21}$$

and setting this to zero yields

$$\mathrm{Re}\left[s_m^*\left(\sum_{i=0}^{2} w_i s_i \phi_i\right)\right] = -\sigma^2 \,\mathrm{Re}\,(w_m)$$

or

$$\mathrm{Re}\,(w_m) = -\mathrm{Re}\left[s_m^*\left(\sum_{i=0}^{2} w_i s_i \phi_i\right)\right]\Big/\sigma^2 = w_m^r \tag{6.3.22}$$

$$dp/dw_m^j = -\left(\sum_{i=0}^{2} w_i s_i \phi_i\right) s_m^* \phi_m + \left(\sum_{i=0}^{2} w_i^* s_i^* \phi_i\right) s_m \phi_m$$

$$+ \sigma^2(w_m^*\phi_m - w_m \phi_m) \tag{6.3.23}$$

whence

$$\mathrm{Im}\left[s_m\left(\sum_{i=0}^{2} w_i^* s_i^* \phi_i\right)\right] + \sigma^2\, \mathrm{Im}\,(w_m^*) = 0$$

$$\mathrm{Im}\,(w_m^*) = -\mathrm{Im}\left[s_m\left(\sum_{i=0}^{2} w_i^* s_i^* \phi_i\right)\right]\bigg/\sigma^2$$

or

$$\mathrm{Im}\,(w_m) = -\mathrm{Im}\left[s_m^*\left(\sum_{i=0}^{2} w_i s_i \phi_i\right)\right]\bigg/\sigma^2 = w_m^j \tag{6.3.24}$$

Adding eqns. (6.3.22) and (6.3.24) provides the complex solution for w_m

$$w_m = w_m^r + jw_m^j = -s_m^*\left(\sum_{i=0}^{2} w_i s_i \phi_i\right)\bigg/\sigma^2 \tag{6.3.25}$$

Effect of constraints: If the minimisation above is carried out with some linear equality constraints

$$w_i = a_i \text{ for some } i \tag{6.3.26}$$

then the Lagrange method for solution would be to minimise the power p augmented by additive terms:

$$p = \left(\sum_{i=0}^{N-1} w_i s_i \phi_i\right)\left(\sum_{i=0}^{N-1} w_i^* s_i^* \phi_i\right) + \sigma^2\left(\sum_{i=0}^{N-1} w_i w_i^* \phi_i\right) + \sum_{\text{some } i} \lambda_i(w_i - a_i) \tag{6.3.27}$$

The partial derivatives then contain terms involving λ_i for those coefficients w_i which are constrained. However, as the coefficients w_i in eqn. (6.3.25) are noninteracting due to the selection of an orthogonal basis vector representation, it is merely necessary to set $w_i = a_i$ for the constrained coefficients.

6.4 Linear main-lobe constraints[22, 70, 109]

In this Section, a variety of main-lobe constraints will be analysed using the power minimisation above. Linear constraint will be applied to the pattern derivatives in the vicinity of the look direction, a technique described by

Appelbaum.[22] The use of real-only weighting coefficients will also be considered.

Look direction constraint only: If it is assumed that w_0 in eqn. (6.3.17) is fixed then because of the definition of $\mathbf{S}_0$, which is equivalent to $\mathbf{S}(0)$ of eqn. (6.3.1), the amplitude gain of the antenna in the broadside direction is fixed. This follows directly from the expression for broadside response $g(0)$:

$$g(0) = \mathbf{W}^t\mathbf{S}_0 = \left(\sum_{i=0}^{N-1} w_i \mathbf{S}_i^*\right)^t \mathbf{S}_0 = w_0\,\phi_0 \tag{6.4.1}$$

For small variations of the signal bearing about the look direction the second-order term $s_2\mathbf{S}_2$ in eqn. (6.3.7) is small, since $s_2 = \frac{1}{2}\theta^2$ and in the optimal solution eqn. (6.3.25) it is reasonable to consider only w_0 and w_1 as having significant amplitudes. The solutions are obtained from

$$w_1 = -s_1^*(w_0 s_0 \phi_0 + w_1 s_1 \phi_1)/\sigma^2 \tag{6.4.2}$$

which, on solving for w_1, yields

$$w_1 = \frac{-s_1^* w_0 s_0 \phi_0}{\sigma^2 + |s_1|^2\phi_1} = \frac{-s_1^* c(\theta)}{\sigma^2 + |s_1|^2\phi_1} \tag{6.4.3}$$

Substituting into eqn. (6.3.17) we find the optimal weighting vector as

$$\mathbf{W}_{\text{opt}} \simeq w_0\mathbf{S}_0^* + w_1\mathbf{S}_1^* = w_0\mathbf{S}_0^* - \frac{w_0 s_0 \phi_0}{\sigma^2 + |s_1|^2\phi_1}\, s_1^*\mathbf{S}_1^* + \cdots \tag{6.4.4}$$

The amplitude gain of the antenna for the signal, $g(\theta)$, is

$$\begin{aligned} g(\theta) &= \mathbf{S}(\theta)^t\mathbf{W}_{\text{opt}} = s_0 w_0 \phi_0 + w_1 s_1 \phi_1 + \cdots \\ &\simeq w_0 s_0 \phi_0\left(1 - \frac{|s_1|^2\phi_1}{\sigma^2 + |s_1|^2\phi_1}\right) = \frac{\sigma^2 c(\theta)}{\sigma^2 + |s_1|^2\phi_1} \end{aligned} \tag{6.4.5}$$

This result demonstrates immediately the sensitivity of the optimal solution to desired signal mismatch. For any finite s_1, i.e. if the signal does not lie exactly at the look direction, then when the uncorrelated noise level σ^2 vanishes so too does the amplitude gain of the antenna; in other words a main-lobe null is formed arbitarily close to the look direction under this condition.

To determine the output signal-to-noise ratio it is necessary to find the norm of the weighting vector since the output noise power is given by $\sigma^2\mathbf{W}^H\mathbf{W}$. From eqn. (6.3.17) with (6.3.14) and 6.3.15),

$$\mathbf{W}^H\mathbf{W} = |w_0|^2\phi_0 + |w_1|^2\phi_1 + \cdots \tag{6.4.6}$$

and substituting from eqn. (6.4.3) yields

$$\mathbf{W}^H\mathbf{W} \simeq |w_0|^2\phi_0 + \frac{|s_1|^2\phi_1|c(\theta)|^2}{(\sigma^2 + |s_1|^2\phi_1)^2} \tag{6.4.7}$$

where $c(\theta)$ is the conventionally weighted antenna response

$$c(\theta) = \mathbf{S}(\theta)^t \mathbf{W}_c = s_0 w_0 \phi_0 \tag{6.4.8}$$

Expressing $\mathbf{W}^H\mathbf{W}$ as

$$\mathbf{W}^H\mathbf{W} = \frac{|w_0|^2 \phi_0 (\sigma^2 + |s_1|^2 \phi_1)^2 + |s_1|^2 \phi_1 |c(\theta)|^2}{(\sigma^2 + |s_1|^2 \phi_1)^2} \tag{6.4.9}$$

we form the ratio

$$\begin{aligned} Q &= \frac{|g(\theta)|^2}{\sigma^2 \mathbf{W}^H\mathbf{W}} = \text{output SNR} \\ &= \frac{|(w_0 s_0 \phi_0 \sigma^2)|^2}{\sigma^2[|w_0|^2 \phi_0 (\sigma^2 + |s_1|^2 \phi_1)^2 + |s_1|^2 \phi_1 |c(\theta)|^2]} \\ &= \frac{\sigma^2 |c(\theta)|^2}{N(\sigma^2 + |s_1|^2 \phi_1)^2 + |s_1|^2 \phi_1 c^2(\theta)} \end{aligned} \tag{6.4.10}$$

By inspection, if s_1 (i.e. θ) $\rightarrow 0$ or $\sigma^2 \rightarrow \infty$ the output SNR tends to the conventional level as might be expected. If s_1 is finite, the array having a pointing error, then the output SNR is heavily dependent upon input SNR in a very nonlinear fashion showing a peak value on either side of which output SNR falls away. Fig. 6.4.1 shows signal output power, Q (output SNR), and output uncorrelated noise power as a function of source bearing in a five element array with $\lambda/2$ spaced elements with a signal power of 10 and σ^2 of 1. Results are shown by an exact numerical technique and also by the approximate equations above. The behaviour of the approximation is excellent in the main-lobe region. Fig. 6.4.2 shows how output SNR and noise power vary as a function of signal power when the direction of arrival is fixed with a bearing error of 5·85° corresponding to $c(\theta) = 4{\cdot}5$. Again the approximate theory is in good agreement with the exact result. This problem is tackled numerically below.

Example 6.4.1 : A five-element array has adaptive weights, $\lambda/2$ spacing, and receives a signal of 10 dB SNR from a source with a bearing of 5·85° relative to broadside. Calculate the amplitude gain of the antenna, the noise output power and the output SNR.

The conventional pattern gain is found as $c(\theta) = 4{\cdot}5$. We have $\sigma^2 = 0{\cdot}1$ (the theory assumes a unit power signal), $\phi_0 = 5$ [eqn. (6.3.14)], $\phi_1 = 10\pi^2$ [eqn. (6.3.15)], $s_1 = \theta = \sin(5{\cdot}85°) = 0{\cdot}102$.

Substitution into eqn. (6.4.5) gives

$$g(\theta) = \frac{4{\cdot}5 \times 0{\cdot}1}{0{\cdot}1 + 1{\cdot}027} = 0{\cdot}399(-7{\cdot}98 \text{ dB}) \tag{6.4.11}$$

yielding an output signal power of 2·02 dB.

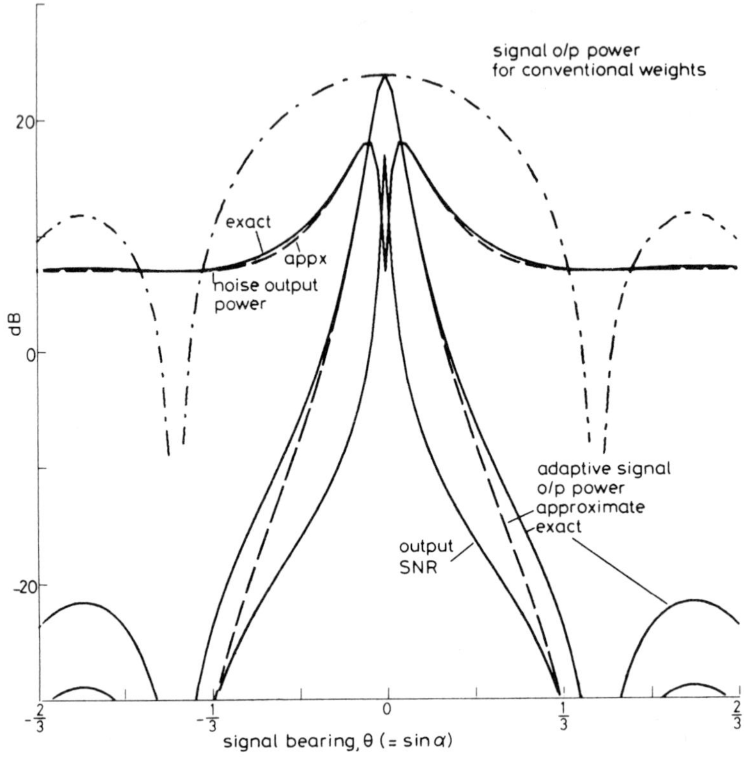

Fig. 6.4.1 *Main-lobe performance of adaptive antenna with look direction constraint at 10 dB input SNR*

We have $w_0 = 1$ [broadside gain of 5 in eqn. (6.4.1)] and substitution into eqn. (6.4.7) yields

$$\mathbf{W}^H\mathbf{W} = 5 + \frac{(0{\cdot}102 \times 4{\cdot}5)^2 10\pi^2}{(0{\cdot}1 + 1{\cdot}027)^2} = 21{\cdot}37 \qquad (6.4.12)$$

yielding an output noise power of $\sigma^2\mathbf{W}^H\mathbf{W} = 21{\cdot}37(13{\cdot}3 \text{ dB})$. The output SNR is thus $2{\cdot}02 - 13{\cdot}3 = -11{\cdot}28$ dB.

Look-direction constraint plus derivative constraint : From eqn (6.3.5) we observe that for small θ the variation of the pattern in the vicinity of the look direction is governed mostly by the first-order term c^i in the Taylor expansion and the associated basis vector is the $\mathbf{S}_1$ in (6.3.7). Using eqn. (6.3.7 & 17) we can write, for a single source,

$$g(\theta) = \mathbf{S}(\theta)^t\mathbf{W} = \left(\sum_{i=0}^{N-1} s_i \mathbf{S}_i\right)^t \left(\sum_{i=0}^{N-1} w_i \mathbf{S}_i^*\right)$$

$$= \sum_{i=0}^{N-1} s_i w_i \phi_i \qquad (6.4.13)$$

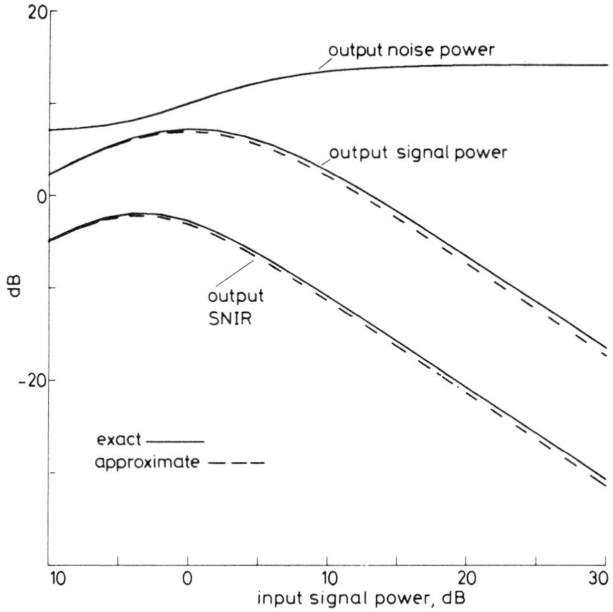

Fig. 6.4.2 *Performance of array with mismatched signal* $\sigma^2 = 1$

and substituting the values for the s_i from eqns. (6.3.11) to (6.3.13) we get

$$g(\theta) = c(\theta) + \theta w_1 \phi_1 + \tfrac{1}{2}\theta^2 w_2 \phi_2 \tag{6.4.14}$$

Clearly if the coefficient w_1 in the spectrum of $\mathbf{W}$ is held at zero, the first derivative of the beam amplitude must be zero at the look direction. Actually we are not only constraining the derivative of the modulus of the pattern but also the phase, but since little phase shift in the beam pattern is expected anyway [e.g. eqn. (3.4.13)] the net effect will be much the same as a constraint on the modulus derivative. If the beam has zero tilt, then one would intuitively expect that the ability of the array to steer a null at a source close to the look direction would be reduced, and this is the case, but the situation is far from simple and there is a complicated interaction with the uncorrelated noise level.

The constraint on the derivative is applied by methods similar to those used for the look direction constraint eqn. (4.5.10) and an algorithm could be based on two projections:

$$\mathbf{W}(k+1) = \mathbf{W}(k) - 2\mu \mathbf{P}_0 \mathbf{P}_1 \mathbf{R}\mathbf{W}(k) \tag{6.4.15}$$

where $\mathbf{P}_0$ projects the gradient into a subspace orthogonal to the steering vector $\mathbf{S}_0$

$$\mathbf{P}_0 = \mathbf{I} - \frac{\mathbf{S}_0^* \mathbf{S}_0^t}{\mathbf{S}_0^t \mathbf{S}_0^*} \tag{6.4.16}$$

and $\mathbf{P}_1$ projects the gradient into a subspace orthogonal to the 'tilt' vector $\mathbf{S}_1$

$$\mathbf{P}_1 = \mathbf{I} - \frac{\mathbf{S}_1^* \mathbf{S}_1^t}{\mathbf{S}_1^t \mathbf{S}_1^*} \tag{6.4.17}$$

Alternatively the 'tilt' component of the input data vector $\mathbf{X}(t)$ can be isolated by a matrix transformation and applied to a hardware coefficient w_1. A suitable transformation matrix would have N rows with the orthogonal basis vectors $\mathbf{S}_i^H$ and would output a vector $\mathbf{Y}(t)$ which would be weighted and summed.

$$\mathbf{Y}(t) = \begin{bmatrix} \mathbf{S}_0^H \\ \mathbf{S}_1^H \\ \mathbf{S}_2^H \\ \cdots \\ \cdots \end{bmatrix} \mathbf{X}(t) \tag{6.4.18}$$

$$y(t) = \mathbf{w}^t \mathbf{Y}(t) \tag{6.4.19}$$

The coefficients in the vector $\mathbf{w}$ now correspond to the w_i coefficients in eqn. (6.3.17). This is the approach used by Appelbaum[22] and is most suited for analogue implementations. Fig. 6.4.3 shows the block diagram of the processor. Since the coefficient w_1 is to be held at zero it is clearly not necessary to provide such a coefficient at all and the corresponding row of the transformation could be dispensed with.

Whichever implementation is used the net result is to constraint $w_1 = 0$ in eqn. (6.4.14). For the output power of the antenna, eqn. (6.3.20) we now assume that w_2 is the only significant variable in the region where, in the Taylor expansion eqn. (6.3.3), $\theta^3/3!$, $\theta^4/4!$ etc., are still small. Therefore, in eqn. (6.3.25) we have the result

$$w_2 = -s_2^* \left(\sum_{i=0}^{2} w_i s_i \phi_i \right) \Big/ \sigma^2 \tag{6.4.20}$$

and, since w_1 is zero,

$$w_2 = -s_2^*(w_0 s_0 \phi_0 + w_2 s_2 \phi_2)/\sigma^2 \tag{6.4.21}$$

whence

$$w_2 = \frac{-s_2^* w_0 s_0 \phi_0}{\sigma^2 + |s_2|^2 \phi_2} \tag{6.4.22}$$

Thus

$$\mathbf{W} = w_0 \mathbf{S}_0^* - s_2^* \mathbf{S}_2^* \frac{w_0 s_0 \phi_0}{\sigma^2 + |s_2|^2 \phi_2} \tag{6.4.23}$$

The signal amplitude gain is

$$g(\theta) = \mathbf{S}(\theta)^t \mathbf{W} = c(\theta) \left[1 - \frac{|s_2|^2 \phi_2}{\sigma^2 + |s_2|^2 \phi_2} \right] = \frac{c(\theta)\sigma^2}{\sigma^2 + |s_2|^2 \phi_2} \tag{6.4.24}$$

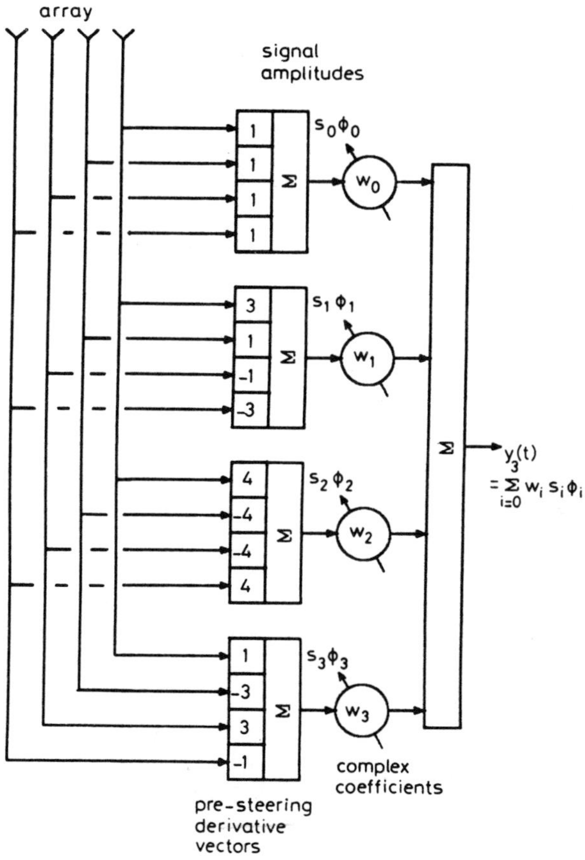

Fig. 6.4.3 *Preprocessing to isolate derivative components in the desired signal*

and

$$\|\mathbf{W}\|^2 = \|\mathbf{W}_c\|^2 + \frac{|s_2|^2\phi_2 |c(\theta)|^2}{(\sigma^2 + |s_2|^2\phi_2)^2} \tag{6.4.25}$$

which has a similar form to eqn. (6.4.7). By analogy with eqn. (6.4.10) we have for the output SNR

$$Q = \frac{\sigma^2 |c(\theta)|^2}{N(\sigma^2 + |s_2|^2\phi_2)^2 + |s_2|^2\phi_2 |c(\theta)|^2} \tag{6.4.26}$$

in which s_2 is given by eqn. (6.3.13) and ϕ_2 by (6.3.16). As the smallest power of θ in the SNR equation above is θ^4 it is expected that for small pointing errors the behaviour will be closer to conventional than for the look direction constrained array. Fig. 6.4.4 shows the variation of signal output power, output SNR, and output noise for a 5-element $\lambda/2$ spaced array as a 10 dB source is scanned

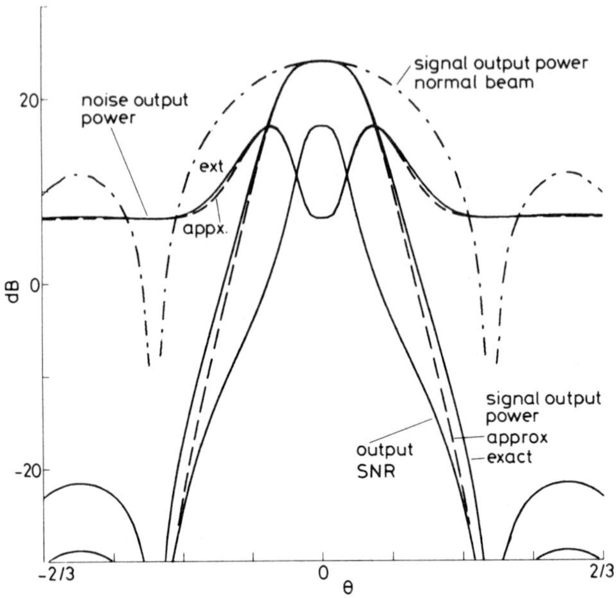

Fig. 6.4.4 *Main-lobe performance of adaptive antenna with look direction and derivative constraints at 10 dB input SNR*

across the broadside direction in the presence of unit power uncorrelated noise. The approximate theory above is presented along with an exact solution and shows good agreement. The output SNR remains high over a much greater angular zone than is the case with the simple look-direction constraint shown in Fig. 6.4.1. In Fig. 6.4.5 the signal has a fixed pointing error but its power is varied, the conditions being the same as those for Fig. 6.4.2. There is a useful improvement in output SNR.

Example 6.4.2: For the five-element array used in example 6.4.1, impose a derivative constraint and calculate the output SNR for a pointing error of 5·85° at 10 dB input SNR. In eqn. (6.4.26) above we have

$$\phi_2 = \sum_{i=1}^{N} k^4 x_i^4 - \frac{1}{N}\left\{\sum_{i=1}^{N} k^2 x_i^2\right\}^2 = 14\pi^4 = 1364 \qquad (6.4.27)$$

from eqn. (6.3.16) and

$$s_2 = \tfrac{1}{2}\theta^2 = \tfrac{1}{2}\sin^2(5{\cdot}85^\circ) = 0{\cdot}0052 \qquad (6.4.28)$$

and obtain

$$Q = \frac{4{\cdot}5^2 \times 0{\cdot}1}{5(0{\cdot}1 + 0{\cdot}0052^2 \times 14 \times \pi^4)^2 + 0{\cdot}0052^2 \times 14 \times \pi^4 \times 4{\cdot}5^2}$$

$$= 2{\cdot}409 \quad \text{or} \quad +3{\cdot}82 \text{ dB}$$

An exact solution yields 4·09 dB, so agreement is tolerable.

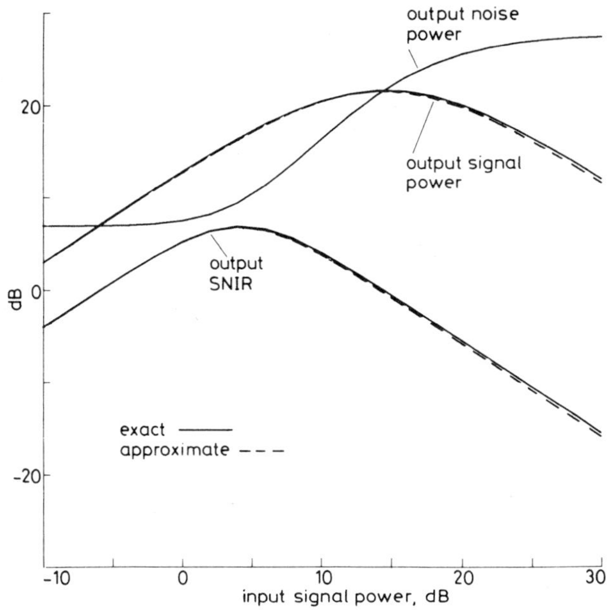

Fig. 6.4.5 *Performance of array with mismatched signal*
Exact and approximate solutions
$\sigma^2 = 1$

Constraints on higher derivatives: One can, in principle, apply a constraint to the second derivative of the pattern and if, in eqn. (6.3.17), both w_1 and w_2 were held at zero then the w_3 coefficient would play the dominant role in main-lobe distortion and signal suppression effects. However, since this constraint lies outside the range of the validity of the approximations used so far it will not be developed further; moreover, the w_3 coefficient only becomes significant when the conventional main lobe is already falling away and it is likely that a constraint on w_1 will be adequate.

A slightly different approach to main lobe constraints was expounded by Wang[110] who simply constrained the weighting coefficients to take on real values only (for a broadside pointing direction). This constraint, on the surface, is very different from derivative constraints yet a little investigation shows that this is not the case. If the expansion of the derivatives of $\mathbf{S}(\theta)$ at $\theta = 0$ are examined in eqn. (6.3.2) then it will be observed that all the odd derivatives are wholly imaginary and all the even derivatives wholly real. Moreover, the first $N-1$ derivatives produce linearly independent vectors and, in conjunction with the original vector $\mathbf{S}(0)$, are a complete set. Thus we may represent an arbitrary weighting vector $\mathbf{W}$ by a linear combination of the derivative vectors up to $\mathbf{S}^{N-1}(0)$:

$$\mathbf{W}^* = \sum_{i=0}^{N-1} b_i \mathbf{S}^i(0) = \sum_{\substack{i \\ \text{even}}} b_i \mathbf{S}^i(0) + j \sum_{\substack{i \\ \text{odd}}} b_i \mathbf{S}^i(0) \tag{6.4.29}$$

and if the weighting vector is constrained to be wholly real it may be assumed that all odd derivatives have zero coefficients in eqn. (6.4.29), if all the b_i are taken as real-valued. Therefore for a source in the main beam, where the $\mathbf{S}_0$ and $\mathbf{S}_1$ vectors in eqn. (6.3.7) predominate, both in $\mathbf{S}(\theta)$ and in the adaptive weight $\mathbf{W}$, it may be taken that the real weight constraint will produce similar results to the beam tilt constraint.

Since the even derivatives of $\mathbf{S}(\theta)$ at $\theta = 0$ are even vectors in the sense that coefficients equidistant from the ends have the same values then a linear combination has the same property, i.e. $\mathbf{W}$ is an even vector.

The solution for the linearly constrained minimum output power condition with real weights is found from Appendix 2 and eqn. (3.4.17). For the complex case, we have with $\mathbf{W}_j = 0$

$$p = \mathbf{W}_r^t \mathbf{R}_r \mathbf{W}_r + \lambda(\mathbf{W}_r^t \mathbf{S}_r - g) \tag{6.4.30}$$

noting that, for a broadside look direction $\mathbf{S}_j$ is zero also. The weighting vector is thus

$$\mathbf{W}_r = \frac{g\mathbf{R}_r^{-1}\mathbf{S}_r}{\mathbf{S}_r^t \mathbf{R}_r^{-1}\mathbf{S}_r} \tag{6.4.31}$$

Figure 6.4.6 shows the 5-element antenna with real coefficients as the source is scanned through the main lobe. The performance is almost indistinguishable from Fig. 6.4.4 but there is greater output power when the source lies in the

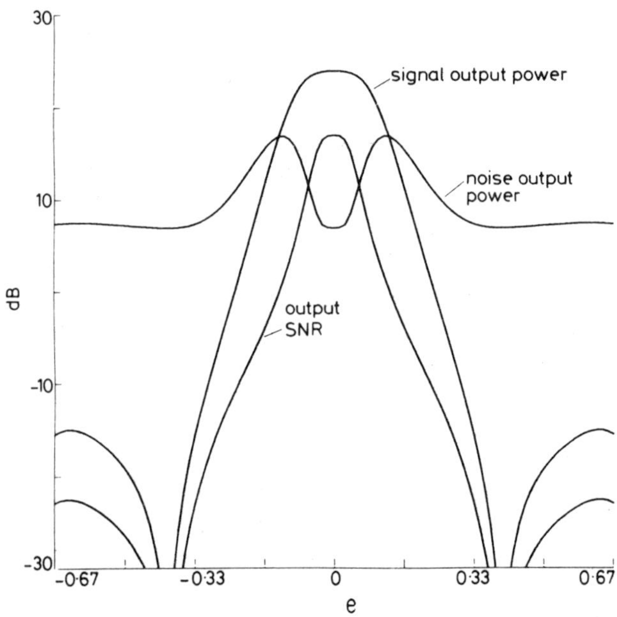

Fig. 6.4.6 *Performance of antenna with real-only weighting coefficients*

sidelobes indicating that interference cancellation has suffered by the loss of degrees of freedom.

6.5 Norm-bounded optimal solutions[100, 105, 120]

Although the optimal weighting vectors obtained in the last Section with linear constraints are effective if the input signal-to-noise ratio is small, the protected zone of the main lobe becomes so narrow at high input signal-to-noise ratios that serious pointing accuracy difficulties are likely to reappear. An effective method for reducing the signal-to-noise ratio dependence of the linear constraints is to combine them with a norm bound on the weighting vector of the form

$$\mathbf{W}^H\mathbf{W} \le B \tag{6.5.1}$$

The general effect of such a constraint is first investigated. Suppose we minimise the output power of an array by variation of $\mathbf{W}$ subject to M linear equality constraints

$$\mathbf{C}_k^t \mathbf{W} = c_k, \qquad k = 1, \ldots, M \tag{6.5.2}$$

and the norm inequality constraint (6.5.1). Using the Lagrange method we minimise the function

$$p = \mathbf{W}^H\mathbf{R}\mathbf{W} + \lambda_0 \mathbf{W}^H\mathbf{W} + \sum_{k=1}^{M} \lambda_k(\mathbf{W}^t\mathbf{C}_k - c_k) \tag{6.5.3}$$

where the λ_k are assumed to be real, as in eqn. (3.4.23), and the partial derivatives of p with respect to the nth element of $\mathbf{W}$ are

$$dp/d\mathbf{W}_r^n = 2 \text{ Re } (\mathbf{RW})^n + 2\lambda_0 \text{ Re } (\mathbf{W}^n) + \sum_{k=1}^{M} \lambda_k \text{ Re } (\mathbf{C}_k)^n$$

$$dp/d\mathbf{W}_j^n = 2 \text{ Im } (\mathbf{RW})^n + 2\lambda_0 \text{ Im } (\mathbf{W})^n - \sum_{k=1}^{M} \lambda_k \text{ Im } (\mathbf{C}_k)^n \tag{6.5.4}$$

and setting these derivatives to zero gives

$$\mathbf{RW} + \lambda_0 \mathbf{W} = \frac{1}{2} \sum_{k=1}^{M} \lambda_k \mathbf{C}_k^* \tag{6.5.5}$$

This equation is equivalent to

$$(\mathbf{R} + \lambda_0 \mathbf{I})\mathbf{W}_{\text{opt}} = \frac{1}{2} \sum_{k=1}^{M} \lambda_k \mathbf{C}_k^* \tag{6.5.6}$$

and in this form we have the interpretation that the norm-bounded weight vector is the same as the optimal vector obtained without the norm bound but with the covariance matrix augmented by the $\lambda_0 \mathbf{I}$ matrix. The value of λ_0 must be determined so that the norm bound is just not exceeded. Intuitively the weight vector is expected to converge smoothly to the conventional one as λ_0 is increased from 0 to ∞ and the norm of the linearly constrained optimal solution will fall to its minimum value $\|\mathbf{W}_c\|$. Somewhere within this range a value may be found such that $\|\mathbf{W}\| = B$. However, should the norm limit not be exceeded even for zero λ_0 then the linearly constrained solution is satisfactory; no attempt is made to use negative λ_0s. Numerical solutions are discussed in Appendix 4.

The output power of the array is not directly changed by the added identity matrix and remains as

$$p = \mathbf{W}_{\text{opt}}^H \mathbf{R} \mathbf{W}_{\text{opt}} \tag{6.5.7}$$

but in general p will be greater due to displacement of $\mathbf{W}_{\text{opt}}$. This is easily seen since if there is a norm-bounded solution with smaller power than the linear solution then the linear solution should have assumed the same weight vector as the norm-bounded one. In a complex interference environment the assumption is that the norm bound will preferentially increase the output power of sources lying within the main lobe, although it is very difficult to prove that this is actually the case. The best that can be done, as in the analysis below is to demonstrate the robustness of the main lobe for signals applied one at a time. In the presence of multiple signals this main-lobe robustness will continue to be effective but the norm-bound may have slight deleterious effects on the depth of nulls achieved in the directions of interferences lying outside the main lobe. The extent of this effect will depend upon the size of λ_0 in eqn. (6.5.6) because the nonlinear optimal weight vector is the same one as would be obtained by injecting additional uncorrelated noise.

Norm-bound plus look-direction constraint: For analytic purposes the Lagrange solution eqn. (6.5.6) is rather intractible though it is useful for numerical work and an eigenvector analysis is therefore described in Appendix 4.[120] The alternative approach is to use the orthogonal basis expansion of the weight vector of Section 6.4. Recall from eqn. (6.4.2) that if the single applied signal is very close to the look direction the weight vector may be approximated as

$$\mathbf{W} \simeq w_0 \mathbf{S}_0^* + w_1 \mathbf{S}_1^* \tag{6.5.8}$$

The value of w_0 is fixed by the look-direction constraint and previously w_1 was varied to minimise the output power. The value of w_1 will now be chosen to minimise the output power while satisfying the norm bound (6.5.1) and we assume that, in place of eqn. (6.4.3) we have

$$w_1 = -x s_1^* \tag{6.5.9}$$

The norm of $\mathbf{W}$ is now

$$\|\mathbf{W}\|^2 = |w_0|^2 \phi_0 + |x|^2 |s_1|^2 \phi_1 \tag{6.5.10}$$

In eqn. (6.5.1) let

$$B = \|\mathbf{W}_c\|^2(1 + \Delta) = |w_0|^2\phi_0(1 + \Delta) \tag{6.5.11}$$

where $\mathbf{W}_c$ is the conventional weighting vector; then, equating eqns. (6.5.10) and (6.5.11) yields

$$|x|^2 = \frac{|w_0|^2\phi_0}{|s_1|^2\phi_1}\Delta \tag{6.5.12}$$

The sign of x in eqn. (6.5.12) is chosen to get the smallest signal output amplitude. The antenna amplitude gain for the signal is

$$\begin{aligned} g(\theta) &= w_0\, s_0\, \phi_0 + w_1\, s_1\, \phi_1 \\ &= c(\theta) - |x||s_1|^2\phi_1 \end{aligned} \tag{6.5.13}$$

and substitution of $|x|$ from eqn. (6.5.12), choosing the phase which minimises $g(\theta)$, yields

$$g(\theta) = c(\theta) - w_0\, \phi_0^{1/2}\Delta^{1/2}|s_1|\phi_1^{1/2} \tag{6.5.14}$$

As $|s_1|$ increases linearly as pointing error $|\theta|$ increases we see that the divergence of the constrained pattern from the conventional one increases linearly with $|\theta|$. However, this is only true if the norm bound is activated and this is not the case for very small $|\theta|$.

The output signal-to-noise ratio of the norm-bounded system is given by

$$Q = \frac{|g(\theta)|^2}{\sigma^2\mathbf{W}^H\mathbf{W}} = \frac{|g(\theta)|^2}{\sigma^2 B} \tag{6.5.15}$$

if the norm limit is active. This SNR is greater than the linearly constrained system output SNR for $|\theta|$ values which activate the norm bound.

Figure 6.5.1 shows signal and noise output power and output SNR for a five-element antenna with a signal input power of 10 dB, $\sigma^2 = 0$ dB, $\Delta = 1$, as the source is scanned through the main lobe. The spike in SNR in the centre corresponds to nonactivation of the norm bound and performance within this region is the same as in Fig. 6.4.1. Outside this the norm bound is active and prevents further deterioration of signal response and output SNR. At $|\theta| \gg 0{\cdot}166$, the norm bound becomes inactive and the performance reverts to the linearly constrained antenna. Uncorrelated noise output is strictly limited by the bound at +10 dB.

In Fig. 6.5.2, the pointing error is fixed at 5·85° as in Fig. 6.4.2 and the source power is varied for $\sigma^2 = 1$. When the norm bound becomes active at about 0 dB signal power the fall in signal response $g(\theta)$ is halted and output SNR improves as signal power increases unlike the nonbounded norm case.

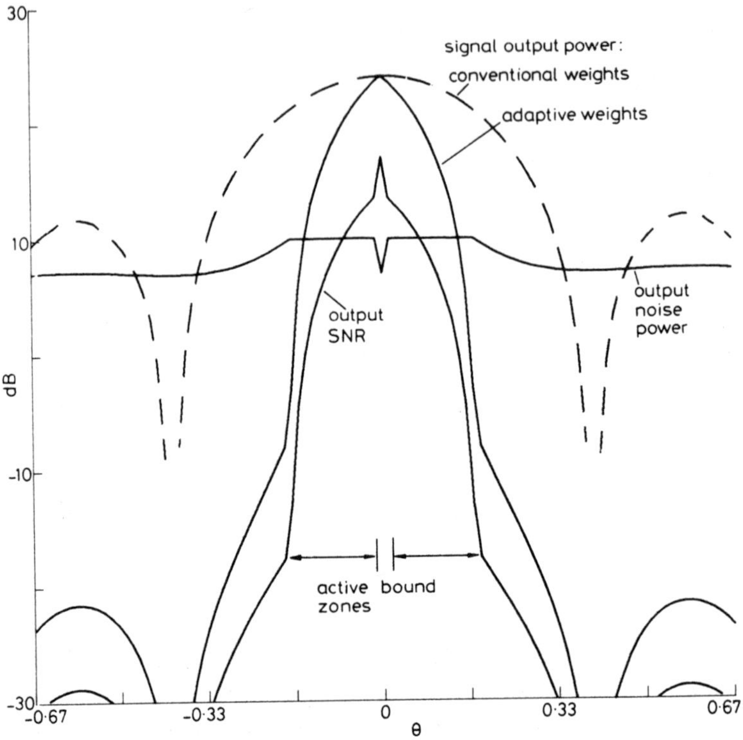

Fig. 6.5.1 *Main-lobe performance of norm-bounded antenna*

Example 6.5.1 : For the antenna above, calculate the desired signal amplitude gain and SNR at 10 dB input SNR. Using eqn. (6.5.14) we set $w_0 = 1$, $\phi_0 = 5$, $\Delta = 1$, $s_1 = \sin(5{\cdot}85°) = 0{\cdot}102$, $\phi_1 = 10\pi^2$ (from example 6.4.1) and obtain, with $c(\theta) = 4{\cdot}5$

$$g(\theta) = 4{\cdot}5 - 2{\cdot}26 = 2{\cdot}23$$

The noise output power is $\sigma^2 B = 1 \times 2 \times 5 = 10$. Thus

$$Q = \frac{2{\cdot}23^2}{10} = 4{\cdot}98 \quad \text{or} \quad +6{\cdot}97 \text{ dB}$$

The exact result is 7·33 dB.

Norm bound plus look direction and derivative constraint : The solution for this case is very similar to the simple look direction constraint above. In place of eqn. (6.4.22) we write

$$w_2 = -xs_2^* \tag{6.5.16}$$

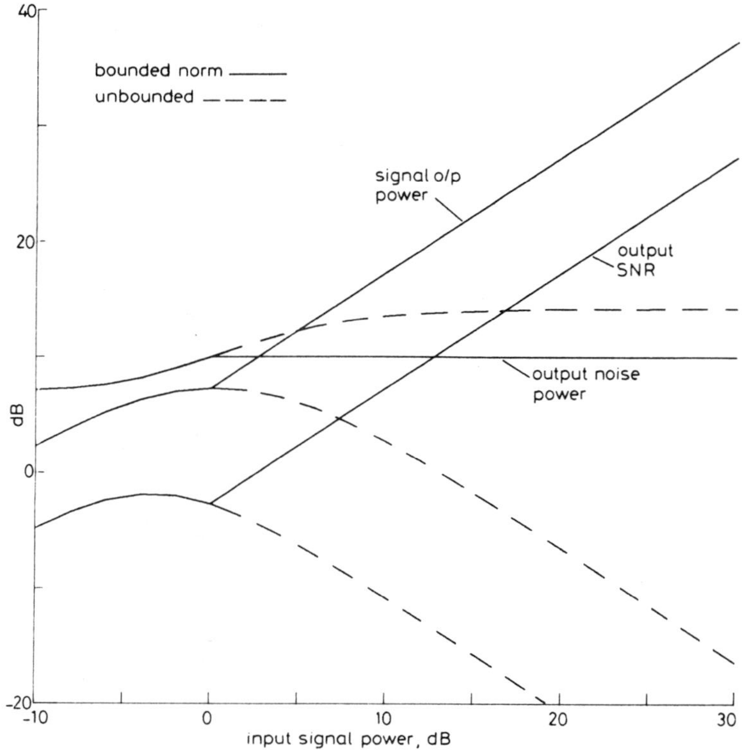

Fig. 6.5.2 *Effect of varying signal power at constant pointing error*

and obtain

$$\|\mathbf{W}\|^2 = |w_0|^2\phi_0 + |x|^2|s_2|^2\phi_2 \tag{6.5.17}$$

and solve for x as before

$$|x|^2 = \frac{|w_0|^2\phi_0}{|s_2|^2\phi_2}\Delta \tag{6.5.18}$$

The signal amplitude gain is

$$g(\theta) = w_0 s_0 \phi_0 + w_2 s_2 \phi_2$$

$$= c(\theta) - |x||s_2|^2\phi_2 \tag{6.5.19}$$

$$= c(\theta) - |w_0||s_2|\phi_0^{1/2}\phi_2^{1/2}\Delta^{1/2} \tag{6.5.20}$$

in which the signal of $(|x|^2)^{1/2}$ giving the smallest amplitude has been taken. The result shows that the adaptive pattern has a quadratic form for small θ if the norm bound is active, since $s_2 = \frac{1}{2}\theta^2$, and better control over distortion of the lobe is exercised. Figure 6.5.3 shows signal and noise output power and SNR as

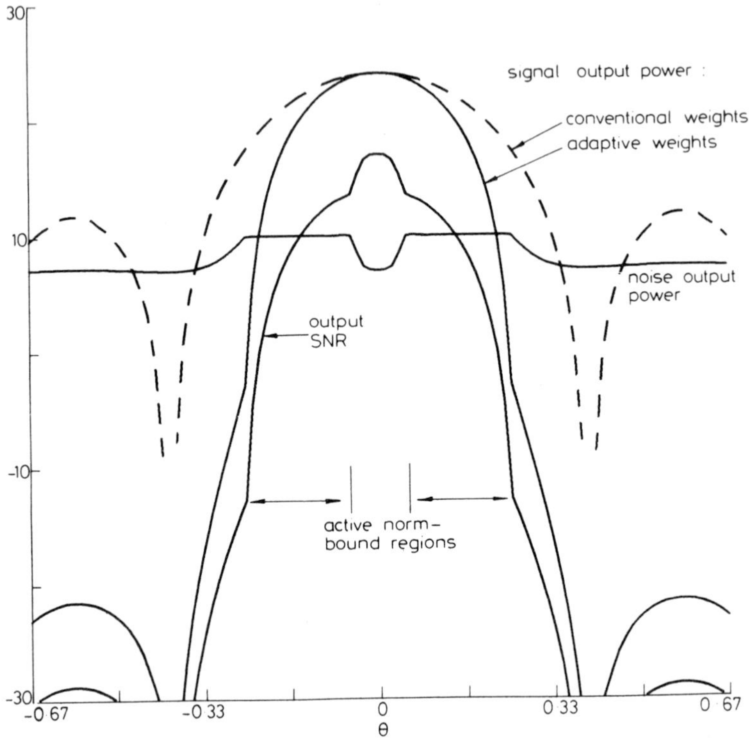

Fig. 6.5.3 *Antenna performance with derivative and norm constraints as a source is swept through the main lobe*

a 10 dB input SNR source is scanned through the lobe the array being the same one used in Fig. 6.4.4. A small zone in the centre fails to produce a limited norm condition and again the bound is slack for $|\theta| > 0{\cdot}25$. Here the pattern is identical to Fig. 6.4.4. The width of the region where an acceptable output SNR is observed is much greater than for the antenna without the derivative constraint. In Fig. 6.5.4, the pointing error is fixed at 5·85° as for Fig. 6.4.5 and there is now little loss in SNR performance relative to the conventional antenna.

When real-only coefficients are used, the performance is virtually indistinguishable from the derivative constrained system except for a poorer rejection of the source when it lies in the sidelobes.

Example 6.5.2 : Calculate the desired signal amplitude gain and output SNR for the antenna above at an input SNR of 10 dB. In eqn. (6.5.20) we have $c(\theta) = 4{\cdot}5$, $w_0 = 1$, $s_2 = \frac{1}{2}\theta^2 = 5{\cdot}2 \times 10^{-3}$, $\phi_0 = 5$, $\phi_2 = 14\pi^4$ (from example 6.4.2). Thus

$$g(\theta) = 4{\cdot}5 - 0{\cdot}43 = 4{\cdot}07$$

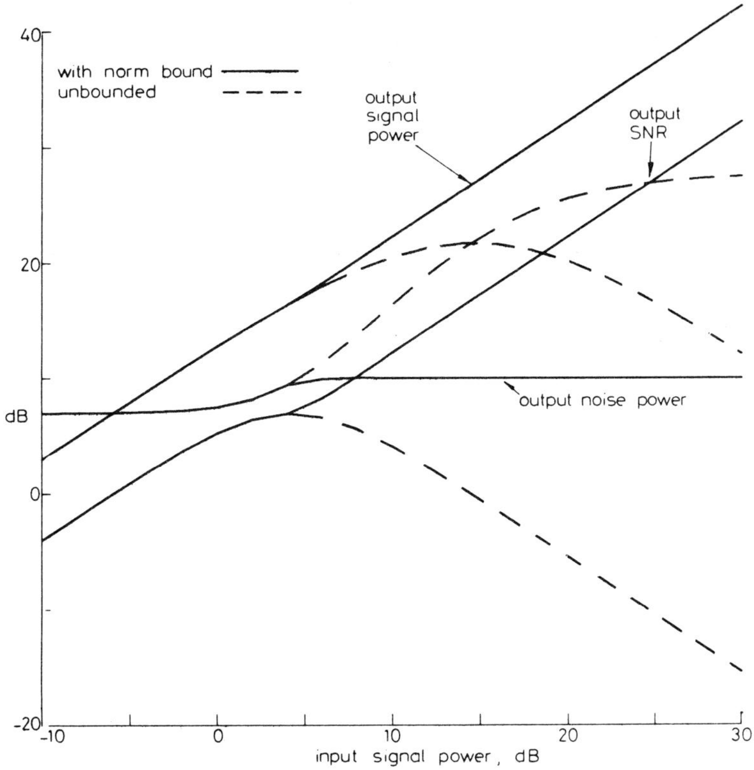

Fig. 6.5.4 *Performance for a fixed pointing error of 5·85°*

The output noise power is $\sigma^2 B = \sigma^2 N(1 + \Delta) = 10$ and the output SNR is

$$Q = \frac{4{\cdot}07^2}{10} = 1{\cdot}66 \quad \text{or} \quad +2{\cdot}2 \text{ dB}.$$

Exact analysis yields $g(\theta) = 4{\cdot}09$ and $Q = +2{\cdot}24$ dB.

6.6 Random channel gain errors

As well as pointing errors, the effects of random channel gain errors discussed in Section 6.2 are very significant at high input desired SNRs.[97] The implementation of a norm bound can effectively control desired signal suppression.[100]

If we write the desired signal as

$$\mathbf{S} = \mathbf{C} + \sigma\mathbf{E} \tag{6.6.1}$$

where **C** is the steering vector and **E** a vector of random errors such that

$$\mathbf{E}^t\mathbf{C}^* = 0 \quad \text{and} \quad \mathbf{E}^t\mathbf{E}^* = \mathbf{C}^t\mathbf{C}^* = N \tag{6.6.2}$$

σ corresponding to the sample standard deviation of the errors, then if the weighting vector is written, quite generally, as

$$\mathbf{W} = \mathbf{C}^* + \mathbf{F}^* \tag{6.6.3}$$

where $\mathbf{F}^t\mathbf{C}^* = 0$, the desired signal amplitude gain is

$$g = \mathbf{S}^t\mathbf{W} = \mathbf{C}^t\mathbf{C}^* + \sigma\mathbf{E}^t\mathbf{F}^* = N + \sigma\mathbf{E}^t\mathbf{F}^* \tag{6.6.4}$$

Now, from the triangle inequality for complex numbers,

$$N - |\sigma\mathbf{E}^t\mathbf{F}^*| \leq |g| \leq N + |\sigma\mathbf{E}^t\mathbf{F}^*| \tag{6.6.5}$$

and from the Cauchy inequality (Section 2.4) we have further

$$|\sigma\mathbf{E}^t\mathbf{F}^*| \leq \sigma(\mathbf{E}^t\mathbf{E}^*)^{1/2}(\mathbf{F}^t\mathbf{F}^*)^{1/2} = \sigma N^{1/2}(\mathbf{F}^t\mathbf{F}^*)^{1/2} \tag{6.6.6}$$

Substituting into eqn. (6.6.5) yields

$$N - \sigma N^{1/2}(\mathbf{F}^t\mathbf{F}^*)^{1/2} \leq |g| \leq N + \sigma N^{1/2}(\mathbf{F}^t\mathbf{F}^*)^{1/2} \tag{6.6.7}$$

We now limit the norm of the weighting vector:

$$\mathbf{W}^H\mathbf{W} = \mathbf{C}^t\mathbf{C}^* + \mathbf{F}^t\mathbf{F}^* \leq N(1 + \Delta) \tag{6.6.8}$$

where Δ is a positive number and thus have

$$\mathbf{F}^t\mathbf{F}^* \leq N(1 + \Delta) - N = N\Delta \tag{6.6.9}$$

If we take the lower limit in eqn. (6.6.7) as a worst case then substituting eqn. (6.6.9) into (6.6.7) yields

$$N(1 - \sigma\Delta^{1/2}) \leq |g| \tag{6.6.10}$$

and by limiting the value of Δ the effect of a finite channel gain standard deviation σ can be limited.

Fig. 6.6.1 shows the main-lobe performance of a five element array with $\lambda/2$ spacing for channel gain errors of 10% ($\sigma \simeq 0{\cdot}1$). In the absence of a norm bound, there is a signal suppression of about 25 dB even when it arrives from the pointing direction and, associated with a high weighting coefficient norm, there is a high output of uncorrelated noise resulting in an output SNR of −6 dB. With the norm bound ($\Delta = 1$) the signal loss is reduced to 11 dB, the output noise power is reduced and the output SNR is +23 dB.

Channel failures: The effect of complete loss of signal in one or more channels due to amplifier failure is often negligible in the conventional antenna and even in the LMS adaptive algorithm. However, linearly constrained minimum output power arrays may be rendered totally useless by such an event. To see this consider the simple case of a broadside look direction constraint

$$\sum_{n=1}^{N} W_n = N \tag{6.6.11}$$

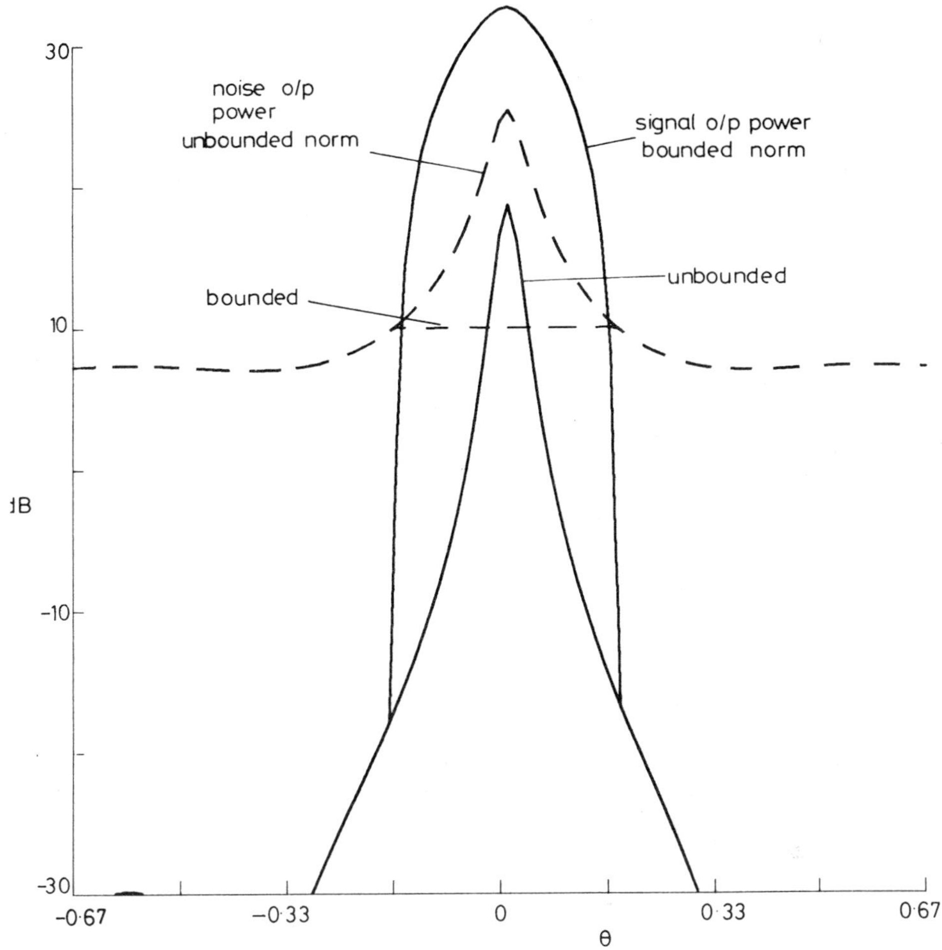

Fig. 6.6.1 *Effect of norm-bound on signal suppression in the presence of random channel gain errors*
Input SNR 20 dB $\sigma_n^2 = 1$, $\Delta = 1$

If there is zero signal in one channel, say channel M, then if W_M is set to a value of N while all other coefficient are set to zero then zero output power results while the look direction constraint is satisfied. An adaptive algorithm will find this trivial solution very efficiently.

Using a norm bound may be effective in partially inhibiting such effects as the norm of the weighting vector is N which may be much greater than the norm of the conventional weighting vector, $N^{1/2}$.

If this is not enough, i.e. there is a high probability of several channel failing without service attention, more active measures may be taken which normally would require monitoring of the channel powers and causing the algorithm to set coefficients in defunct channels to zero.

A similar though perhaps less expected effect occurs if two channels get crossed such that $X_i(t) = X_j(t)$, say, and the look direction is other than broadside. If we solve the simultaneous equations

$$W_i + W_j = 0$$

$$W_i e^{j\phi_i} + W_j e^{j\phi_j} = N \tag{6.6.12}$$

then we obtain

$$W_i = \frac{N}{e^{j\phi_i} - e^{j\phi_j}} \qquad W_j = \frac{N}{e^{j\phi_j} - e^{j\phi_i}} \tag{6.6.13}$$

and using these coefficients while setting all the rest to zero will cause zero power to be output $(W_i X_i(t) + W_j X_j(t) = 0)$ while satisfying the constraint which is of the form

$$\sum_{n=1}^{N} W_n e^{j\phi_n} = N$$

6.7 Effect of norm bound on interference cancellation

This is the most important aspect of the functioning of a norm-bounded antenna, since adaptation and the possibility of main lobe distortion, which invokes the norm bound in the first place, is only necessary in the presence of strong interference. We wish to consider the performance of the array when there is a signal present in the main lobe whose direction is slightly different from the look direction together with an interference somewhere in the conventional pattern sidelobes.

It was observed in eqn. (6.5.6) that the effect of a norm bound is equivalent to augmenting the covariance matrix by an identity matrix $\lambda_0 \mathbf{I}$ with the positive variable λ_0 being increased until the norm limit is just reached from above. If the norm is below the limit in the unmodified optimal solution ($\lambda_0 = 0$) then nothing is done. Whether the optimal solution is modified by imposing a norm bound or by actually injecting uncorrelated noise is immaterial so far as the final weighting vector is concerned, although there would be a problem with the latter approach as the noise would appear at the output and might interfere with signal detectibility. In gradient methods it is normally possible to get the effect of injected noise by modification of the algorithm.

If the main-lobe desired signal is presented in isolation to the antenna, which uses a norm bound to prevent its cancellation, then in principle the constant λ_0 can be calculated. The approximation is now made that, when an additional interference is presented in a direction remote from the main beam and is subsequently nulled by the antenna, the extra displacement of the weighting coefficients is so small that the value of λ_0 is hardly changed. In other words, cancellation of a source in the sidelobes is a process which interacts very little with the main lobe. This assumption will be demonstrated experimentally in due course but first a theoretical solution for the weighting vector will be derived which makes use of it.

Theoretical solution: For the main-lobe signal only, using a simple look direction constraint and a norm bound we proceed as in Section 6.4, obtaining

$$w_1 = \frac{-s_1^* w_0 s_0 \phi_0}{\lambda_0 + |s_1|^2 \phi_1} \tag{6.7.1}$$

where, in place of the true uncorrelated noise variance σ^2 of eqn. (6.4.3), we have used the variable λ_0 which is adjusted to satisfy the norm bound. A new orthogonal basis vector $\mathbf{S}_2$ is now introduced and defined as the projection of the space vector of the sidelobe interference 'U' onto a subspace orthogonal to $\mathbf{S}_0$ and $\mathbf{S}_1$. The latter retain their meanings defined in Section 6.3

$$\mathbf{S}_2 = \left(I - \frac{\mathbf{S}_0^* \mathbf{S}_0^t}{\mathbf{S}_0^t \mathbf{S}_0^*}\right)\left(I - \frac{\mathbf{S}_1^* \mathbf{S}_1^t}{\mathbf{S}_1^t \mathbf{S}_1^*}\right)\mathbf{U} \tag{6.7.2}$$

where $\mathbf{U} = \mathbf{S}(\theta_u)$, θ_u being the bearing of the interference. We can now define $\mathbf{U}$ in terms of the orthogonal basis vectors:

$$\mathbf{U} = u_0 \mathbf{S}_0 + u_1 \mathbf{S}_1 + u_2 \mathbf{S}_2 \tag{6.7.3}$$

In order to allow cancellation of this interference a new component $w_2 \mathbf{S}_2^*$ is introduced into $\mathbf{W}$:

$$\mathbf{W} = w_0 \mathbf{S}_0^* + w_1 \mathbf{S}_1^* + w_2 \mathbf{S}_2^* \tag{6.7.4}$$

and, mathematically, the approximation made earlier that the introduction of the interference does not significantly change the weight vector is equivalent to assuming that w_1 of eqn. (6.7.4) is the same as w_1 in (6.7.1). This would be exactly true if $\mathbf{U}$ were orthogonal to $\mathbf{S}_0$ and $\mathbf{S}_1$ and the validity of the analysis relies upon small inner products. The new coefficient w_2 is found from eqn. (6.3.25) as

$$w_2 = \frac{-u_2^* \sum_{i=0}^{2} w_i u_i \phi_i \pi_u}{\lambda_0} \tag{6.7.5}$$

where π_u is the interference power; a multiplier of $\mathbf{U}^*\mathbf{U}^t$ in the covariance matrix. For small w_2 and quasiorthogonality of $\mathbf{S}_2$ and $\mathbf{S}_0$ the look direction amplitude gain will not be significantly changed. We solve for w_2:

$$w_2 = \frac{-u_2^*}{\lambda_0}(w_0 u_0 \phi_0 + w_1 u_1 \phi_1 + w_2 u_2 \theta_2)\pi_u \tag{6.7.6}$$

whence

$$w_2 = \frac{-u_2^* \rho_u \pi_u}{\lambda_0 + |u_2|^2 \phi_2 \pi_u} \tag{6.7.7}$$

where $\rho_u = w_0 u_0 \phi_0 + w_1 u_1 \phi_1$ is the absolute response of the adaptive antenna to interference $\mathbf{U}$ in its original condition, i.e. when only the main-lobe desired signal and the norm bound determine the solution and no attempt is made to cancel the new interference. Eqn. (6.7.7) is clearly analogous to (6.7.1) which determined w_1.

Interference output power: The amplitude gain of the adapted antenna in the direction of the sidelobe interference is

$$g_u = \mathbf{U}'\mathbf{W} = \sum_{i=0}^{2} w_i u_i \phi_i = \rho_u \left[1 - \frac{|u_2|^2 \phi_2 \pi_u}{\lambda_0 + |u_2|^2 \phi_2 \pi_u}\right]$$

$$= \frac{\rho_u \lambda_0}{\lambda_0 + |u_2|^2 \phi_2 \pi_u} \tag{6.7.8}$$

If $\mathbf{U}$ is almost orthogonal to $\mathbf{S}_0$ and $\mathbf{S}_1$ then

$$\phi_2 = \|\mathbf{S}_2\|^2 \simeq \|\mathbf{U}\|^2 = N \tag{6.7.9}$$

and, as u_0 and u_1 are small in eqn. (6.7.3), we can take

$$|u_2| \simeq 1 \tag{6.7.10}$$

The interference output power is

$$p_u = \pi_u |g_u|^2 = \frac{\pi_u |\rho_u|^2 \lambda_0^2}{(\lambda_0 + \pi_u N)^2} \tag{6.7.11}$$

Output SNIR: The signal output power is approximated by the equivalent result obtained in the absence of the sidelobe interference, eqn. (6.5.14) and the output signal-to-interference ratio is just

$$Q \simeq \frac{\pi_s |g_s|^2}{\pi_u |g_u|^2} \tag{6.7.12}$$

and all variables are calculable.

The trend of the behaviour can be seen from eqn. (6.7.11) which has an easily found maximum as π_u is varied of

$$p_u(\max) = \frac{\lambda_0 |\rho_u|^2}{4N} \tag{6.7.13}$$

which depends directly upon the value of λ_0. This can be found as follows. From eqn. (6.5.11) we have

$$\|W\|^2 = |w_0|^2 \phi_0 (1 + \Delta) \tag{6.7.14}$$

and equating with (6.4.7) we get

$$\frac{|s_1|^2\phi_1|c(\theta)|^2}{(\lambda_0+|s_1|^2\phi_1)^2}=|w_0|^2\phi_0\Delta$$

whence

$$\lambda_0=\frac{|s_1|\phi_1^{1/2}|c(\theta)|}{|w_0|\phi_0^{1/2}\Delta^{1/2}}-|s_1|^2\phi_1 \tag{6.7.15}$$

which is of the form

$$\lambda_0=\frac{x|c(\theta)|}{(N\Delta)^{1/2}}-x^2$$

having a maximum value at

$$x=\frac{|c(\theta)|}{2(N\Delta)^{1/2}}$$

of

$$\lambda_0(\max)=\frac{|c(\theta)|^2}{2N\Delta}-\frac{|c(\theta)|^2}{4N\Delta}=\frac{|c(\theta)|^2}{4N\Delta}\simeq\frac{N}{4\Delta} \tag{6.7.16}$$

Thus,

$$p_u(\max)\simeq|\rho_u|^2/16\Delta \tag{6.7.17}$$

As unity signal power is assumed and $g_s \simeq N$ we have in eqn. (6.7.12)

$$Q_{\min}\simeq\frac{16N^2\Delta}{|\rho_u|^2} \tag{6.7.18}$$

which is likely to be much greater than unity. Thus, even with the worst possible combination of desired signal bearing and interference power the output SIR will be quite good, though perhaps not of high fidelity quality.

Norm bound with derivative constraint : The solution for this problem is exactly equivalent to the above with the exception that vector $\mathbf{S}_2$, defined in eqn. (6.3.8) replaces $\mathbf{S}_1$ everywhere and s_2 replaces s_1. Result (6.7.18) remains unchanged. Vector $\mathbf{S}_2$ above, defined in (6.7.2), also stays the same. Thus the difference is a transformation of θ to $\frac{1}{2}\theta^2$.

Simulation results : The adaptive beamformer has been solved numerically to illustrate some of the results above. Various parameters in the theory are somewhat inaccessible, particularly ρ_u, the amplitude gain for the interference when there is a main lobe desired signal. Thus although the theory gives a model for system behaviour it is not entirely satisfactory for precise quantitative results. Fig. 6.7.1 shows interference output power against interference input power in a

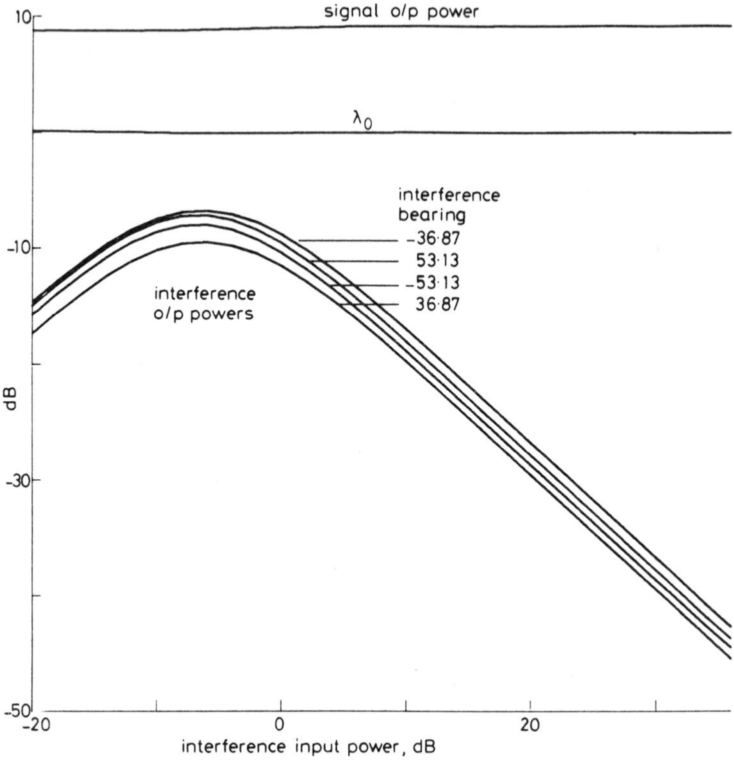

Fig. 6.7.1 *Breakthrough of a side lobe interference for norm-bounded system*

five element array with $\lambda/2$ spacing. The interference is placed in the sidelobes at four different bearings, as shown while the desired signal is placed 5° from the pointing direction (broadside). The maximum interference output power of about -8 dB agrees well with eqn. (6.7.17) and the lowest output SNIR of 16 dB agrees favourably with eqn. (6.7.18), noting that $|\rho_u| \simeq 2$ from Fig. 6.7.2 and there is a 4 dB loss of desired signal to be taken into account, evident in this beam pattern. The desired signal output power and the λ_0 coefficient are seen to have insignificant variation which supports the initial theoretical assumption. The value of Δ is 1.

Fig. 6.7.2 shows the shapes of the optimal beam patterns when the interference is placed at $-53{\cdot}13°$ and its power is varied. The shape of the main lobe hardly changes, supporting the assumption that sidelobe interference influences the main lobe only slightly. It is interesting that the interference bearing is actually a null of the conventional pattern and its amplitude gain at small powers, $\simeq \rho_u$ is due entirely to the presence of the main lobe desired signal which causes considerable disturbance in the sidelobes.

Fig. 6.7.3 shows how desired signal output power varies together with inter-

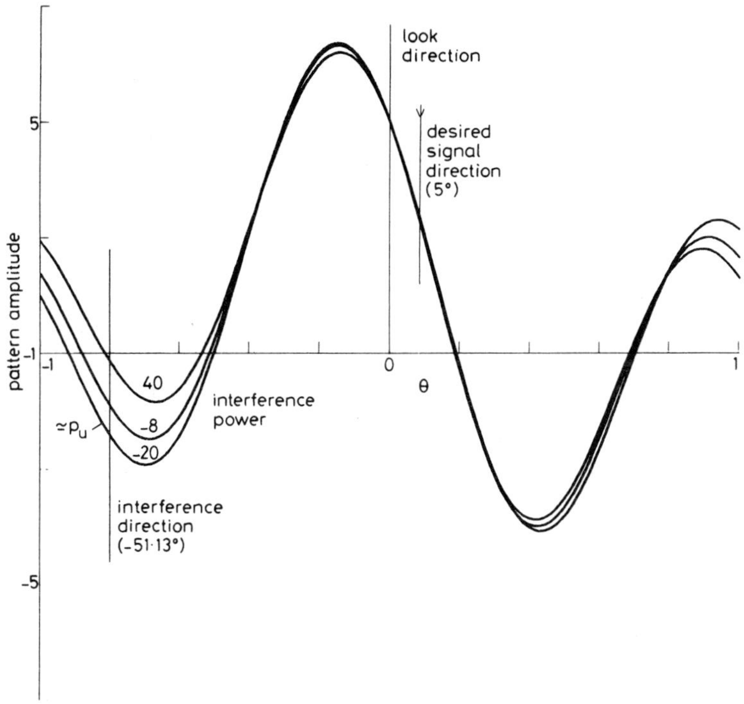

Fig. 6.7.2 *Array patterns as a function of interference input power*

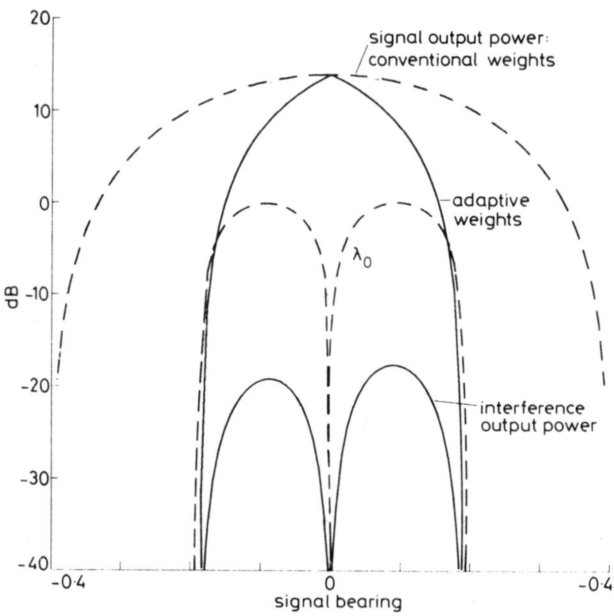

Fig. 6.7.3 *Main-lobe performance of norm-bounded array*
Desired signal in main lobe and interference in sidelobes

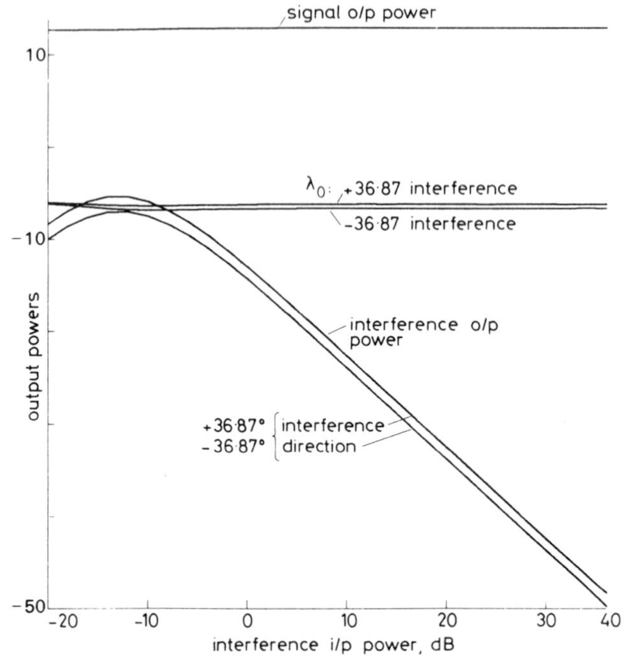

Fig. 6.7.4 *Breakthrough of sidelobe interference for derivative constraint*

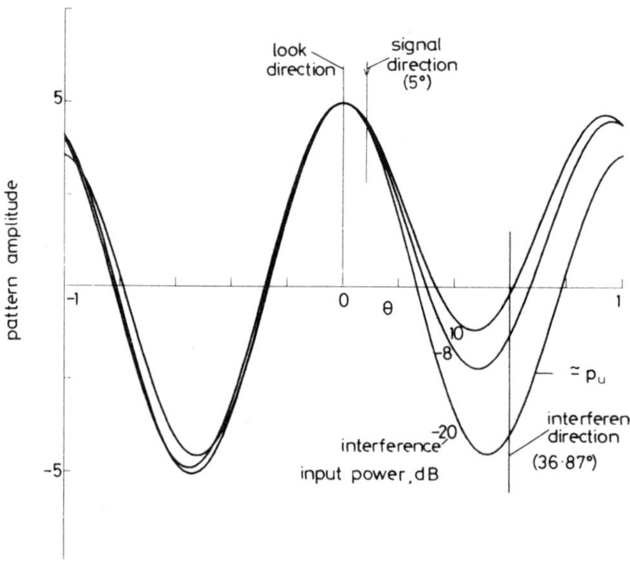

Fig. 6.7.5 *Beam patterns for array with norm-bound and derivative constraint against side-lobe interference power*

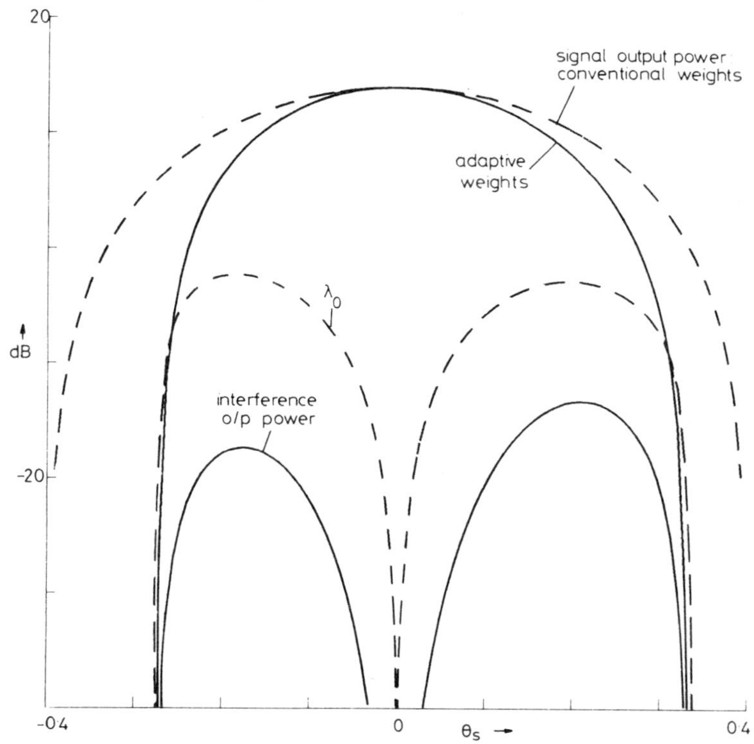

Fig. 6.7.6 *Array with derivative and norm constraints*
Performance against desired signal direction
10 dB interference at 36·9°

ference output power and λ_0 as the desired signal is scanned through the main lobe region. A maximum of λ_0 and interference output power occurs for a particular pointing error, a result expected from eqns. (6.7.16) and (6.7.11). At the interference power of $+10$ dB chosen the interference output power is less than its maximum value indicated by eqn. (6.7.17). The norm-squared increase Δ is fixed at unity in all cases.

Figs. 6.7.4, 6.7.5, and 6.7.6 show the same results but with the addition of a derivative constraint. The output SNIRs are quite similar apart from occuring at different ranges of θ_s in the scanned desired signal case.

Main-lobe interference: When the interference enters the main lobe and activates the norm bound itself then the situation is not described at all by the above technique. If the interference is stronger than the desired signal it will exert the predominant control over the beam shape and the desired signal will ride through without affecting it very much. The extent of cancellation of the interference could be predicted by the methods discussed in Section 6.4.

6.8 Hardware approximations to norm bounds

The norm bound is expensive to calculate in terms of computation load and, in hardware systems, almost impossible. Some approximation to the exact result is allowable.

Limitation on individual element moduli: Let W_n be the nth element of weight vector $\mathbf{W}$. Constraining the modulus of W_n to be less than a certain limit

$$|W_n| \leq b \qquad n = 1, \ldots, N \tag{6.8.1}$$

gives a norm bound since

$$\mathbf{W}^H\mathbf{W} = \sum_{n=1}^{N} |W_n|^2 \leq Nb \tag{6.8.2}$$

If the elements are complex they are constrained to lie in the circle shown in Fig. 6.8.1(a). The same thing could be done with quadrature real coefficients.

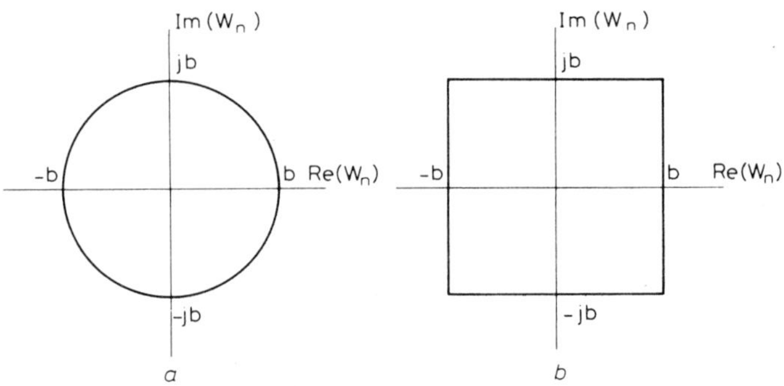

Fig. 6.8.1 *Individual moduli bounds*

The arithmetic may be simplified if the real and imaginary parts of the coefficients are separately constrained:

$$|\mathrm{Re}\,(W_n)| \leq b \qquad |\mathrm{Im}\,(W_n)| \leq b \tag{6.8.3}$$

which gives the norm limit

$$\mathbf{W}^H\mathbf{W} \leq 2b \tag{6.8.4}$$

The feasible region for the coefficients is then as shown in Fig. 6.8.1(*b*).

Limit on coefficient excursion: The conventional weighting vector is $\mathbf{C}^*$ where $\mathbf{C}$ is the steering vector. We can limit each coefficient of $\mathbf{W}$ to lie within a certain distance of the corresponding element in $\mathbf{C}^*$:

$$W_n = C_n^* + \Delta_n \qquad |\Delta_n| \leq b \tag{6.8.5}$$

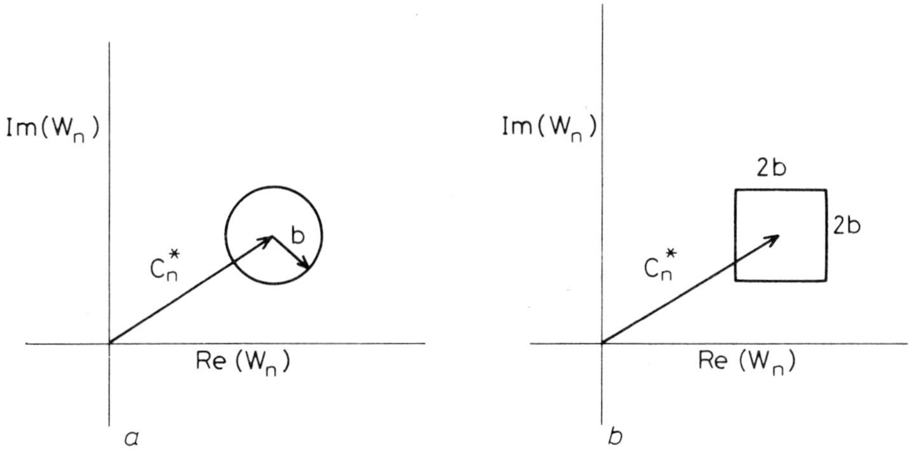

Fig. 6.8.2 *Coefficient excursion bounds*

which gives a norm limit of

$$\mathbf{W}^H\mathbf{W} \leq N(1 + b)^2 \tag{6.8.6}$$

The feasible region is shown in Fig. 6.8.2(*a*). The real and imaginary part may be separately constrained giving the region shown in Fig. 6.8.2(*b*).

This system is particularly suitable for hardware applications where the variable part of the element is in a shunt path as shown in Fig. 6.8.3.

This scheme has a further advantage which is that if the feasible region is fairly small as suggested in the diagrams then it is not possible for the coefficients in the weighting vector to vanish and the constraint vector in the algorithm may be dispensed with. The coefficients will now tend to fall to the

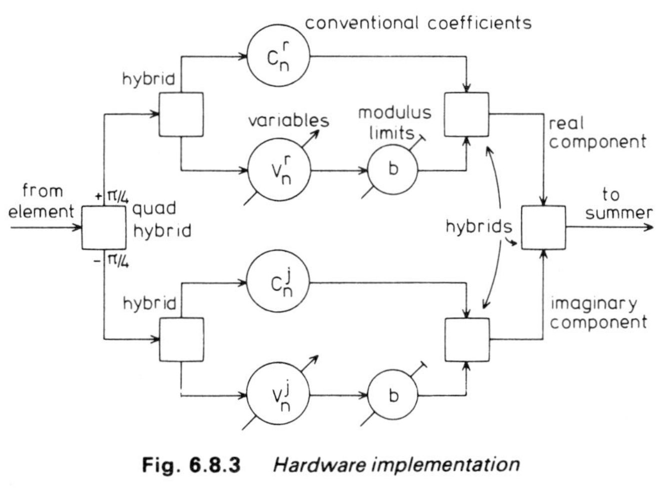

Fig. 6.8.3 *Hardware implementation*

smallest amplitude end of the feasible region which will involve some loss of desired signal amplitude but if there is a related decrease in uncorrelated noise then this is no disadvantage.

The actual shape of the main lobe when a desired signal is scanned through it is rather difficult to determine mathematically when such nonlinear constraints are in use and must be determined by simulation.

6.9 Effect of data transformation on norm bound

If the antenna data vectors are passed through a beamformer or other matrix transformation **H** the equivalent optimal solution in **H**-space is given by eqn. (3.7.9) as

$$\mathbf{W}_H = (\mathbf{H}^t)^{-1}\mathbf{W}_x \tag{6.9.1}$$

if **H** is an orthogonal transformation then

$$\mathbf{W}_H^H \mathbf{W}_H = \mathbf{W}_x^H \mathbf{H}^t(\mathbf{H}^t)^H \mathbf{W}_x = \mathbf{W}_x^H \mathbf{W}_x \tag{6.9.2}$$

and the norm bound is unaffected in value. Otherwise the equivalent norm in x-space must be found:

$$\mathbf{W}_x^H \mathbf{W}_x = \mathbf{W}_H^H(\mathbf{H}^t)^H \mathbf{H}^t \mathbf{W}_H \tag{6.9.3}$$

and the problem of constraining $\mathbf{W}_H$ such that

$$\mathbf{W}_x^H \mathbf{W}_x \leq B \tag{6.9.4}$$

is greatly complicated.

Chapter 7

Suboptimal arrays and other modified systems

Adaptive antennas of the type discussed in earlier Chapters tend to be expensive to construct if the number of elements in the array is large since a pair of variable quadrature attenuators and associated control must be used in every channel. Much of this expense is wasted if the antenna is only to cancel one or two interferences as, in principle, it is only necessary to provide one quadrature attenuator for each interference. However, the task of finding the best way to deploy a limited number of degrees of freedom in an antenna is far from trivial. A number of techniques for doing this are discussed in this chapter and their performances are compared, mostly on their ability to cancel single interferences. The assumption is made that best single interference cancellation coincides with best multiple interference cancellation (provided sufficient degrees of freedom are provided) though simulation tests should be performed on a realistic model of the system as a final check.

7.1 Arrays with thinned adaptive elements

Fig. 7.1.1 shows a suboptimal system with N elements out of which M are connected for variable weighting by the shunt coefficients V_i. The N-vector $\mathbf{S}(\theta)$ is the space vector of a signal at the whole array:

$$\mathbf{S}(\theta) = (S_1, S_2, \ldots, S_N)^t \tag{7.1.1}$$

The constraint vector $\mathbf{C}$ appropriate for the whole array is

$$\mathbf{C} = \mathbf{S}(\theta_d) \tag{7.1.2}$$

where θ_d is the direction of the desired signal. If the elements of $\mathbf{C}$ not connected to variable coefficients are set to zero we get a 'thinned' or sparse vector $\mathbf{C}_v$ of the form

$$\mathbf{C}_v = (0, 0, 0, 0, 0, C_6, C_7, C_8, 0)^t \tag{7.1.3}$$

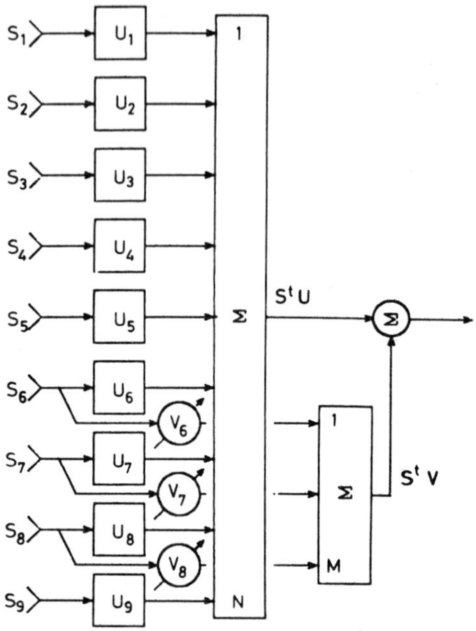

Fig. 7.1.1 *Adaptive antenna with thinned variable coefficients*

The conventional, phase-compensated, weighting vector with the adaptive coefficients set to zero is

$$\mathbf{U} = \mathbf{C}^* \tag{7.1.4}$$

Variable vector $\mathbf{V}$ is an N-vector with elements not associated with adaptive channels set to zero. Basically the total weighting vector is of the form

$$\mathbf{W} = \mathbf{U} + \mathbf{V} \tag{7.1.5}$$

but to keep track of the constraints on $\mathbf{V}$ a projection matrix $\mathbf{P}$ is defined which is null except for an M^2 element submatrix $\mathbf{P}_v$

$$\mathbf{P} = \left[\begin{array}{c|c|c} 0 & & \\ \hline & P_v & \\ \hline & & \end{array}\right] \tag{7.1.6}$$

and the form used in minimisation will be

$$\mathbf{W} = \mathbf{U} + \mathbf{PV} \tag{7.1.7}$$

in which $\mathbf{V}$ can then be varied without constraint. For the array of the type shown the projection $\mathbf{P}$ will always be of the form

$$\mathbf{P} = \mathbf{I}_v - \frac{\mathbf{C}_v^* \mathbf{C}_v^t}{\mathbf{C}_v^t \mathbf{C}_v^*} \tag{7.1.8}$$

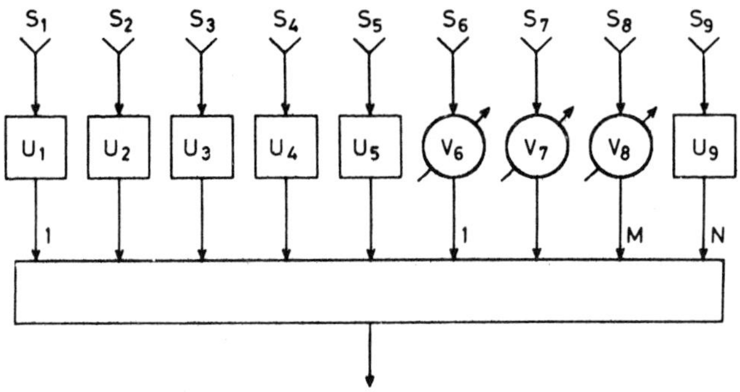

Fig. 7.1.2 *'Thinned' suboptimal antenna with unconstrained variable coefficients*

Total elements = *N*
Fixed elements = *K*
Variable elements = *M*

For example, if the pointing direction is broadside, we have for $\mathbf{P}_v$ when $M = 3$

$$\mathbf{P}_v = \begin{bmatrix} \frac{2}{3} & -\frac{1}{3} & -\frac{1}{3} \\ -\frac{1}{3} & \frac{2}{3} & -\frac{1}{3} \\ -\frac{1}{3} & -\frac{1}{3} & \frac{2}{3} \end{bmatrix} \tag{7.1.9}$$

Matrix $\mathbf{I}_v$ is defined in eqn. (7.1.14).

This constraint prevents the amplitude gain of the adaptive subarray varying in the direction of the desired signal since the space vector of the desired signal, if there is no pointing error, is just **C** and

$$\mathbf{C}^t\mathbf{W} = \mathbf{C}^t\mathbf{U} + \mathbf{C}^t\mathbf{P}\mathbf{V} = \mathbf{C}^t\mathbf{U} = N \tag{7.1.10}$$

Fig. 7.1.2 shows a slightly different arrangement in which the adaptive subarray is separate from the fixed one. The total number of elements remains at N of which K are fixed at the phase compensation values and M are variable. As before an N-vector **S** represents the signal vector at the whole aperture but subsidiary vectors $\mathbf{S}_u$ and $\mathbf{S}_v$ represent the signals present at the fixed and variable subarrays respectively with the rest set to zero:

$$\mathbf{S}_u = (S_1, S_2, S_3, S_4, S_5, 0, 0, 0, S_9)^t \tag{7.1.11}$$

$$\mathbf{S}_v = (0, 0, 0, 0, 0, 0, S_6, S_7, S_8, 0)^t \tag{7.1.12}$$

If $\mathbf{S} = \mathbf{S}(\theta)$ is set to the pointing direction then we obtain the constant vectors $\mathbf{C}$, $\mathbf{C}_u$, and $\mathbf{C}_v$ which are defined in the same way as **S**. The variable coefficient vector **V** is defined as an N-vector whose elements not connected to variable coefficients are set to zero.

The conventional weighting vector **U** consists of the vector $\mathbf{C}^*$ with elements connected to variable coefficients set to zero:

$$\mathbf{U} = (C_1^*, C_2^*, C_3^*, C_4^*, C_5^*, 0, 0, 0, C_9^*)^t \tag{7.1.13}$$

The total weighting vector is $\mathbf{W} = \mathbf{U} + \mathbf{PV}$ as in eqn. (7.1.7) but this time the projection matrix $\mathbf{P}$ may take one of two forms, either the null constraint of eqn. (7.1.8) in which case the adaptive subarray has a null in the steering direction of the fixed subarray or otherwise the elements of $\mathbf{V}$ are unconstrained and the projection $\mathbf{P}$ consists of an N by N identity matrix with the diagonal elements corresponding to the fixed channels set to zero, for example,

$$\mathbf{P} = \text{diag}\,(0, 0, 0, 0, 0, 1, 1, 1, 0)^t \tag{7.1.14}$$

Power minimisation: The antenna will be optimised by minimisation of its output power subject to the constraints described. Unlike the fully adaptive aperture this will not lead to the optimum output SNR condition for the given number of adaptive channels. It does not appear to be always easy to converge to the best output SNR unless a reference signal is available and the LMS algorithm is used. Covariance matrix $\mathbf{R}$ is the N by N complex covariance across the whole array and the output power is

$$\begin{aligned} p &= \mathbf{W}^H\mathbf{R}\mathbf{W} = (\mathbf{U} + \mathbf{PV})^H\mathbf{R}(\mathbf{U} + \mathbf{PV}) \\ &= \mathbf{U}^H\mathbf{RU} + \mathbf{V}^H\mathbf{PRPV} + \mathbf{U}^H\mathbf{RPV} + \mathbf{V}^H\mathbf{PRU} \end{aligned} \tag{7.1.15}$$

Let $(\ )^n$ be the nth element of a complex vector $(\ )$. The partial derivatives of p with respect to the real and imaginary parts of V^n are

$$\begin{aligned} \partial p/\partial\, \text{Re}\,(\mathbf{V}^n) &= 2\,\text{Re}\,(\mathbf{PRPV})^n + 2\,\text{Re}\,(\mathbf{PRU})^n \\ \partial p/\partial\, \text{Im}\,(\mathbf{V}^n) &= 2\,\text{Im}\,(\mathbf{PRPV})^n + 2\,\text{Im}\,(\mathbf{PRU})^n \end{aligned} \tag{7.1.16}$$

and equating both to zero yields

$$\mathbf{PRPV} = -\mathbf{PRU} \tag{7.1.17}$$

This equation cannot be solved by matrix inversion as was the case for fully optimal antennas since $\mathbf{PRP}$ is singular having at least one eigenvalue of zero with eigenvector $\mathbf{C}_v$. However, we can use the pseudoinverse of $\mathbf{PRP}$ to get the general solution[132, 136]

$$\hat{\mathbf{V}} = -(\mathbf{PRP})^+\mathbf{PRU} + \mathbf{Z} \tag{7.1.18}$$

by premultiplying throughout by the pseudoinverse $(\mathbf{PRP})^+$ defined in Appendix A1.4). The vector $\mathbf{Z}$ represents the unknown part of the solution and is any vector outside the space spanned by the eigenvectors of $\mathbf{PRP}$.[132] In fact, in the present case, if $\mathbf{R}$ is invertible then there is only one such vector

$$\mathbf{Z} = k\mathbf{C}_u \tag{7.1.19}$$

and as we are using $\mathbf{W} = \mathbf{U} + \mathbf{PV}$ as the weight vector we have $\mathbf{PZ} = 0$ and it makes no difference what value is assigned to k. Therefore $\mathbf{Z}$ is neglected in the present problem. Thus the optimal weighting vector is taken as

$$\hat{\mathbf{W}} = \mathbf{U} - (\mathbf{PRP})^+\mathbf{PRU} \tag{7.1.20}$$

An alternative approach, preferable for numerical work, but otherwise intractible, is to minimise the function

$$p = \overline{|u(t) + v(t)|^2} = \overline{|\mathbf{X}_u(t)^t\mathbf{U} + \mathbf{X}_v(t)^t\mathbf{V}|^2} \tag{7.1.21}$$

with the constraint $\mathbf{C}^t\mathbf{V} = 0$ as appropriate if the adaptive aperture has a null look direction constraint. In this equation we think of $\mathbf{X}_u$ and $\mathbf{U}$ as being K-vectors and $\mathbf{X}_v$ and $\mathbf{V}$ as M-vectors to avoid the use of pseudoinverses. The solution to this minimisation problem is

$$\hat{\mathbf{V}} = -\mathbf{R}_v^{-1}\boldsymbol{\phi}^* \tag{7.1.22}$$

when the adaptive aperture is unconstrained, where $\boldsymbol{\phi}$ is the vector of correlations between $\mathbf{X}_v(t)$ and $u(t)^*$ and $\mathbf{R}_v$ is the cross-covariance matrix of the data $\mathbf{X}_v(t)$ at the adaptive elements. When the adaptive aperture has a null pointing-direction constraint, the solution is

$$\hat{\mathbf{V}} = \mathbf{R}_v^{-1}\left\{\mathbf{C}_v^* \frac{\mathbf{C}_v^t\mathbf{R}_v^{-1}\boldsymbol{\phi}^*}{\mathbf{C}_v^t\mathbf{R}_v^{-1}\mathbf{C}_v^*} - \boldsymbol{\phi}^*\right\} \tag{7.1.23}$$

which clearly satisfies $\mathbf{C}_v^t\hat{\mathbf{V}} = 0$.

The divergence of these solutions is significant, neither yielding maximum output SNIR. If SNIR or mean square error is optimised the resulting vector is a linear combination of eqns. (7.1.22) and (7.1.23) with a relative weighting which depends upon the power of the desired signal.

Single-source solution: The output SNR will be investigated extensively for the case that there is a single interference present. For the system with a look-direction gain constraint, the desired signal response is constant and need not be explicitly considered while, in the case for unconstrained coefficient variation, the signal will be assumed to be so weak as not to affect the solution. The main-lobe performance of the antenna is of greatest interest as most suboptimal systems give tolerable sidelobe interference cancellation. Good main-lobe performance tends to be indicative of good performance in multisource scenarios although, as the degrees of freedom of a suboptimal system are reduced, some sacrifice is inevitable in comparison with a fully adaptive antenna.

Let N by N complex matrix $\mathbf{R}$ be the covariance of the single source data at the full aperture:

$$\mathbf{R} = \sigma^2\mathbf{I} + \pi_s\mathbf{S}^*\mathbf{S}^t \tag{7.1.24}$$

where uncorrelated noise of power σ^2 is assumed. We have

$$\mathbf{PRP} = \mathbf{P}(\mathbf{I} + \pi_s\mathbf{S}^*\mathbf{S}^t)\mathbf{P} = \mathbf{P} + \pi_s\mathbf{PS}^*\mathbf{S}^t\mathbf{P} \tag{7.1.25}$$

where σ^2 is set to unity for convenience. From a pseudoinverse formula (Appendix A1.4),

$$(\mathbf{PRP})^+ = \mathbf{P} - \frac{\pi_s\mathbf{PS}^*\mathbf{S}^t\mathbf{P}}{1 + \pi_s\mathbf{S}^t\mathbf{PS}^*} \tag{7.1.26}$$

In eqn. (7.1.17) we have

$$\mathbf{PRU} = \mathbf{P}(\mathbf{I} + \pi_s \mathbf{S}^*\mathbf{S}')\mathbf{U} = \mathbf{PU} + \pi_s \mathbf{PS}^*\mathbf{S}'\mathbf{U} \tag{7.1.27}$$

But $\mathbf{PU} = 0$ by definition of $\mathbf{P}$ and $\mathbf{U}$ and we define

$$\mathbf{S}'\mathbf{U} = c \tag{7.1.28}$$

the conventional aperture amplitude gain for the source. Substitution in eqn. (7.1.18) gives

$$\begin{aligned}\hat{\mathbf{V}} &= -\left(\mathbf{P} - \frac{\pi_s \mathbf{PS}^*\mathbf{S}'\mathbf{P}}{1 + \pi_s \mathbf{S}'\mathbf{PS}^*}\right)\mathbf{PS}^*\pi_s c \\ &= -\pi_s c\mathbf{PS}^* + \frac{\pi_s^2 c\mathbf{S}'\mathbf{PS}^*}{1 + \pi_s \mathbf{S}'\mathbf{PS}^*}\mathbf{PS}^* \\ &= -\frac{\pi_s c}{1 + \pi_s \mathbf{S}'\mathbf{PS}^*}\mathbf{PS}^* \end{aligned} \tag{7.1.29}$$

Source output power: The amplitude gain for source is

$$\begin{aligned} g_s &= \mathbf{S}'\mathbf{W}_{\text{opt}} = \mathbf{S}'\mathbf{U} + \mathbf{S}'\mathbf{PV} = \mathbf{S}'\mathbf{U} - \frac{c\pi_s \mathbf{S}'\mathbf{PS}^*}{1 + \pi_s \mathbf{S}'\mathbf{PS}^*} \\ &= c\left(1 - \frac{\pi_s \mathbf{S}'\mathbf{PS}^*}{1 + \pi_s \mathbf{S}'\mathbf{PS}^*}\right) = \frac{c}{1 + \pi_s \mathbf{S}'\mathbf{PS}^*} \end{aligned} \tag{7.1.30}$$

Thus the source output power is

$$p_s = \pi_s |g_s|^2 = \frac{\pi_s |c|^2}{(1 + \pi_s \mathbf{S}'\mathbf{PS}^*)^2} \tag{7.1.31}$$

Uncorrelated noise output power: This is just the squared norm of the weighting vector.

$$p_n = \mathbf{U}^H\mathbf{U} + \hat{\mathbf{V}}^H\hat{\mathbf{V}} = \mathbf{U}^H\mathbf{U} + \frac{\pi_s^2 |c|^2 \mathbf{S}'\mathbf{PS}^*}{(1 + \pi_s \mathbf{S}'\mathbf{PS}^*)^2} \tag{7.1.32}$$

Note that orthogonality of $\mathbf{U}$ and $\hat{\mathbf{V}}$ is necessary to get this result.

Total output power: Adding eqns. (7.1.31) and (7.1.32) gives the total noise plus source output power:

$$\begin{aligned} p_{\text{tot}} &= \mathbf{U}^H\mathbf{U} + |c|^2 \frac{\pi_s + \pi_s^2 \mathbf{S}'\mathbf{PS}^*}{[1 + \pi_s \mathbf{S}'\mathbf{PS}^*]^2} \\ &= \mathbf{U}^H\mathbf{U} + \frac{\pi_s |c|^2}{1 + \pi_s \mathbf{S}'\mathbf{PS}^*} \end{aligned} \tag{7.1.33}$$

Up to this point the solution has been perfectly general except that c varies with the length of the conventional aperture. However, it is now necessary to consider the two choices of **P** matrix defined in eqns. (7.1.8) and (7.1.14) separately.

7.2 Linearly constrained minimum output power suboptimal system

To get this system, the $\mathbf{P}_v$ matrix is set to the projection (7.1.8). The array may correspond to either Fig. 7.1.1 or Fig. 7.1.2. The desired signal space vector will be taken as **C** across the whole array and $\mathbf{C}_u$ and $\mathbf{C}_v$ at the two subarrays. The amplitude gain of the antenna for the desired signal is

$$g_d = \mathbf{C}^t\mathbf{W} = \mathbf{C}^t\mathbf{U} + \mathbf{C}_v^t\mathbf{P}_v\mathbf{V} = \mathbf{C}^t\mathbf{U} = \mathbf{U}^H\mathbf{U} = \mathbf{K} \tag{7.2.1}$$

since

$$\mathbf{C}_v^t\mathbf{P}_v = \mathbf{0}^t \tag{7.2.2}$$

Vector **S** is now the space vector of the interference source and the quadratic form $\mathbf{S}^t\mathbf{P}\mathbf{S}^* = \mathbf{S}_v^t\mathbf{P}_v\mathbf{S}_v^*$ is equal to

$$\mathbf{S}_v^t\left(\mathbf{I}_v - \frac{\mathbf{C}_v^*\mathbf{C}_v^t}{\mathbf{C}_v^t\mathbf{C}_v^*}\right)\mathbf{S}_v^* = \mathbf{S}_v^t\mathbf{S}_v^* - \frac{|\mathbf{S}_v^t\mathbf{C}_v^*|^2}{\mathbf{C}_v^t\mathbf{C}_v^*} \tag{7.2.3}$$

The quantity $\mathbf{S}_v^t\mathbf{C}_v$ is the amplitude gain of the adaptive subarray if it were set up with unity modulus conventional weights phase compensated for the desired signal but actually receives the interference. Since $|\mathbf{S}_v^t\mathbf{C}_v^*| \le M$ we can define a normalised amplitude gain ρ_v for the interference $(0 \le |\rho_v| \le 1)$ such that

$$\mathbf{S}_v^t\mathbf{C}_v^* = M\rho_v \tag{7.2.4}$$

Since

$$\mathbf{S}_v^t\mathbf{S}_v^* = \mathbf{C}_v^t\mathbf{C}_v^* = M \tag{7.2.5}$$

we have

$$\mathbf{S}_v^t\mathbf{P}_v\mathbf{S}_v^* = M(1 - |\rho_v|^2) \tag{7.2.6}$$

The variable c is the response of the conventionally weighted antenna ($\mathbf{W} = \mathbf{U}$) for the interference. Let K be the number of fixed elements and, since $|\mathbf{S}^t\mathbf{U}| \le K$, we can define a normalised interference amplitude gain ρ_u such that $0 \le |\rho_u| \le 1$ and

$$c = K\rho_u \tag{7.2.7}$$

Noting that $\mathbf{U}^H\mathbf{U} = K$, eqn. (7.1.33) becomes finally

$$p_{n+i} = K + \frac{\pi_i K^2 |\rho_u|^2}{1 + \pi_i M(1 - |\rho_v|^2)} \tag{7.2.8}$$

where π_i is the interference input power.

Sidelobe interference: The first case to consider is that there is interference present some distance from the pointing direction such that $|\rho_v|$ and $|\rho_u|$ are much less than unity. Using the former approximation, eqn. (7.2.8) reduces to

$$p_{n+i} \simeq K + \frac{K^2\pi_i|\rho_u|^2}{1+\pi_i M} \tag{7.2.9}$$

which indicates that the interference plus noise output power increases monotonically as interference power π_i increases, to a maximum level of

$$p_{n+i}(\max) = K + K^2|\rho_u|^2/M \tag{7.2.10}$$

Clearly total output is minimised by maximising M, the number of adaptive elements and this can be taken to its limit of N, the total number of elements.

In order to choose K it is necessary to consider the signal output power as K is the number of conventional channels. Assuming that the desired signal lies in the look direction, with space vector $\mathbf{C}$, then the desired signal output power is $\pi_d K^2$ and the output SNIR is

$$Q_{\text{cons}} = K^2\pi_d\left[K + \frac{K^2\pi_i|\rho_u|^2}{1+\pi_i M}\right]^{-1} \tag{7.2.11}$$

which for large π_i tends to

$$\frac{\pi_d}{K^{-1} + M^{-1}|\rho_u|^2} \tag{7.2.12}$$

and is clearly maximised by finding a small value for $K^{-1} + M^{-1}|\rho_u|^2$. Again one would tend to increase both K and M up to N for the best results. However, if $|\rho_u|^2$ is small enough that $M^{-1}|\rho_u|^2$ is small in comparison to K^{-1} then the exact value of M may not matter too much and this is the basis for the use of suboptimal arrays. Thus suppose we take $K = N$ and $\overline{|\rho_u|}$ as $(6K)^{-1/2}$, an average result for the sidelobe level derived in eqn. 5.1.21:

$$Q_{\text{cons}} \simeq \pi_d K\left(1 + \frac{1}{6M}\right) \tag{7.2.13}$$

which suggests that M may be very small without noticeable deterioration in performance. However, if M is small then the approximation $|\rho_v| \ll 1$ used in eqn. (7.2.9) may cease to be reasonable.

If an average value of $\overline{|\rho_v|} \simeq (6M)^{-1/2}$ is taken then substitution of this and the approximation for ρ_u into eqn. (7.2.8) yields

$$Q_{\text{cons}} \simeq K\pi_d\left(1 + \frac{1}{5+6M}\right)^{-1} \tag{7.2.14}$$

which again suggests that K, the number of conventional channels, rather than M dominates the output SNIR.

Main-lobe interference: If the interference strays into the main lobe then the conclusions are rather different. If we let $\pi_i \to \infty$ in eqn. (7.2.8), then we obtain

$$p_{n+i} \simeq K + \frac{K^2 |\rho_u|^2}{M(1 - |\rho_v|^2)} \tag{7.2.15}$$

If the interference is very close to the look direction then it is reasonable to set $|\rho_u|^2$ in the numerator to 1 and consider that only the denominator varies. Thus, if

$$|\rho_v|^2 = 1 - \epsilon^2 \tag{7.2.16}$$

then

$$p_{n+i} \simeq K + K^2/M\epsilon^2 \tag{7.2.17}$$

which may even approximate to $K^2/M\epsilon^2$ if ϵ^2 is very small. Thus the performance in terms of output SNIR is

$$Q_{\text{cons}} = \frac{\pi_d K^2}{p_{n+i}} \simeq \frac{\pi_d}{K^{-1} + (M\epsilon^2)^{-1}} \simeq \pi_d M\epsilon^2 \tag{7.2.18}$$

This result is somewhat surprising, indicating that performance depends only on the number of adaptive channels M and not at all on K the number of conventional channels. If the adaptive channels have a uniform spacing of D, then ρ_v can be found via eqn. (5.1.23) and Q_{cons} approximates to

$$Q_{\text{cons}} \simeq \tfrac{1}{12}\pi_d M(M^2 - 1)k^2D^2\theta^2 \simeq \tfrac{1}{12}\pi_d M^3k^2D^2\theta^2 \tag{7.2.19}$$

If they are spread uniformly throughout the length of the aperture then $D \simeq Nd/M$ where d is the interelement spacing and

$$Q_{\text{cons}} \simeq \tfrac{1}{12}\pi_d MN^2k^2d^2\theta^2 \tag{7.2.20}$$

Comparison with eqn. (5.1.24) shows that the output SNIR is inferior to the fully adaptive antenna by a factor of M/N, N being the total number of elements in the array. This is perhaps a reasonable trade off of performance against number of adaptive channels but more efficient structures can in fact be found and are derived later.

Owsley[107] has simulated the performance of this suboptimal antenna for a single main lobe interference source and has come to roughly the same conclusions experimentally that the theory above leads to.

7.3 Power inversion solution

In the power inversion array, as discussed, for example by Compton[39] and Morgan,[68] the adaptive subarray is unconstrained and must be separated from the fixed aperture. The antenna will tend to steer nulls not only at interferences but also at the desired signal if this becomes appreciably strong. The power inversion system relies for its operation on either (*a*) provision of only enough degrees of freedom to cancel interferences stronger than the desired signal, or (*b*) adaptation in a zero signal environment. This latter may sound unreasonable but many applications provide a suitable opportunity for power inversion. These include frequency-hopping systems, where adaption is done at frequencies not currently in use, low power spread spectrum systems where broadband input SNR may be very small, and finally radar systems where the average power of echo pulses, if clutter can be ignored, is very weak in comparison to CW jammers.

To get the PI (power inversion) solution, the matrix $\mathbf{P}_v$ of eqn. (7.1.6) is set to an M by M identity matrix as defined in (7.1.14). In (7.1.33) we now have

$$\mathbf{S}^t\mathbf{P}\mathbf{S} = \mathbf{S}_v^t\mathbf{P}_v\mathbf{S}_v^* = \mathbf{S}_v^t\mathbf{S}_v^* = M \tag{7.3.1}$$

and if K elements are used for conventional weighting the total noise plus interference output power is

$$p_{n+i} = K + \frac{\pi_i K^2 |\rho_u|^2}{1 + M\pi_i} \tag{7.3.2}$$

Desired signal response: The optimal weighting vector is given by eqns. (7.1.14) and 7.1.29) as

$$\hat{\mathbf{W}} = \mathbf{U} - \mathbf{P}\mathbf{S}^*\frac{\pi_i c}{1 + \pi_i M} \tag{7.3.3}$$

If a weak desired signal arrives exactly from the look direction so that $\mathbf{S}_d = \mathbf{C}$, its amplitude gain is

$$g_d = \mathbf{C}^t\hat{\mathbf{W}} = \mathbf{C}^t\mathbf{U} - \frac{\pi_i K\rho_u \mathbf{C}_v^t\mathbf{S}_v^*}{1 + \pi_i M} \tag{7.3.4}$$

in which c has been replaced by its alternative $K\rho_u$ [eqn. (7.2.7)]. From eqn. (7.2.4) we have $\mathbf{C}_v^t\mathbf{S}_v^* = M\rho_v^*$ so that eqn. (7.3.4) reduces finally to

$$g_d = K\left(1 - \frac{M\pi_i\rho_v^*\rho_u}{1 + \pi_i M}\right) \tag{7.3.5}$$

For small ρ_v and ρ_u while it is difficult to predict the exact value of this expression which is a function of the relative phases of the interferences received in the fixed and adaptive subarrays, it will stay close to K.

An interesting condition occurs when the interference arrives from the same direction as the desired signal, for here we have $\rho_v = \rho_u = 1$ and eqn. (7.3.5) becomes

$$g_d = K(1 + \pi_i M)^{-1} \tag{7.3.6}$$

and there is significant loss of desired signal response as soon as $\pi_i M$ approaches unity (the uncorrelated noise power). The same is true if a desired signal only is present of course, there could be substantial self-cancellation. However, from eqn. (7.3.2) it can be seen that there is no reduction in overall noise output of the array under this condition and if a desired signal is initially just detectible in the presence of uncorrelated noise it is possible that an interference of power

$$\frac{1}{M} < \pi_i \ll 1 \tag{7.3.7}$$

could substantially suppress the antenna desired signal response if it arrives from the same direction. However, because of eqn. (7.3.7) the input SNIR is hardly changed and an omnidirectional element or a conventional antenna would not be disturbed by the interference. Thus it is possible for a power inversion array to give subconventional and even subomni performance for weak interferences lying in the same direction as the desired signal. This phenomenon can be otherwise explained by the sensitive self-cohering action of the PI antenna which detects correlations due to interferences below the uncorrelated noise level and actively steers nulls at them. An example of this is discussed later.

This effect could be eliminated by reducing M, the number of adaptive elements, to a small value—only enough to cancel the strongest interferences though this may not be too easy to program into an adaptive algorithm.

Interference close to signal: If there is a strong interference close to the desired signal the output SNIR can be estimated by setting $\pi_i \to \infty$ in eqn. (7.3.5) and obtaining

$$g_d \simeq K(1 - \rho_v^* \rho_u) \tag{7.3.8}$$

then allowing ρ_v^* and ρ_u^i to approach unity. This yields the output desired signal power. The noise power, from eqn. (7.3.2), approaches $K + K^2 |\rho_u|^2/M \simeq K + K^2/M$. Finally,

$$Q_{PI} = \frac{\pi_d |g_d|^2}{p_{n+i}} \simeq \frac{\pi_d |1 - \rho_v^* \rho_u|^2}{K^{-1} + M^{-1}} \tag{7.3.9}$$

This result is most sensitive to the term $1 - \rho_v^* \rho_u$ which is a function of the relative phase of the fixed and adaptive subarrays' gain for the interference. Thus for given K and M best performance on the whole will be obtained if the subarrays are as far apart as possible. This is in contrast to the steered result eqn. (7.2.19) in which the output SNR is independent of the phase of ρ_v and is only sensitive to its magnitude. Therefore in the steered system it would not matter where in the array the adaptive subarray was placed; indeed to minimise ρ_v it would be best to spread the elements throughout the whole length of the antenna to give the longest base line and smallest adaptive subarray bandwidth.

By way of illustration the performance of an array in which the fixed and adaptive subarrays occupy the two ends of an array of N elements is investigated in more detail. Defining the centre of the K element fixed subarray as the phase centre we have, from eqn. (5.1.23),

$$\rho_u \simeq 1 - \tfrac{1}{24}(K^2 - 1)k^2d^2\theta^2 \qquad |\theta| \ll 1 \tag{7.3.10}$$

and

$$\rho_v = \frac{\sin \frac{1}{2}Mkd\theta}{M \sin \frac{1}{2}kd\theta} \exp\{-\tfrac{1}{2}j(K + M)kd\theta\} \simeq 1 - \tfrac{1}{2}jNkd\theta \tag{7.3.11}$$

For small θ, eqn. (7.3.10) varies much more slowly than eqn. (7.3.11) and it is reasonable to put $\rho_v^* \rho_u \simeq 1 - \frac{1}{2}jNkd\theta$, and eqn. (7.3.9) reduces to

$$Q_{PI} \simeq \frac{\frac{1}{4}\pi_d N^2k^2d^2\theta^2}{K^{-1} + M^{-1}} \tag{7.3.12}$$

For a given even number of elements this expression is maximised when $K = M = \frac{1}{2}N$ indicating that the array should be split in the centre. For this case, illustrated in Fig. 7.3.1, the output SNIR approximates to

$$\underset{M=K}{Q_{PI}(\text{opt})} \simeq \tfrac{1}{2}\pi_d M^3k^2d^2\theta^2 \tag{7.3.13}$$

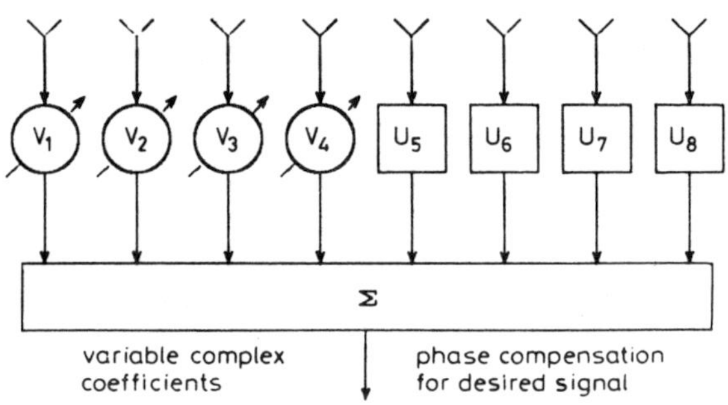

Fig. 7.3.1 *'Optimised' power-inversion array*

For comparison, the equivalent constrained suboptimal array is found to have an output SNIR, given by eqn. (7.2.19), of

$$Q_{\text{cons}} \simeq \tfrac{1}{12}\pi_d M^3k^2d^2\theta^2 \tag{7.3.14}$$

whose performance is 8 dB worse. Spreading the variable coefficients uniformly throughout the aperture would put $D = 2d$ and improve the SNIR by 6 dB.

The fully adaptive antenna gives an output SNIR of $\frac{2}{3}M^3\pi_d k^2d^2\theta^2$ and is thus 1·25 dB better than the power inversion array [eqn. (5.1.24)].

The reason for the very poor performance of the constrained suboptimal

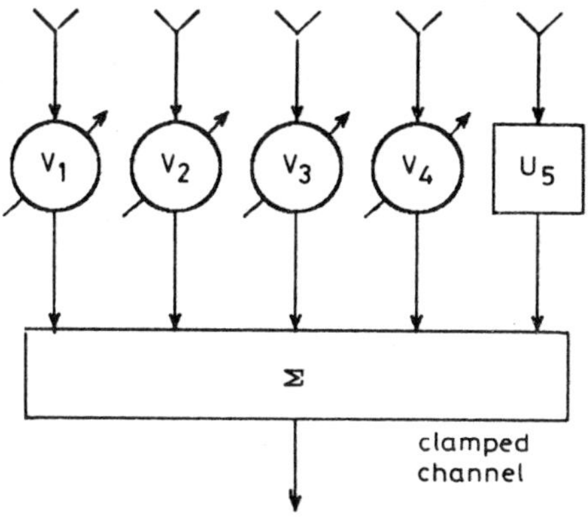

Fig. 7.3.2 *Power-inversion array with one clamped channel*

system is that it is not an optimal SNR system for the degrees of freedom; the look direction null constraint effectively prevents the phase response of its adaptive subaperture varying relative to the fixed subaperture and it is unable to effectively cancel interference in the main lobe.

It is common to operate the PI system with $K = 1$,[54, 72] Fig. 7.3.2, as no prior knowledge of desired signal direction is required. We now have $\rho_u = 1$ while ρ_v remains the same as in eqn. (7.3.11). The output SNR for main-lobe jamming them approximates to

$$\underset{M=1}{Q_{PI}} \simeq \tfrac{1}{4}\pi_d N(N-1)k^2 d^2 \theta^2 \tag{7.3.15}$$

which is worse by a factor $4(N-1)/N^2$ than the best result, eqn. (7.3.13). This main-lobe performance is consistent with the sidelobe performance which is worse by a factor of K because there is no beam to improve desired signal amplitude.

Fig. 7.3.3 shows the output SNIR for a five-element $\lambda/2$ array as a 40 dB interference is scanned from $-90°$ to $+90°$. The desired signal and the uncorrelated noise both have unit power. The power inversion result (with $K = 1$) is not much inferior to the result for the fully adaptive case when the interference is close to the desired signal but is about 7 dB worse when the interference lies in the sidelobes. In Fig. 7.3.4, the interference power is reduced to 0 dB. For this case both the fully adaptive and conventional arrays give a worst result of $-0{\cdot}8$ dB when the interference lies in the signal direction but the power inversion array gives a worst SNIR of $-14{\cdot}7$ dB from eqns. (7.3.2) and (7.3.6) which is $11{\cdot}7$ dB below an omnidirectional element performance. Thus if an interference, or even the desired signal, is present during adaptation at this level a serious loss of performance results.

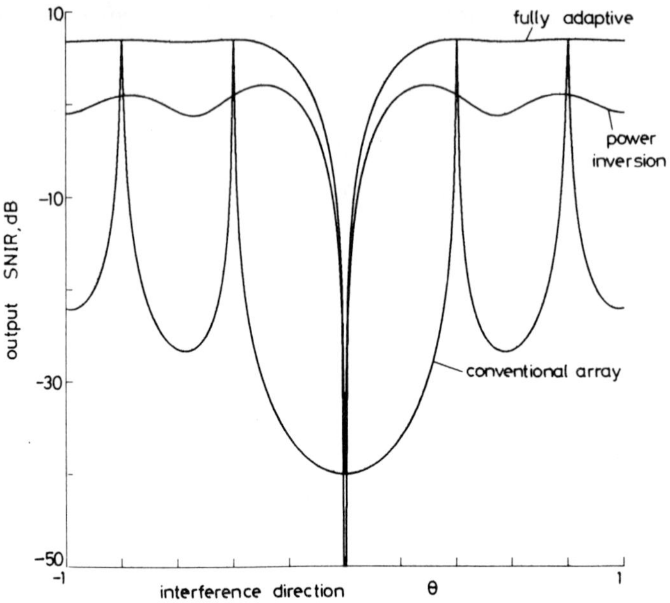

Fig. 7.3.3 *Output SNIR of power-inversion antenna with one clamped channel as direction of a 40 dB interference is varied*

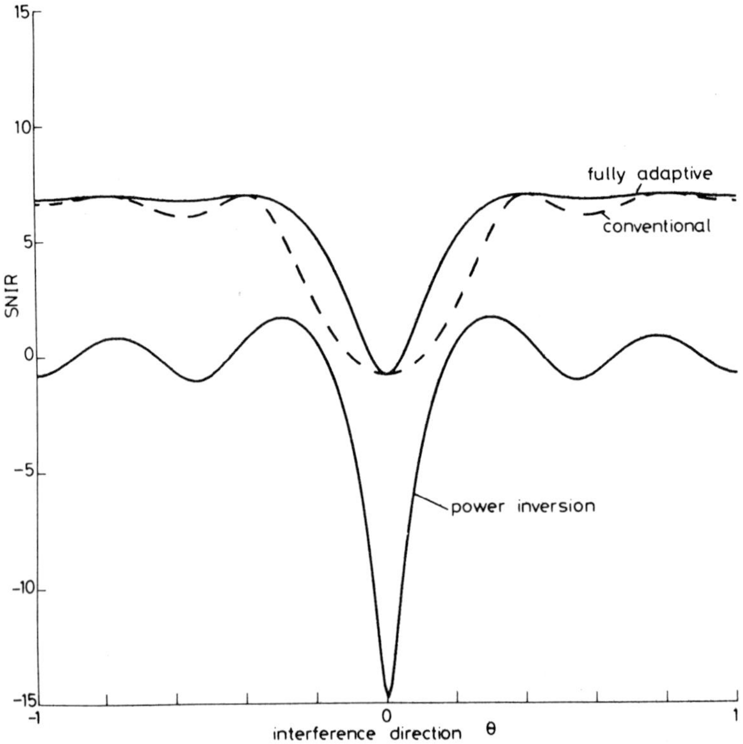

Fig. 7.3.4 *Output SNIR of power-inversion antenna with 0 dB interference*

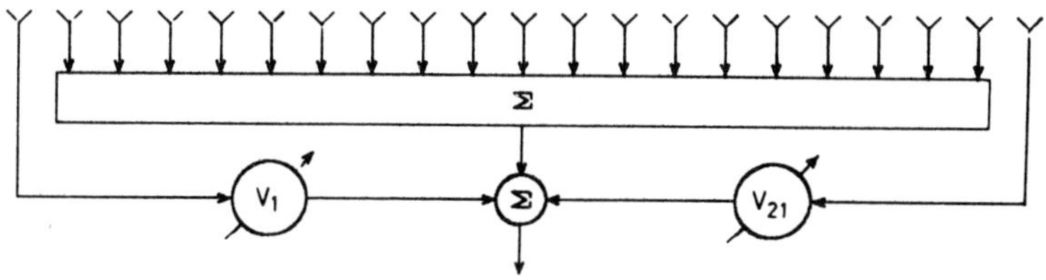

Fig. 7.4.1 *Representation of a sidelobe canceller as a power inversion array* ($M = 2$)

7.4 The sidelobe canceller

A simple version of the sidelobe canceller was shown in Fig. 1.5.1 and it is proposed to model the system as a power inversion antenna shown in Fig. 7.4.1 where two auxilliary elements are shown. The fixed aperture would have to be two dimensional.

Sidelobe interference: For an interference in the sidelobes of the main aperture, we can use eqn. (7.3.2) to find the total noise plus interference output power of the adaptive system

$$p_{n+i} = K + \frac{\pi_i K^2 |\rho_u|^2}{1 + M\pi_i} \tag{7.4.1}$$

where K is the total number of elements which represent the dish, e.g. for a 50λ diameter dish of area $1963\lambda^2$ we could use about 7854 elements on a $\lambda/2$ grid to represent the dish as an array. To get an idea of average performance we can take the RMS sidelobe level as

$$|\rho_u|(\text{RMS}) = (6K)^{-1/2} \tag{7.4.2}$$

based on eqn. (5.1.21). Substitution of eqn. (7.4.2) into (7.4.1) gives

$$p_{n+i} \simeq K + \frac{\pi_i K/6}{1 + M\pi_i} = K\left(1 + \frac{\pi_i/6}{1 + M\pi_i}\right) \tag{7.4.3}$$

which increases monotonically from K to $K(1 + 1/6M)$ as the input interference power increases from zero to infinity.

As the desired signal amplitude gain is not varied very much for sidelobe interference cancellation [eqn. (7.3.5)] the p_{n+i} results above can be taken as a true measure of the output SNIR performance which is tolerably good.

The situation is different if the interference falls on a large sidelobe of the main aperture, for example on the first sidelobe, which may have a $|\rho_u|$ value of -25 dB or 0·056. For a strong interference the maximum value of p_{n+i} is then

$$p_{n+i} = K\left(1 + \frac{0{\cdot}0032K}{M}\right) \tag{7.4.4}$$

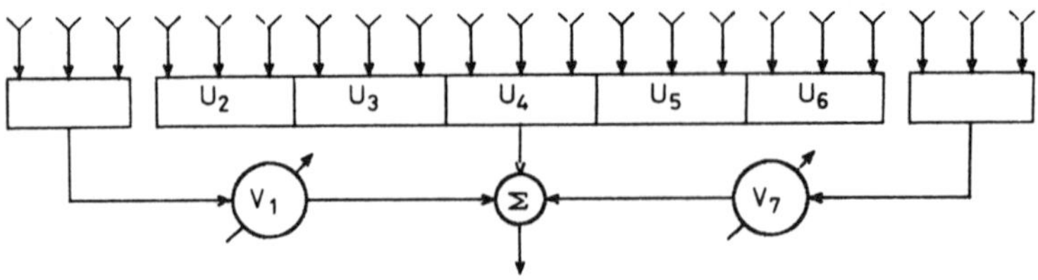

Fig. 7.4.2 *Use of directional auxiliaries*
The main aperture is subdivided into similar units

If K is large, the increase of noise and interference may be considerable. Taking the case above, of $K = 7854$ and $M = 2$, we have

$$p_{n+i} = K(1 + 12{\cdot}56) \tag{7.4.5}$$

an increase of 11·3 dB.

Directional auxilliaries: This increase in uncorrelated noise output can be overcome by improving the directivity of the auxilliaries which can be represented by using a group of elements as a fixed subarray as in Fig. 7.4.2. In order to avoid a complete reanalysis of the configuration these new subarrays can be considered as single elements provided the main aperture is divided into similar units. The effects of the increase in interference and noise from these subarrays is summarised in the following changes if the subarrays have N elements:

(*a*) Uncorrelated noise is increased by N times.

(*b*) Interference power is increased by $N^2\rho_N^2$ times where ρ_N is the normalised directional response of a subarray.

(*c*) Desired signal power is increased by N^2 times.

(*d*) The number of elements in the fixed aperture is decreased by N times.

(*e*) Since the theory has assumed unity uncorrelated noise power, all powers are divided by N at the inputs to compensate (*a*) then increased N times at the output.

Putting all these changes into eqn. (7.4.6) yields

$$p_{n+i}/N = \frac{K}{N} + \frac{\dfrac{\pi_i N^2 \rho_N^2}{N} |\rho_u|^2 K^2/N^2}{1 + \dfrac{M\pi_i N^2 \rho_N^2}{N}} \tag{7.4.6}$$

which, in the large π_i limit, tends to

$$p_{n+i} \simeq K + \frac{|\rho_u|^2 K^2}{NM} \tag{7.4.7}$$

whose variable part is about N times smaller than the initial result, eqn. (7.4.1). The desired signal output power is unchanged by the reformulation above since actions (*c*) and (*d*) cancel. If $N = 9$ were used in the calculation of eqn. (7.4.5) the result would be

$$p_{n+i} = K\left(1 + \frac{25{\cdot}13}{18}\right) \tag{7.4.8}$$

an increase of only 3·8 dB.

Main-lobe interference : Eqn. (7.3.2) indicates that if the interference enters the main lobe and $\rho_u \rightarrow 1$ the total noise and interference output power increases to a maximum of

$$p_{n+i} = K + K^2/M \tag{7.4.9}$$

a value much greater than the uncorrelated noise output in the absence of any interference. This alone would inhibit the development of a good output SNIR. However, at the same time the desired signal response g_d becomes small as indicated by eqn. (7.3.5). What has happened here is that the auxilliary channel weights are being used to form a main-lobe null and, because of the small aperture of the auxilliaries, very large weights—on the order of K/M—are needed to do this. The enhanced noise output in eqn. (7.4.9) is due almost entirely to the uncorrelated noise produced by the auxilliary elements.

In practice, it is unlikely that the coefficients in the auxilliary channels would have sufficient dynamic range to achieve this effect and main-lobe interference would remain uncancelled. It is concluded that the sidelobe canceller, as its name might imply, is purely a sidelobe system.

7.5 Adaptively combined steered subarrays[35, 51]

This technique is illustrated in Fig. 7.5.1. The M front end subarrays each of K elements are conventionally steered to receive the desired signal and their outputs are then combined through M complex coefficients which are variable. The

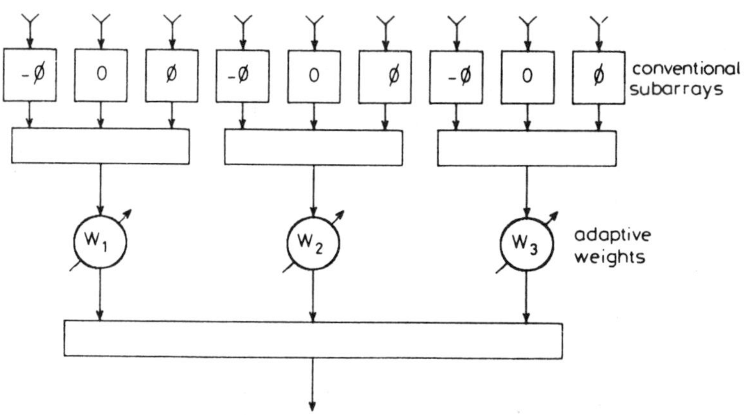

Fig. 7.5.1 *Adaptively combined subarrays*

M variable coefficients satisfy a look direction constraint to prevent cancellation of the desired signal. The performance of the system is easily predictable from the modifications introduced in the last section to deal with multielement subarrays. If these changes are made to eqn. (5.2.19) with the exception of (d) which is inapplicable we obtain

$$Q = \frac{KM\pi_d[1 + |\rho_K|^2 KM\pi_i(1 - |\rho_u|^2)]}{1 + |\rho_K|^2 KM\pi_i} \underset{\pi_i \to \infty}{\simeq} KM\pi_d(1 - |\rho_u|^2) \qquad (7.5.1)$$

in which π_d and π_i are the desired signal and interference powers, ρ_K is the normalised directional response of a subarray for the interference after phasing for the desired signal, and ρ_u is the normalised directional response for the interference of the M element array formed by replacing each subarray by an omnidirectional element and setting the variable weighting coefficients to conventional values for receiving the desired signal.

For uniformly spaced subarrays the pattern ρ_u is given by the usual formula, eqn. (5.1.23), if the desired signal is broadside,

$$\rho_u = \frac{\sin \frac{1}{2}MkD\theta}{M \sin \frac{1}{2}kD\theta} \qquad -1 \leq \theta \leq 1 \qquad (7.5.2)$$

where D is the interarray spacing.

Sidelobe interference: For an interference remote from the main lobe directed at the desired signal a reasonable mean-square estimate of sidelobe level is $|\rho_u|^2 \simeq (6M)^{-1}$ from eqn. (5.1.21) and substitution into (7.5.1) yields an average output SNIR of

$$Q = KM\pi_d\left(1 - \frac{1}{6M}\right) \qquad (7.5.3)$$

which, for large M, will not be much inferior to the performance of a fully adaptive antenna.

Main-lobe interference and diffraction secondaries: For small θ the pattern ρ_u in eqn. (7.5.2) approximates to

$$|\rho_u|^2 \simeq 1 - \tfrac{1}{12}(M^2 - 1)k^2D^2\theta^2 \simeq 1 - \tfrac{1}{12}M^2k^2D^2\theta^2 \qquad (7.5.3)$$

But $MD \simeq$ aperture length and is the same as for a fully adaptive antenna. Substituting $N = KM$ in eqn. (7.5.1) and $MD = Nd$ in (7.5.3) yields

$$Q \simeq \tfrac{1}{12}\pi_d N^3k^2d^2\theta^2 \qquad (7.5.4)$$

the same result as for a fully adaptive array, eqn. (5.1.25). However, due to the large subarray spacing of Kd, the adaptive aperture will have repetitive grating lobes at which interferences, although weakened by the shading offered by the subarrays, will have desired signal type phasing and $\rho_u = 1$. In principle, an interference lying exactly at the centre of one of these lobes would lie in a null of the subarray pattern (if a overall uniform element spacing is assumed) and would be removed by 'conventional' methods. However, it is unlikely that the accuracy of construction of the subarrays will produce a conventional null better than 20–30 dB and considerable interference power will leak through. The adaptive weights will be unable to cancel this interference because of the desired signal steering constraint. Thus periodic patches of poor interference cancellation are expected at the grating lobe periodicity of the adaptive aperture, a point noted by Chapman.[35] These effects can be avoided to some extent by breaking up the regular spacings on which the secondary lobes depend. Thus if the subarrays have differing numbers of elements and are irregularly spaced it is unlikely that ρ_u will get too close to unity for any direction except that of the desired signal and good interference cancellation will be obtained for all interference directions.

In conclusion, provided enough adaptive weights are provided to allow null formation for the expected number of jammers this suboptimal adaptive array offers a good all-round performance and, most importantly, is comparable to a fully adaptive array for main-lobe jamming where adaptive performance is most required.

7.6 Conventionally combined adaptive subarrays

This approach is illustrated in Fig. 7.6.1 in which K subarrays each have M elements, all adaptive, and are combined through K phase compensation units. As there are a total of $N = MK$ adaptive channels this system may not appear to offer much advantage in cost over a fully adaptive system. Its potential advantages become apparent below when we consider matrix inversion techniques.

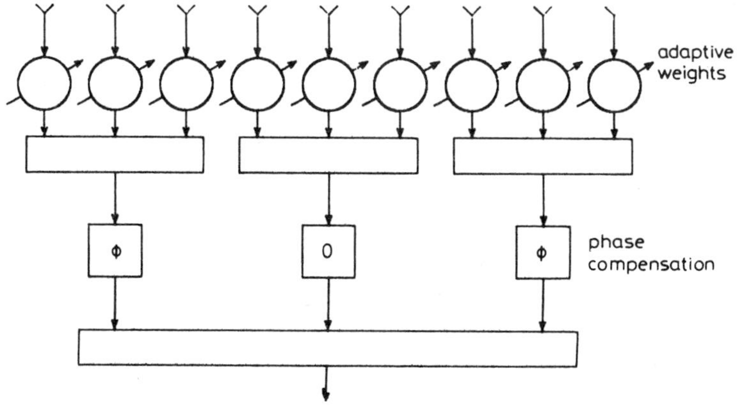

Fig. 7.6.1 *Adaptive subarrays combined with conventional weights*

The performance of the whole system is easily predicted by first of all finding the output SNIR of each subarray then, on the assumption that the interference outputs of the subarray are uncorrelated, increasing the SNIR by a factor of K when they are added. This uncorrelated assumption is certainly valid at high interference power levels where the interference is heavily cancelled and uncorrelated noise originating at the element level dominates the output.

The SNIR of the output of a single subarray is given by eqn. (5.1.19) and multiplying by K yields:

$$Q = KM\pi_d\left(1 - \frac{M\pi_i|\rho_u|^2}{1 + M\pi_i}\right) \underset{\pi_i \to \infty}{\longrightarrow} KM\pi_d(1 - |\rho_u|^2) \tag{7.6.1}$$

where ρ_u is the normalised directional pattern of an individual subarray, using conventional weights, for the interference.

Sidelobe interference: If the interference lies in the sidelobes of an individual subarray conventional pattern then $\rho_u \ll 1$ and (7.6.1) approximates to $KM\pi_d$ which is the output SNR of a fully adaptive antenna. Thus the sidelobe performance of this technique is promising. However as the main lobes of the subarrays are relatively broad one might expect that main-lobe jamming may have a greater importance.

Main-lobe jamming: Using the approximation (5.1.23) for the sin $(x)/x$ pattern of a subarray gives

$$1 - |\rho_u|^2 \simeq \tfrac{1}{12}(M^2 - 1)k^2d^2\theta^2 \tag{7.6.2}$$

and for large M the output SNIR approximates to

$$Q \simeq \tfrac{1}{12}\pi_d KM^3k^2d^2\theta^2 \simeq \tfrac{1}{12}\pi_d NM^2k^2d^2\theta^2 \tag{7.6.3}$$

The output SNIR of this system is seen to be much inferior to that of the fully adaptive antenna and the suboptimal system of Section 7.5, the difference being a factor of M^2/N^2.

The interest in the system lies in the computation reduction afforded by the small subarrays. If the computation load used to calculate a sample matrix inversion weighting vector is, say, N^3 for an N element array, then the total load to optimise all K subarrays is KM^3 which is much less than the $(KM)^3$ required for a fully adaptive array. However, it would be better to use sample matrix inversion in the system of Section 7.5 unless there is some problem which prevents its use. Such a case is examined by Wirt[88] who has used power inversion subarrays without steering vectors and can steer the phase compensation to received desired signals quite independently. This system will give an interference reduction capability but the overall SNR gain will only be K rather than the KM of which the array is capable.

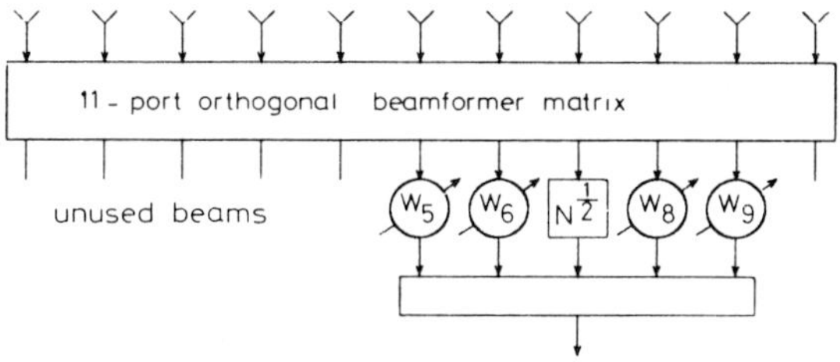

Fig. 7.7.1 *Adaptive array 'thinned' in beam space*

7.7 Beam-space suboptimal systems[21, 49, 65, 66]

Figure 7.7.1 shows diagrammatically a suboptimal processor operating in beam space in which the beams are thinned before adaptive combination. Such a system may use a hardware beamformer like a Butler matrix or may comprise a cluster of dipoles or feeds at the focus of a parabolic dish or a microwave lens.[21] The beams to be used could be based on a ranking of output powers or, as indicated, the beams closest to the desired signal may be taken.

Optimal solution—all beams used: Assume that the beamformer is a unitary matrix transformation, i.e. the beams are orthogonal and there is no loss of power overall. A unit power input signal feeds a total power N into all the input ports and distributes power N among the output ports which might appear all at one port if the signal lies in a beam direction. In general the unit amplitude signal from direction θ is represented by a beam-space vector $\mathbf{S}(\theta)$ which contains the complex amplitudes at the output ports and is scaled so that

$$\|\mathbf{S}(\theta)\|^2 = N \tag{7.7.1}$$

In the case of a uniformly spaced line array and a DFT beamformer it is not difficult to show that $S_n(\theta)$, the nth element of $\mathbf{S}(\theta)$ is given by

$$S_n(\theta) = N^{-1/2} \frac{\sin \frac{1}{2}Nkd(\theta - \theta_n)}{\sin \frac{1}{2}kd(\theta - \theta_n)} \tag{7.7.2}$$

where the variables correspond to those in eqn. (5.1.23) and the θ_n are a set of directions uniformly distributed on the circular interval $(-1, 1)$. The 'steering vector' at the beam outputs corresponds to the elementary vector

$$\mathbf{C} = (0, 0, 0, 0, 0, 0, 0, N^{1/2}, 0, 0, 0)^t \tag{7.7.3}$$

where the nonzero dth element defines the pointing direction of the array. This example refers to Fig. 7.7.1. The conventional weighting vector is thus

$$\mathbf{U} = \mathbf{C}^* \tag{7.7.4}$$

Unit power uncorrelated noise at the beamformer inputs is transformed into statistically equivalent noise at its outputs.

Equation (7.1.33) can be used to calculate total noise and interference output power if there is only a single source present. Variable c is the amplitude gain of the interference in the desired signal beam

$$c = \mathbf{S}(\theta_i)^t\mathbf{U} = N^{1/2}S_d(\theta_i) = N\rho_u \tag{7.7.5}$$

and the **P** matrix is a diagonal matrix with unit entries except for the element corresponding to the desired signal beam which is zero. The **P** matrix which corresponds with the diagram, when all beams are used, is

$$\mathbf{P} = \text{diag}\,(1, 1, 1, 1, 1, 1, 1, 0, 1, 1, 1) \tag{7.7.6}$$

and $\mathbf{S}^t\mathbf{P}\mathbf{S}^*$ is the sum of the interference powers at the auxillary beam ports for unit input powers:

$$\mathbf{S}^t\mathbf{P}\mathbf{S}^* = \sum_{n \neq d} \|S_n(\theta_i)\|^2 = \Sigma_{\text{aux}} \tag{7.7.6}$$

Because of the unitariness of the beamformer the sum of the interference powers over all ports is N:

$$\Sigma_{\text{aux}} + |S_d(\theta_i)|^2 = \Sigma_{\text{aux}} + N|\rho_u|^2 = N \tag{7.7.7}$$

Thus

$$\Sigma_{\text{aux}} = N(1 - |\rho_u|^2) \tag{7.7.7}$$

and substituting in eqn. (7.1.33) yields

$$p_{\text{tot}} = N + \frac{\pi_i N^2 |\rho_u|^2}{1 + \pi_i N(1 - |\rho_u|^2} \tag{7.7.8}$$

which is in agreement with the fully adaptive result for direct element weighting obtainable from eqn. (7.2.8) by setting $K = M = N$ and $\rho_u = \rho_v$.

The desired signal response is

$$p_s = \pi_d |\mathbf{C}^t\mathbf{U}|^2 = \pi_d N^2 \tag{7.7.9}$$

and the output SNIR corresponds to eqn. (5.1.19).

Suboptimal system: Suppose that some number M of the beams are used for cancellation with $M \leq N - 1$. This constraint on degrees of freedom can be represented by appropriately setting elements of **P** to zero. The figure corresponds to

$$\mathbf{P} = (0, 0, 0, 0, 0, 1, 1, 0, 1, 1, 0)^t \tag{7.7.10}$$

and, in eqn. (7.1.33), we now have

$$\mathbf{S}^t\mathbf{P}\mathbf{S}^* = \sum_{\text{some } n} |S_n(\theta_i)|^2 = \sigma_{\text{aux}} \tag{7.7.11}$$

and since $\sigma_{\text{aux}} \leq \Sigma_{\text{aux}}$ it is immediately seen that there is an increase of output noise and interference. If we let $\pi_i \to \infty$ in eqn. (7.1.33) we obtain

$$p_{n+i} \to N + |c|^2/\mathbf{S}^t\mathbf{P}\mathbf{S}^* = N + N^2|\rho_u|^2/\sigma_{\text{aux}} \tag{7.7.12}$$

If the interference happens to arrive in one of the auxillary beams connected to a variable coefficient we have $\sigma_{\text{aux}} \to N$ (since the appropriate element of $\mathbf{S}(\theta_i)$ in eqn. (7.7.11) goes close to $N^{1/2}$) and

$$p_{n+i} \simeq N(1 + |\rho_u|^2) \tag{7.7.13}$$

a result which agrees with eqn. (7.2.10) for the fully adaptive case. However, should the interference lie in a direction not covered by a variable beam, then it is only detected through sidelobe leakage and σ_{aux} could become very small causing an increase in output power from eqn. (7.7.12). If a cluster of beams surrounds the desired signal beam as suggested in the diagram then it is likely that the response of the interference throughout the cluster will be more or less uniform since it will be equidistant from the nulls of every beam pattern. If we set

$$\sigma_{\text{aux}} \simeq NM|\rho_u|^2 \tag{7.7.14}$$

to describe this condition then

$$p_{n+i} \simeq N\left(1 + \frac{1}{M}\right) \tag{7.7.15}$$

which is quite an acceptable result comparable to eqn. (7.2.10) for example.

Main-lobe interference: For main-lobe jamming, σ_{aux} becomes small as most of the interference power is output into the signal beam. In eqn. (7.7.12) we obtain

$$p_{n+i} \simeq N^2|\rho_u|^2/\sigma_{\text{aux}} \tag{7.7.16}$$

and the output SNIR becomes

$$Q = \frac{\pi_d|g_d|^2}{p_{n+i}} \simeq \frac{\pi_d\sigma_{\text{aux}}}{N|\rho_u|^2} \tag{7.7.17}$$

clearly proportional to σ_{aux}. Since it is known that the output SNIR has the fully adaptive value, eqn. (5.1.19), if $\sigma_{\text{aux}} = \Sigma_{\text{aux}}$ we may write

$$\frac{Q}{Q_{\text{opt}}} = \frac{\sigma_{\text{aux}}}{\Sigma_{\text{aux}}} \tag{7.7.18}$$

and attention is concentrated on collecting as much of the interference power as possible, a conclusion reached by Widrow *et al.*[142] in a related problem. If the beam patterns are described by eqn. (7.7.2), an interference close to the desired signal bearing has a space vector approximating to

$$\mathbf{S}(\theta_i) = \mathbf{S}(\theta_d + \Delta) \simeq \mathbf{S}(\theta_d) + \Delta\mathbf{S}'(\theta_d) \tag{7.7.19}$$

where $\mathbf{S}^i$ is a vector of derivatives with respect to θ (as used in Chapter 6). Thus

$$S_n(\theta_i) \simeq C_n + N^{-1/2}\Delta\delta_{nd}(-1)^{n-d}[\sin \tfrac{1}{2}kd(\theta_n - \theta_d)]^{-1} \quad n = 0, \ldots, N-1 \tag{7.7.20}$$

$$\simeq \frac{2\Delta(-1)^{n-d}}{N^{1/2}kd}(\theta_n - \theta_d)^{-1} \qquad n \neq d \tag{7.7.21}$$

and

$$\sigma_{\text{aux}} = \sum_{\text{some } n} |S_n(\theta_i)|^2 \simeq \frac{4\Delta^2}{Nk^2d^2} \sum_{\text{some } n} (\theta_n - \theta_d)^{-2} \tag{7.7.22}$$

This sum is a rapidly converging sequence and 64% of the available power is contained in the two beams straddling the signal beam and 80% if the next two are included. Thus it is not difficult for σ_{aux} to approach Σ_{aux} and main-lobe jamming performance can approach the fully adaptive result. In practice, more beams than this might be provided since enough degrees of freedom must be provided to deal with multiple interference sources.

7.8 Comparison of suboptimal systems

The performances of the circuits discussed up to this point are summarised. The most interesting feature is main-lobe performance since all the system give some sort of sidelobe protection. In all cases the number of adaptive channels is M, there are N channels altogether.

System		SNIR	Relative SNIR
1	Fully adaptive	$\frac{1}{12}N(N^2-1)k^2d^2\theta^2$	1
2	Thinned adaptive (7.1.19)	$\frac{1}{12}(M^2-1)N^2d^2\theta^2/M$	$(M^2-1)/MN$
3	Conventional subarrays (7.5.4)	$\frac{1}{12}(M^2-1)N^3k^2d^2\theta^2/M^2$	$(M^2-1)/M^2$
4	Adaptive half-aperture (7.3.14)	$\frac{1}{12}M(M^2-1)k^2d^2\theta^2$	$\frac{1}{2}(\frac{1}{4}-1/N^2)$
5	Power inversion $\frac{1}{2}$ aperture (7.3.13)	$\frac{1}{2}M^3k^2d^2\theta^2$	$\frac{3}{4}$
6	Power inversion 1 channel fixed (7.3.15)	$\frac{1}{4}N(N-1)k^2d^2\theta^2$	$3/(N+1)$
7	Adaptive subarrays (7.6.3)	$\frac{1}{12}N(M^2-1)k^2d^2\theta^2$	$(M^2-1)/(N^2-1)$
8	Beam space (7.7.18)	$\frac{1}{12}N(N^2-1)k^2d^2\theta^2\sigma/\Sigma$	σ/Σ

$k = 2\pi/\lambda$, d = interelement spacing, θ = sine of interference bearing

7.9 Amplitude or phase-only weighting

Amplitude weighting:[110] Amplitude-only weighting requires only an attenuator in each element output instead of the quadrature system. The effects of using such weighting is best illustrated by its effect on a line array with a broadside pointing direction. If w_n is the sequence of real weighting coefficients the directional pattern is of the form

$$g(\theta) = \sum_{n=1}^{N} w_n e^{jkx_n\theta} \qquad -1 \leq \theta \leq +1 \tag{7.9.1}$$

where $k = 2\pi/\lambda$, x_n are the element locations, and θ is the sine of true bearing. Clearly

$$\begin{aligned} g(\theta) &= \sum_{n=1}^{N} w_n \cos(kx_n\theta) + j\sum_{n=1}^{N} w_n \sin(kx_n\theta) \\ &= g^e(\theta) + jg^0(\theta) \end{aligned} \tag{7.9.2}$$

where g^e is an even function of θ: $g^e(\theta) = g^e(-\theta)$ and g^0 is an odd function: $g^0(\theta) = -g^0(-\theta)$. Taking the modulus of (7.9.2) yields

$$|g(\theta)|^2 = |g^e(\theta)|^2 + |g^0(\theta)|^2 \tag{7.9.3}$$

which is an even function. Thus real-only weights for a broadside steered line array will produce a symmetrical modulus pattern. If the array is steered to an angle other than broadside by phase-shifting before weighting (see Fig. 7.10.1) then the same conclusion applies but the pattern is symmetrical about its new look direction in θ-space.

The use of real only weights is equivalent to the constraint system discussed in Section 6.4 and the main lobe will therefore be more robust than if fully variable coefficients are used.

Phase-only weighting:[24, 64, 78] This procedure is discussed fully by Baird and Rassweiler.[24] For very small perturbations of the coefficients from their nominal settings the phase shift exp $(j\phi)$ may be approximated by $1 + j\phi$. Thus if the sequence of coefficients in the elements is exp $(j\phi_n)$ the directional pattern approximates to

$$g(\theta) \simeq \sum_{n=1}^{N} (1 + j\phi_n) e^{jkx_n\theta} \tag{7.9.4}$$

$$= c(\theta) + j\sum_{n=1}^{N} \phi_n^e \cos(kx_n\theta) + \sum_{n=1}^{N} \phi_n^0 \sin(kx_n\theta) \tag{7.9.5}$$

where $c(\theta)$ is the conventional pattern without adaptation, and ϕ^e and ϕ^0 are the even and odd parts respectively of the phase function. Thus $g(\theta)$ is the sum of an even real conventional part, an odd real adaptive part, and an imaginary even adaptive part. If a null is steered in some direction θ_i then the sum of the real parts must cancel and the imaginary part must be zero. At bearing $-\theta_i$ the real parts will add rather than cancel while the imaginary part will still be zero. Thus the sidelobe level will increase by 6 dB at the mirror bearing.

The algorithm for adjusting the phase shift is somewhat nonlinear[64] in operation and a gradient descent procedure cannot guarantee to find the minimum output power or mean square error point as multiple minima may be present.

The use of phase-only coefficients is attractive as many operational systems already have phase control as part of a beam steering system and an economical adaptive system could result, especially if a perturbation algorithm were used to determine the gradients. Unfortunately, many beam steering phase shifters are rather coarsely quantised at 22·5° or 45° steps and the null depths suffer accordingly.

7.10 Broad-band arrays[45, 47, 50, 51, 60, 74, 84, 106]

Quite good cancellation of broad-band interference sources is obtainable in essentially narrow-band arrays with complex weights or quadrature attenuation but the penalty is that increased degrees of freedom are required to cancel each source (Section 5.3). Performance of the narrow-band array is improved if the quadrature attenuators are specifically designed to produce phase shifts which are proportional to frequency, for example by using true time delays to get the 90° split between the quadrature channels. However, the best results come from using tapped delay line processing in the element outputs. This scheme was used in the early papers by Widrow[84] and Frost[47] though there was little discussion of the performance and no information which enabled design calculations to be carried out. However, Rodgers[74] showed that a simple system with only a few taps could give excellent nulling characteristic over fractional bandwidths of the order of 50%. Such a system is shown in Fig. 7.10.1.

If a power inversion result is required one tap in any line can be clamped at unity gain (all taps are real). However, if a steered system is desired then it is necessary to put broad band time delays into the element outputs to bring the desired signal into time coincidence at the processor input. This is a simple modification of the preprocessing discussed in the last Section. The look direction constraint then consists of constraining the sums of taps in each column to a set of constants since for the desired signal the adaptive antenna looks like a single channel discrete time filter with weighting coefficients given by column sums of weighting coefficients. This technique is discussed in detail by Frost.[47]

The improvement in cancellation performance improves rapidly at first as the number of taps is increased but beyond four or five taps per channel only small gains are to be had. The reason for this is (*a*) if the interference is broad-band it becomes increasingly decorrelated along each line and adding extra taps simple adds uncorrelated noise to the output; (*b*) if the interference is narrow-band there is nothing to be gained by using more than two taps anyway. The eigen-

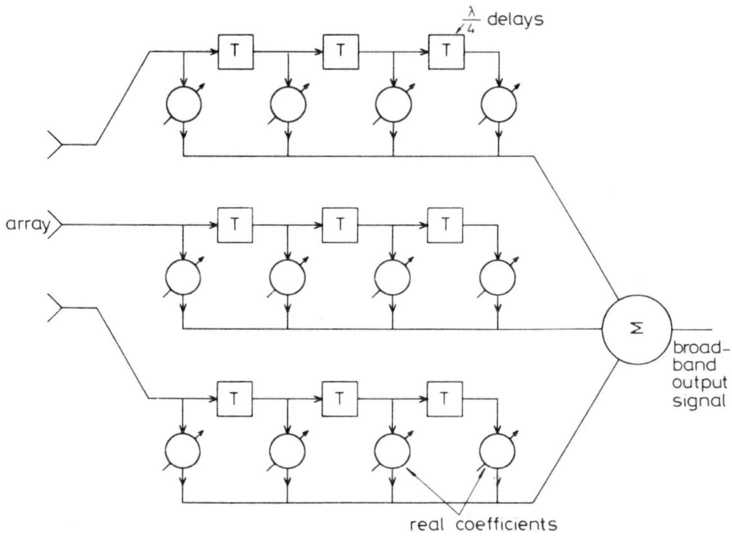

Fig. 7.10.1 *Broad-band antenna with real coefficients*

value analysis of Section 5.3, although done for complex signals, indicates that the number of eigenvalues of appreciable magnitude is not likely to increase by a factor of more than three or four for broad band signals of less than, say, 50% bandwidth and the use of three or four taps presumably provides enough degrees of freedom for the cancellation of the 'eigenvector sources'.

The eigenvalue distribution for the tapped delay line system is not very good and much slower convergence may be anticipated than for an equivalent narrow-band system. Because of the real nature of the signals there are twice as many eigenvalues as one would expect from each source and these occur in pairs of equal values (Section 4.10).

The length of the delay units in the delay lines appears to be best fixed at $\lambda/4$ at the midband frequency.

Multiple narrow-band array:[100, 109] An alternative to tapped delay line processing is to filter the antenna signals into a number of fairly narrow bands, of say 10% fractional bandwidth, and apply these to a number of narrow band processors operating independently. This approach is not much more expensive than the tapped delay line one as the total number of degrees of freedom is likely to be about the same. It has the advantage that narrow receiver bandwidths are advantageous in cancellation of interference suffering from certain multipath effects as discussed in Sections 2.7 and 5.5. However, this is the only condition where there appears to be a cut and dried case for using one system in preference to the other.

The eigenvalues of narrow-band signals were derived in Section 5.3 and, for, say, 10% bandwidth filters, each source may produce two significant eigenvalues. Thus if deep nulls are required it may be necessary to assume that each interference source consumes two degrees of freedom and the number of antenna elements may have to be increased accordingly. An alternative is to provide tapped delay lines with three or four taps for each narrow-band receiver.

7.11 Nonadaptive preprocessing[52, 60, 22, 93]

This is usually done to facilitate the provision of look direction constraints in hardware oriented steered antennas. If the pointing direction is not broadside in a line array the calculations involved in maintaining a certain gain in an arbitrary direction can be quite difficult [e.g. eqn. (4.5.7b)]. The problem is partially overcome by the signal preprocessor shown in Fig. 7.11.1 which simply phases the antenna signals so that when they arrive at the processor the desired signal appears to be coming from the broadside direction. The broadside look direction constraint then is simply a matter of keeping a sum of the coefficients at a constant value, a somewhat simpler task.

Griffiths[52] has taken the process one step further and has eliminated all constraints form the adaptive processor. This system is shown in Fig. 7.11.2 in which a conventional antenna output is formed while the unconstrained adaptive coefficients are fed by signals from which the desired signal has been removed by spatial filtering in transformation **T**. The adaptive weights thus cannot cancel the desired signal and must operate on the interferences only.

If the desired signal is broadside at the antenna elements (if not the simple phasing preprocessor above will correct the problem) then suitable design of **T** is easy. However, if **T** is not an orthogonal matrix the eigenvalues of the covariance matrix may be altered and convergence characteristics also.

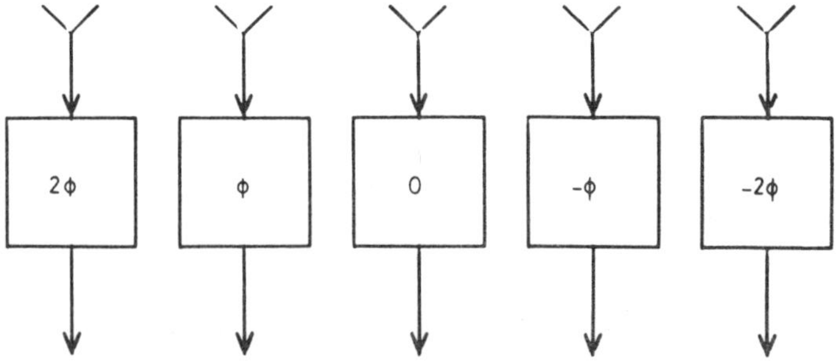

Fig. 7.11.1 *Presteering of antenna signals to swing desired signal to broadside*

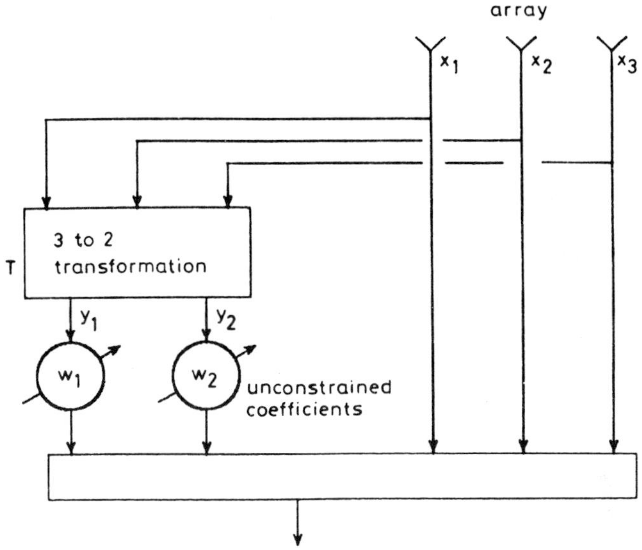

Fig. 7.11.2 *Array preprocessing to allow unconstrained coefficient variation*

For the 3-to-2 channel transformation shown in the diagram the transformed signals y_1, y_2 might be formed by the weightings

$$y_1 = x_1 - x_3$$

$$y_2 = x_1 - 2x_2 + x_3 \tag{7.11.1}$$

so that the two transformation vectors $(1, 0, -1)^t$, $(1, -2, 1)^t$ are mutually orthogonal and orthogonal to the weightings in the conventional part of the antenna $(1, 1, 1)^t$. If the number of elements is a power of two the rows of the Walsh function matrix could be used as weightings. If the transformation consists of an orthogonal beamformer the system reduces to the one shown in Fig. 7.7.1 (with all beam used).

It has been pointed out by Griffiths that the dynamic ranges of the coefficients weighting the y_i signals are very different from those required if the coefficients act directly upon the element signals and this can be used to reduce the effects of coefficient quantisation.

7.12 The adaptive preprocessor

An adaptive preprocessor is a structure which is designed to prefilter the antenna data spatially and improve the eigenvalue distribution of the interferences. A gradient based adaptive antenna then has faster convergence when connected to the output. The ideal preprocessor would be the eigenvector

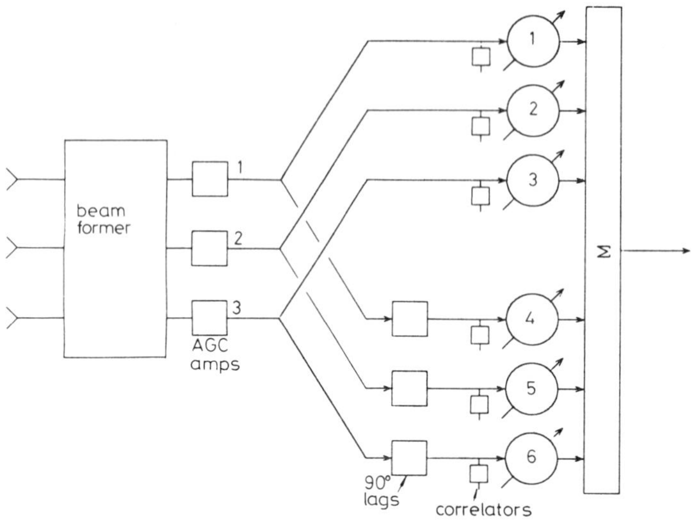

Fig. 7.12.1 *Fixed preprocessor with AGC power equalisation shared between quad channels*

matrix followed by AGC amplifiers to equalise the eigenvalue powers. An alternative is the matrix $\mathbf{R}^{-1/2}$ discussed in Section 3.2. However, estimation of these preprocessor coefficients in not a trivial task and it might be argued that if the amount of information about the data needed to do this is available one might as well simply compute the optimal weighing vector directly $\mathbf{W}_{\text{opt}} = \mathbf{R}^{-1}\mathbf{S}^*$, or use an accelerated descent procedure of the type discussed in Section 4.13. White has described an adaptive eigenvector preprocessor[80] but simulations tend to indicate that the convergence behaviour is not altogether satisfactory in ill-conditioned interference environments and certainly inferior to sample matrix inversion approaches or accelerated descent.

Restricting attention to fixed preprocessors with output AGC action, the aim is to select components of the input data vectors important for convergence by inner product operations with fixed coefficients. The simplest possible such preprocessor would be the beamformer matrix shown in Fig. 7.12.1 with AGC amplifiers connected to the output. It is important to ensure that the desired signal only appears on one output channel so that the steering vector is not altered by the AGC gain variations. The system offers very fast convergence with interferences of very disparate powers so long as the source are well resolved, i.e. well spaced so that there is little leakage of output power between beams to upset the AGC amplifiers. In other words, if a strong source leaks into a weak source's beam it will reduce the gain of its amplifier and suppress the weaker source at the output. This will reduce the convergence rate.

Another preprocessor candidate is the spatial derivative vectors matrix.[77] If one were interested in fast cancellation of main beam interference the derivative signal $\mathbf{X}^t\mathbf{C}^i$ could be formed (Fig. 6.4.3) and amplified up to the same level as the conventional output. Of course it would be essential to ensure that the desired signal is arriving from the assumed direction since otherwise rapid suppression of the desired signal will occur.

Perturbation algorithm: If the perturbation algorithm is used on the coefficients in Fig. 7.12.1 it is clear that changing, say, w_1 is equivalent to perturbation of the beam pattern by a real lobe in the direction of beam 1 and changing w_4 gives a quadrature lobe shift. If a weak interference lies in the direction of beam 1 then good estimates of its output power are obtained as other interference sources are relatively attenuated.

If a digitally controlled weighting vector operates on the signals directly at the elements, in a nontransformed system, it is possible to get the same changes in the beam pattern by shifting all the weighting coefficients simultaneously, each coefficient increment being set to a value proportional to the appropriate beamformer coefficient. Thus a transformed gradient is easily measured. The effects of the AGC amplifiers can be simulated by setting the weighting vector to the sequence of beam coefficients and scanning the beam round in an initial calibration of beam powers. The gradients determined by perturbation can then be scaled by the square roots of these calibration powers. Similar techniques in perturbation arrays can be used to emulate many of the suboptimal processors discussed earlier in this Chapter.

7.13 Convergence properties of suboptimal systems

All of the suboptimal systems discussed earlier in this Chapter are regardable as multiply linearly constrained fully adaptive antennas, the constraints arising through the use of projection $\mathbf{P}$ in eqn. (7.1.15). The convergence theory of this type of antenna was discussed by Frost[47] in connection with multiply constrained time domain processors.

The gradient of the output power in eqn. (7.1.15) in $2\mathbf{PRW}(k)$ and the ensemble averaged convergence equations have the form

$$\mathbf{W}(k+1) = \mathbf{W}(k) - 2\mu\mathbf{PRW}(k) \tag{7.13.1}$$

Subtracting $\hat{\mathbf{W}}$, the optimum value of $\mathbf{W}$, from both sides of this equation yields

$$\mathbf{W}(k+1) - \hat{\mathbf{W}} = \mathbf{W}(k) - \hat{\mathbf{W}} - 2\mu\mathbf{PR}[\mathbf{W}(k) - \hat{\mathbf{W}}] \tag{7.13.2}$$

where the last term is permissable since

$$\begin{aligned}\mathbf{PR}\hat{\mathbf{W}} &= \mathbf{PR}(\mathbf{U} - (\mathbf{PRP})^{+}\mathbf{PRU}) \\ &= \mathbf{PRU} - \mathbf{PRP}(\mathbf{PRP})^{+}\mathbf{PRU} \\ &= \mathbf{PRU} - \mathbf{PRU} = 0\end{aligned} \tag{7.13.3}$$

Setting $\mathbf{W}(k) - \hat{\mathbf{W}} = \mathbf{V}(k)$ we have

$$\mathbf{V}(k+1) = \mathbf{V}(k) - 2\mu\mathbf{PRPV}(k) = (\mathbf{P} - 2\mu\mathbf{PRP})\mathbf{V}(k) \tag{7.13.4}$$

where the idempotence of **P** is invoked, $\mathbf{PPV} = \mathbf{PV}$. The output power of the antenna is

$$p(k) = \hat{\mathbf{W}}(k)^H \mathbf{R} \hat{\mathbf{W}}(k) = [\mathbf{W} + \mathbf{PV}(k)]^H \mathbf{R} [\mathbf{W} + \mathbf{PV}(k)] \tag{7.13.5}$$

and using eqn. (7.13.3) we have

$$p^{(k)} = \hat{\mathbf{W}}^H \mathbf{R} \hat{\mathbf{W}} + \mathbf{V}(k)^H \mathbf{PRPV}(k) \tag{7.13.5a}$$

Substitution of eqn. (7.13.4) into this yields

$$p(k) = \hat{p} + \mathbf{V}(0)^H (\mathbf{P} - 2\mu \mathbf{PRP})^k \mathbf{PRP} (\mathbf{P} - 2\mu \mathbf{PRP})^k \mathbf{V}(0) \tag{7.13.6}$$

where $\hat{p}$ is the power associated with weighting vector $\hat{\mathbf{W}}$ and $\mathbf{V}(0)$ is the initial value of $\mathbf{V}(k)$. Let **Z** be the eigenvector matrix of **PRP** with columns of eigenvectors such that

$$\mathbf{Z}^H \mathbf{PRPZ} = \mathbf{\Lambda} \tag{7.13.7}$$

where $\mathbf{\Lambda}$ is a diagonal eigenvalue matrix with $N - M$ null entries, M being the rank of **P** and equal to the number of adaptive complex coefficients. Inserting identity matrices $\mathbf{ZZ}^H$ into eqn. (7.13.6) results in

$$p(k) = \hat{p} + \mathbf{V}(0)^H \mathbf{ZZ}^H (\mathbf{P} - 2\mu \mathbf{PRP})^k \mathbf{ZZ}^H \mathbf{PRPZZ}^H (\mathbf{P} - 2\mu \mathbf{PRP})^k \mathbf{ZZ}^H \mathbf{V}(0) \tag{7.13.8}$$

and noting that

$$\mathbf{Z}^H \mathbf{A}^k \mathbf{Z} = \mathbf{Z}^H \mathbf{AZZ}^H \mathbf{AZ} \cdots \mathbf{Z}^H \mathbf{AZ} = (\mathbf{Z}^H \mathbf{AZ})^k \tag{7.13.9}$$

we have

$$p(k) = \hat{p} + v(0)^H (\mathbf{P} - 2\mu \mathbf{\Lambda})^k \mathbf{\Lambda} (\mathbf{P} - 2\mu \mathbf{\Lambda})^k v(0) \tag{7.13.9}$$

where v is the transformed vector $\mathbf{Z}^H \mathbf{V}$. Equation (7.13.9) is a set of M independent equations in the elements of vector v:

$$p(k) = \hat{p} + \sum_{n=1}^{M} |v_n(0)|^2 (1 - 2\mu\lambda_n)^{2k} \lambda_n \tag{7.13.10}$$

and these equations represent decaying components of **V**, i.e. components which are, initially, suboptimal. Fig. 7.13.1 shows a two-dimensional example for two coefficients. The initial value of the weighting vector, i.e. **U**, corresponds to $\mathbf{V} = 0$ and the starting point in the **V** plane is at the origin. These ellipses are actually a section of the N-dimensional ellipse set by the $N - M$ dimensional constraint surface as in Fig. 3.4.2. Their contours are level curves of $\mathbf{V}^H \mathbf{PRPV}$ and the diameters are parallel to the eigenvectors of **PRP** with lengths inversely proportional to the square roots of the corresponding eigenvalues (Appendix A1.2).

Fig. 7.13.2 shows the reduction of the output powers of the terms in eqn. (7.13.10) for $\mu = 0{\cdot}1$ and $v_n(0) = 1$. The larger eigenvalue terms decay very rapidly; as time progresses the smaller eigenvalue terms (if they exist) dominate the

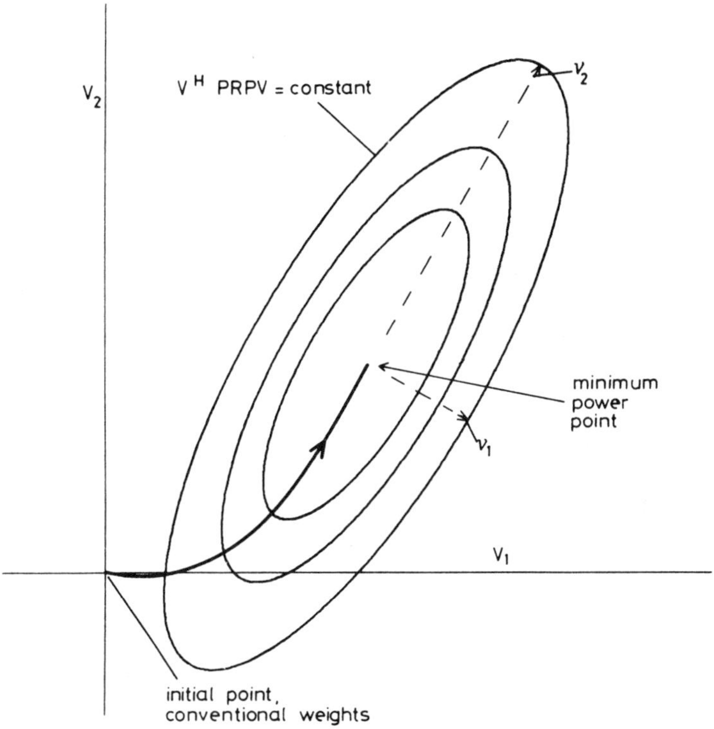

Fig. 7.13.1 *Convergence of a constrained antenna in a subspace*

output. The envelope of the lines is of the form a/k and this might be a good estimate of total power convergence is a complex interference scenario with a diffuse spread of source powers and eigenvalues. If a sample matrix inversion algorithm is used (Section 4.13) the performance is more or less independent of the eigenvalue spread and this is an attractive proposition for suboptimal processors since the dimensionality of the problem is reduced.

Stability in gradient methods: Eqn. (7.13.10) is clearly unstable if, for any λ_n, $|1 - 2\mu\lambda_n| > 1$. Thus for stability we have the limit

$$\mu \leq 1/\mathrm{Tr}\,(\mathbf{PRP}) \leq 1/\lambda_{\max} \tag{7.13.11}$$

Another stability criterion quoted by Frost is

$$\mu \leq 1/\mathrm{Tr}\,(\mathbf{R}) \tag{7.13.12}$$

since

$$\mathrm{Tr}\,(\mathbf{R}) \geqslant \mathrm{Tr}\,(\mathbf{PRP}) \tag{7.13.13}$$

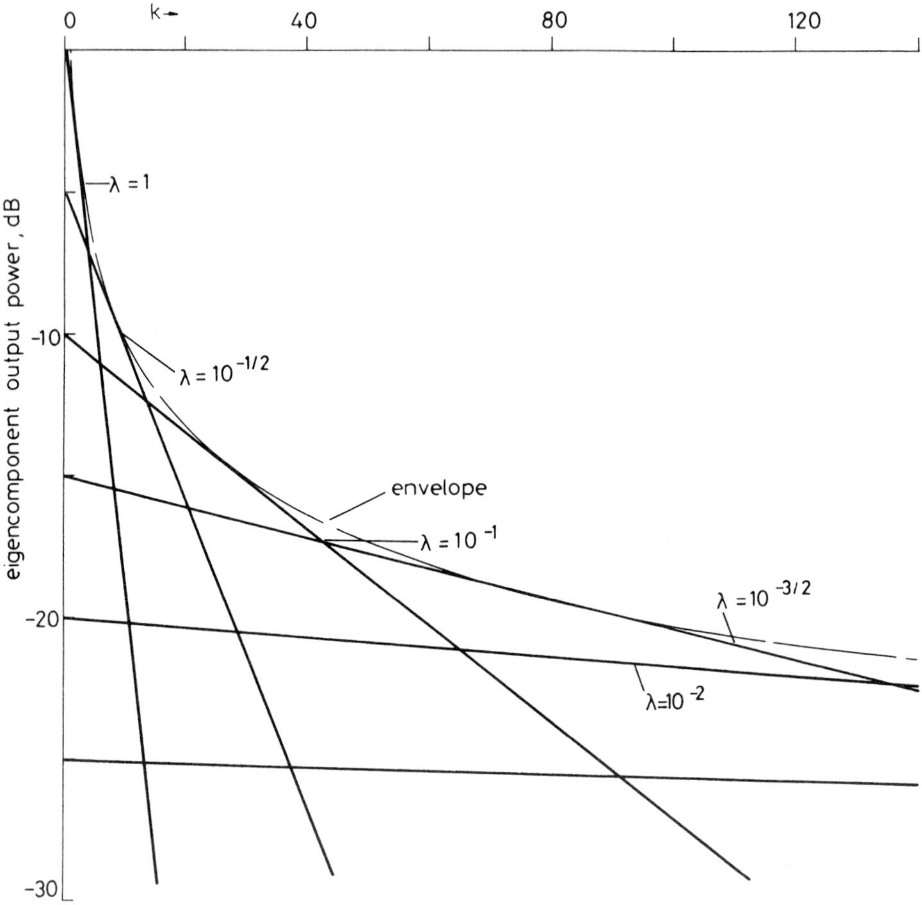

Fig. 7.13.2 *Decay of eigencomponent output powers* $\mu = 0{\cdot}05$

Tr (**PRP**) can be determined for the two cases discussed in Section 7.1 as follows. For eqn. (7.1.8) we have

$$\begin{aligned}
\mathrm{Tr}\,(\mathbf{PRP}) &= \mathrm{Tr}\left(\mathbf{I} - \frac{\mathbf{C}_v^*\mathbf{C}_v^t}{\mathbf{C}_v^t\mathbf{C}_v^*}\right)\mathbf{R}\left(\mathbf{I} - \frac{\mathbf{C}_v^*\mathbf{C}_v^t}{\mathbf{C}_v^t\mathbf{C}_v^*}\right) \\
&= \mathrm{Tr}\,(\mathbf{R}) - 2\,\mathrm{Tr}\left(\frac{\mathbf{C}_v^t\mathbf{R}\mathbf{C}_v^*}{\mathbf{C}_v^t\mathbf{C}_v^*}\right) + \mathrm{Tr}\left(\frac{\mathbf{C}_v^t\mathbf{C}_v^*\mathbf{C}_v^t\mathbf{R}\mathbf{C}_v^*}{(\mathbf{C}_v^t\mathbf{C}_v^*)^2}\right) \\
&= \mathrm{Tr}\,(\mathbf{R}) - \frac{\mathbf{C}_v^t\mathbf{R}\mathbf{C}_v^*}{\mathbf{C}_v^t\mathbf{C}_v^*} = \mathrm{Tr}\,(\mathbf{R}) - \frac{\mathbf{U}^H\mathbf{R}\mathbf{U}}{\mathbf{U}^H\mathbf{U}} \qquad (7.13.14)
\end{aligned}$$

the last term being a function of the array output power with the adaptive coefficients set to zero.

In the case of eqn. (7.1.14), the trace is a sum over a subset of the element outputs.

References

Optimal and null steering arrays

1 BANGS, W. J., and SCHULTHEISS, P. M.: 'Space-time processing for optimal parameter estimation' (part of Ref. 101)

2 BRYNN, F.: 'Optimal signal processing of three-dimensional arrays operating on Gaussian signals and noise', *JASA*, 1962, **34,** pp. 289–297

3 COX, H.: 'Optimum arrays and the Schwartz inequality', *JASA*, 1969, **45,** pp. 228–231

4 CHENG, D. K., and TSENG, F. I.: 'Optimum spatial processing in a noisey environment for arbitrary antenna arrays subject to random errors', *IEEE Trans.*, 1968, **AP-16,** pp. 164–171

5 DAVIDS, N., THURSTON, E. G., and MUESER, R. E.: 'The design of optimum directional acoustic arrays', *JASA*, 1952, **24,** pp. 50–56

6 DAVIES, D. E. N.: 'Independent angular steering of each zero of the directional pattern for a linear array', *IEEE trans.*, 1967, **AP-15,** pp. 296–298

7 DAVIES, D. E. N., and RIZK, Mrs. M.: 'Electronic steering of multiple nulls for circular arrays', *Electron. Lett.*, 1977, **13,** pp. 669–670

8 DRANE, C. Jr., and McILVENNA, J.: 'Gain maximisation and controlled null placement simultaneously achieved in aerial array patterns', *Radio & Electron. Eng.*, 1970, **39,** pp. 49–57

9 GILBERT, E. M., and MORGAN, S. P.: 'Optimum design of directive antenna arrays', *BSTJ*, May 1955, pp. 637–663

10 GLASER, G. C., and WILLIAMS, R. E.: 'Array processors for simultaneous noise and interference suppression or simultaneous signal-to-noise and resolution enhancement', *JASA*, 1976, **60,** pp. 1319–1330

11 HANSEN, W. W., and WOODYARD, J. R.: 'A new principle in directional antenna design', *Proc. IRE*, 1938, **26,** pp. 333–345

12 HODGKISS, W. S., and NOLTE, L. W.: 'A sequential implementation of optimal array processors', *IEEE Trans.*, May 1980, **AES-16,** pp. 349–354

13 LEWIS, J. B., and SCHULTHEISS, P. M.: 'Optimum and conventional detection using a linear array', *JASA*, 1971, **49,** pp. 1083–1091

14 LO, Y. T., LEE, S. W., and LEE, Q. H.: 'Optimisation of directivity and signal-to-noise ratio of an arbitrary antenna array', *Proc. IEEE*, 1966, **54,** pp. 1033–1045

15 MAKSYM, J. N.: 'Directional accuracy of small ring arrays', *JASA*, 1977, **61,** pp. 105–109

16 PASUPATHY, S., and VENETSANPOULOS, A. N.: 'Optimum active array processing structure and space-time factorability', *IEEE Trans.*, 1974, **AES-10,** pp. 770–778

17 PRITCHARD, R. L.: 'Maximum directivity index of a linear point array', *JASA*, 1954, **26,** pp. 1034–1039

18 SCHARF, L. L., and MOOSE, P. H.: 'Information measures and performance bounds for array processors', *IEEE Trans.*, 1976, **IT-22,** pp. 11–21

19 STRAIT, B. J., and HIRASAWA, K.: 'Array design for a specified pattern by matrix methods', *IEEE Trans.*, 1969, **AP-17,** pp. 237–239
20 VANDERKULK, W.: 'Optimum processing for acoustic arrays', *J. Br. IRE.*, Oct 1963, pp. 285–292

Adaptive radio and radar antennas

21 ADAMS, R. N., HOROWITZ, L. L., and SENNE, K. D.: 'Adaptive main-beam nulling for narrow-beam antenna arrays', *IEEE Trans.*, 1980, **AES-16,** pp. 509–516
22 APPLEBAUM, S. P., and CHAPMAN, D. J.: 'Adaptive arrays with main beam constraints', *IEEE Trans.*, 1976, **AP-24,** pp. 650–662
23 BAIRD, C. A.: 'Recursive algorithms for adaptive array antennas'. Final report Contract F30602-72-C-0499, Rome Air Development Centre. (Sep. 1973)
24 BAIRD, C. A., and RASSWEILER, G. G.: 'Adaptive sidelobe nulling using digitally controlled phase shifters', *IEEE Trans.*, 1976, **AP-24,** pp. 638–649
25 BARTON, B. A.: 'Digital beamforming for radar', *IEE Proc. F., Commun., Radar & Signal Process.*, 1980, **127,** pp. 266–277
26 BAUMGARTEN, D.: 'Proposals for simplifying envelope normalisation in adaptive antenna arrays', *IEEE Trans.*, 1980, **AES-16,** pp. 732–734
27 BERNI, A. J.: 'Weight jitter phenomena in adaptive array control loops', *IEEE Trans.*, 1977, **AES-13,** pp. 355–361
28 BORGIOTTI, G. V., and KAPLAN, L. J.: 'Superresolution of uncorrelated interference sources by using adaptive array techniques', *IEEE Trans.*, 1979, **AP-27,** pp. 842–845
29 BOROSON, D. M.: 'Sample size considerations for adaptive arrays', *IEEE Trans.*, 1980, **AES-16,** pp. 446–451
30 BRENNAN, L. E., and MALLETT, J. D.: 'Efficient simulation of external noise incident on arrays', *IEEE Trans.*, 1976, **AP-24,** pp. 740–741
31 BRENNAN, L. E., PUGH, E. L., and REED, I. S.: 'Control loop noise in adaptive array antennas', *IEEE Trans.*, 1971, **AES-7,** pp. 254–263
32 BRENNAN, L. E., and REED, I. S.: 'Theory of adaptive radar', *IEEE Trans.*, 1973, **AES-9,** pp. 237–252
33 BUHRING, W.: 'Adaptive orthogonal projection for rapid converging interference suppression', *Electron. Lett.*, 1978, **14,** pp. 515–516
34 CANTONI, A.: 'Application of orthogonal perturbation sequences to adaptive beamforming', *IEEE Trans.*, 1980, **AP-28,** pp. 191–202
35 CHAPMAN, D. J.: 'Partial adaptivity for the large array', *IEEE Trans.*, 1976, **AP-24,** pp. 685–696
36 COMPTON, R. T.: 'Multiplier offset voltages in adaptive arrays', *IEEE Trans.*, 1976, **AP-24,** pp. 616–627
37 COMPTON, R. T.: 'An experimental four-element adaptive array', *IEEE Trans.*, 1976, **AP-24,** pp. 697–706
38 COMPTON, R. T.: 'An adaptive array in a spread spectrum communication system', *Proc. IEEE*, 1978, **66,** pp. 289–298
39 COMPTON, R. T.: 'The power inversion array: Concept and performance', *IEEE Trans.*, 1979, **AES-15,** pp. 803–814
40 COMPTON, R. T.: 'Power optimisation in adaptive arrays', *IEEE Trans.*, 1980, **AP-28,** pp. 79–85
41 COMPTON, R. T.: 'Pointing accuracy and dynamic range in a steered beam adaptive array', *IEEE Trans.*, 1980, **AES-16,** pp. 280–287
42 COMPTON, R. T., HUFF, R. J., SWARNER, W. G., and KSIENKI, A. A.: 'Adaptive arrays for communication systems', *IEEE Trans.*, 1976, **AP-24,** pp. 599–607
43 DAVIS, R. C., BRENNAN, L. E., and REED, L. S.: 'Angle estimation with adaptive arrays in external noise fields', *IEEE Trans.*, 1976, **AES-12,** pp. 179–186

44 DICARLO, D. M.: 'Reference loop phase shift in an N-element adaptive array', *IEEE Trans.*, 1979, **AES-15,** pp. 576–582

45 DOCTER, R. A., and MASENTEN, W. K.: 'Trends in adaptive antenna circuit design'. Electro. Conference, New York, 1979

46 EGGESTAD, M., and HEIER, S.: 'A combined programmed and adaptive null steering technique', *IEEE Trans.*, 1980, **AES-16,** pp. 639–645

47 FROST, O. L.: 'An algorithm for linearly constrained adaptive array processing', *Proc. IEEE*, 1971, **60,** pp. 661–675

48 GABRIEL, W. F.: 'Adaptive arrays—an introduction', *Proc. IEEE*, 1976, **64,** pp. 239–272

49 GOBERT, J.: 'Adaptive beam weighting', *IEEE Trans.*, 1976, **AP-24,** pp. 744–749

50 GRIFFITHS, L. J.: 'A simple adaptive algorithm for real-time processing in antenna arrays', *Proc. IEEE*, 1969, **57,** pp. 1696–1704

51 GRIFFITHS, L. J.: 'Adaptive array processing experiments at HF' (part of Ref. 108)

52 GRIFFITHS, L. J.: 'An adaptive beamformer which implements constraints using an auxiliary preprocessor' (part of Ref. 108)

53 HICKS, D. S.: 'Null steering in linear arrays by using amplitude control of signals in the feeder network', *Electron. Lett.*, 1977, **13,** pp. 198–199

54 HOROWITZ, L. L., BLATT, H., WESLEY, G. B., and SENNE, K.: 'Controlling adaptive antenna arrays with the sample matrix inversion algorithm', *IEEE Trans.*, 1979, **AES-15,** pp. 840–847

55 HOROWITZ, L. L.: 'Convergence rate of the extended SMI algorithm for narrowband adaptive arrays', *IEEE Trans.*, 1980, **AES-16,** pp. 738–740

56 HOWELLS, P. W.: 'Explorations in fixed and adaptive resolution at GE and SURC', *IEEE Trans.*, 1976, **AP-24,** pp. 575–584

57 HUDSON, J. E.: 'Quantisation effects in adaptive beamformers' (part of Ref. 108)

58 HUDSON, J. E.: 'An adaptive antenna with computer assisted convergence'. IEE conference on antennas and prop., London. IEE Conf. Publ. 169, 1978, pp. 46–50

59 HUDSON, J. E.: 'A Kalman type algorithm for adaptive radar arrays and modelling of non-stationary weights' *in* 'Case studies in advanced signal processing'. IEE Conf. Publ., 180, 1979

60 JIM, C. W.: 'A comparison of two LMS constrained optimal array structures', *Proc. IEEE*, 1977, **65,** pp. 1730–1731

61 KWOK, P. C. K., and BRANDON, P. S.: 'Eigenvalues of the noise covariance matrix of a linear array in the presence of 2 directional interferences', *Electron. Lett.*, 1959, **15,** pp. 50–51

62 KWOK, P. C. K., and BRANDON, P. S.: 'The optimal radiation pattern of an array in the presence of 2 directional interferences', *Electron. Lett.*, 1979, **15,** pp. 251–252

63 LANK, G. W.: 'General effect of envelope normalisation in adaptive array control loops', *IEEE Trans.*, 1973, **AES-9,** pp. 328–329

64 LEAVITT, M. K.: 'A phase adaptation algorithm', *IEEE Trans.*, 1976, **AP-24,** pp. 754–756

65 MAYHAN, J. T.: 'Adaptive nulling with multiple beam antennas', *IEEE Trans.*, 1978, **AP-26,** pp. 267–273

66 MAYHAN, J. T.: 'Some techniques for evaluating the bandwidth characteristics of adaptive nulling systems', *IEEE Trans.*, 1979, **AP-27,** pp. 363–373

67 MONZINGO, R. A., and MILLER, T. W.: 'Introduction to adaptive arrays' (Wiley, 1980)

68 MORGAN, D.: 'Partially adaptive array techniques', *IEEE Trans.*, 1978, **AP-26,** pp. 823–833

69 NITZBERG, R.: 'Computational precision requirements for optimal weights in adaptive processing', *IEEE Trans.*, 1980, **AES-16,** pp. 418–425

70 PRASAD, S.: 'Linear antenna arrays with broad nulls with applications to adaptive arrays', *IEEE Trans.*, 1979, **AP-27,** pp. 185–190

71 PRASAD, S.: 'Generalised array pattern synthesis by the method of alternating orthogonal projections', *IEEE Trans.*, 1980, **AP-28,** pp. 328–332

72 RASSWEILER, G. G., WILLIAMS, M. R., PAYNE, L. M., and MARTIN, G. P.: 'A miniaturised lightweight wideband null steerer', *IEEE Trans.*, 1976, **AP-24,** pp. 749–754

73 REED, I. S., MALLETT, J. D., and BRENNAN, L. E.: 'Rapid convergence rate in adaptive arrays', *IEEE Trans.*, 1974, **AES-10,** pp. 853–863

74 RODGERS, W. E., and COMPTON, R. T.: 'Adaptive array bandwidth with tapped delay-line processing', *IEEE Trans.*, 1979, **AES-15,** pp. 21–28

75 SENSIPER, S.: 'Suggested signal generator-simulator for adaptive antenna test', *IEEE Trans.*, 1976, **AP-24,** pp. 760–764

76 SUSANS, D. E.: 'An adaptive receiving system for UHF television', *BBC Eng.*, June 1975, pp. 39–50

77 TAKAO, K., and KOMIYAMA, K.: 'An adaptive antenna for rejection wideband interference', *IEEE Trans.*, 1980, **AES-16,** pp. 452–459

78 THOMPSON, P. A.: 'Adaptation by direct phase-shift adjustment in narrow-band adaptive antenna systems', *IEEE Trans.*, 1976, **AP-24,** pp. 756–760

79 WASHBURN, T. W., and SWEENEY, L. E. Jun.: 'An on-line adaptive beamforming capability for HF backscatter radar', *IEEE Trans.*, 1976, **AP-24,** pp. 721–732

80 WHITE, W. D.: 'Cascade preprocessors for adaptive antennas', *IEEE Trans.*, 1976, **AP-24,** pp. 670–684

81 WHITE, W. D.: 'Artificial noise in adaptive arrays', *IEEE Trans.*, 1978, **AES-14,** pp. 380–384

82 WHITE, W. D.: 'Adaptive cascade networks for deep nulling', *IEEE Trans.*, 1978, **AP-26,** pp. 396–402

83 WIDROW, B.: 'Adaptive sampled-data systems—a statististical theory of adaptation', *IRE* Wescon Conv. Record. 3 Pt. 4, 1959, pp. 74–85

84 WIDROW, B., MANTEY, P. E., GRIFFITHS, L. J., and GOODE, B. B.: 'Adaptive antenna systems', *Proc. IEEE*, 1967, **55,** pp. 2143–2159

85 WIDROW, B., and MCCOOL, J. M.: 'A comparison of adaptive algorithms based on the methods of steepest descent and random search', *IEEE Trans.*, 1976, **AP-24,** pp. 615–637

86 WINDRAM, M. D.: 'A steerable aerial for UHF reception in Alderney', *Cable Telev. Eng.*, 1976, **10,** pp. 362–369

87 WINDRAM, M. D.: 'Adaptive antennas for UHF broadcast reception', *IEE Proc. F.*, 1980, **127,** pp. 249–256

88 WIRTH, W. D.: 'Suboptimal suppression of directional noise', *IEEE Trans.*, 1976, **AP-24,** pp. 741–744

89 ZAHM, C. L.: 'Effects of errors in the direction of incidence on the performance of an adaptive array', *Proc. IEEE*, 1972, **60,** pp. 1008–1009

90 ZAHM, C. L.: 'Applications of adaptive arrays to suppress strong jammers', *IEEE Trans.*, 1973, **AES-9,** pp. 260–271

Sonar and seismic

91 ANDERSON, V. C.: 'Dicanne, a realisable adaptive process', *JASA*, 1969, **45,** pp. 39–51

92 ANDERSON, V. C., and RUDNICK, P.: 'Rejection of a coherent arrival at an array', *JASA*, 1969, **45,** pp. 406–410

93 BIENVENUE, G.: 'Underwater passive detection and spatial coherence testing', *JASA*, 1979, **65,** pp. 425–437

94 BIENVENUE, G., and VERNET, J. L.: 'Enhancement of antenna performance by adaptive processing' (part of Ref. 101)

95 CAPON, J.: 'High resolution frequency-wavenumber spectrum analysis', *Proc. IEEE*, 1969, **57,** pp. 1408–1418

96 CAPON, J.: 'Probability distributions for extimators of the frequency-wavenumber spectrum', *Proc. IEEE*, 1970, **58,** pp. 1785–1786

97 COX, H.: 'Resolving power and sensitivity to mismatch of optimum array processors', *JASA*, 1973, **54,** pp. 771–785 (see also Ref. 101)

98 CRON, B. F., and SHERMAN, C. H.: 'Spatial-correlation functions for various noise models', *JASA*, 1962, **34,** pp. 1732–1736

99 GIRAUDON, C.: 'Optimum antenna processing: a modular approach' (part of Ref. 108)
100 GRIFFITHS, J. W. R., and HUDSON, J. E.: 'An introduction to adaptive processing in a passive sonar system' (part of Ref. 108)
101 GRIFFITHS, J. W. R. G., STOCKLIN, P. L., and Van SCHOONEVELD, C. (Eds.): 'Signal Processing'. Proceedings of the NATO Advanced Study Institute on signal processing and underwater acoustics, Loughborough, UK, 1972 (Academic Press, 1973)
102 LACOSS, R. T.: 'Adaptive combining of wideband data for optimal reception', *IEEE Trans.*, 1968, **GE-6,** pp. 78–86
103 LACOSS, R. T.: 'Data adaptive spectral analysis methods', *Geophysics*, 1971, **36,** pp. 661–675
104 LUNDE, E. B.: 'The forgotten algorithm in adaptive beamforming' (part of Ref. 108)
105 MAKSYM, J. N.: 'A robust formulation of an optimum cross-spectral beamformer for line arrays', *JASA*, 1979, **65,** pp. 971–975
106 McCOOL, J. M.: 'A constrained adaptive beamformer tolerant of array gain and phase errors' (part of Ref. 108)
107 OWSLEY, N. L.: 'Spatial processing for sensor arrays' (part of Ref. 101)
108 TACCONI, G. (Ed.): 'Aspects of signal processing'. Proceedings of the NATO Advanced Study Institute on signal processing and underwater acoustics, Porto Venere, Italy, 1976 (Reidel, 1977)
109 VURAL, A. M.: 'Effects of perturbations on the performance of optimum/adaptive arrays', *IEEE Trans.*, 1979, **AES-15,** pp. 76–87
110 WANG, H. S. C.: 'Interference reduction by amplitude shading of sonar transducer array', *JASA*, 1977, **61,** pp. 1251–1259
111 WIDROW, B. *et al.*: 'Stationary and non-stationary learning characteristics of the LMS adaptive filter' (part of Ref. 108)
112 WINKLER, L. P., and SCHWARTZ, M.: 'Adaptive non-linear optimisation of the signal to noise ratio of an array subject to a constraint', *JASA*, 1971, **52,** pp. 39–51

Mathematical background

113 ADBY, P. R., DEMPSTER, J. A. H.: 'Introduction to optimisation methods' (Chapman & Hall, 1974)
114 AITKEN, C.: 'Matrices and determinants' (Oliver & Boyd, 1962)
115 ANDERSON, T. W.: 'Asymptotic theory for principle component analysis', *Ann. Math. Stat.*, 1963, **34,** pp. 122–148
116 AYRES, F.: 'Matrices' (Schaum)
117 BARLTETT, M. S.: 'An inverse matrix adjustment arising in discriminant analysis', *Ann. Math. Stat.*, 1951, **22,** pp. 107–111
118 DAVENPORT, W. B., and ROOT, W. L.: 'An introduction to the theory of random signals and noise' (McGraw-Hill, 1958)
119 EISENSCHITZ, R. K.: 'Matrix algebra for physicists' (Heineman, 1966)
120 FORSYTHE, G. E., and GOLUB, G. H.: 'On the stationary values of a second degree polynomial on the unit sphere', *J. Soc. Indus. Appl. Maths.*, 1965, **13,** pp. 1050–1068
121 FOX, L., and MAYERS, D. F.: 'Computing methods for scientists and engineers' (Oxford Univ. Press, 1968)
122 FUKUNAGA, K.: 'Introduction to statistical pattern recognition' (Academic Press, 1972)
123 GOLDBERGER, K.: 'Topics in regression analysis' (Macmillan, 1968)
124 GOLUB, G. H.: 'Stationary values of the ratio of quadratic forms subject to linear constraints', Zeitschrift fur Angewante Mathematika und Physik, 1970, **21,** pp. 318–327
125 GOODMAN, N. R.: 'Statistical analysis based on a certain complex gaussian distribution', *Ann. Math. Stat.*, 1963, **34,** pp. 152–180
126 GREVILLE, T. N. E.: 'Some applications of the pseudoinverse of a matrix', *SIAM Rev.*, 1960, **2,** pp. 15–22
127 KESTEN, H.: 'Accelerated stochastic approximation', *Ann. Math. Stat.*, 1958, **29,** pp. 41–59

128 KOWALIK and OSBORNE: 'Methods for unconstrained optimisation' (Elsevier, 1968)
129 LEE, Y. W.: 'Statistical theory of communication' (Wiley, 1963)
130 LEVINSON, N.: 'The Wiener RMS error criterion in filter design and prediction', *J. Math. Phys.*, 1947, **XXV,** pp. 261-278
131 LUENBERGER, D. G.: 'An introduction to linear and non-linear programming' (Adison Wesley, 1973)
132 PENROSE, R.: 'A generalised inverse for matrices', *Proc. Camb. Phil. Soc.*, 1955, **51,** pp. 406–413
133 PENROSE, R.: 'On best approximate solutions of linear matrix equations', *Proc. Camb. Phil. Soc.*, 1956, **52,** pp. 17–19
134 SMITH, BOYLE, GARBOW, IKEBE, KLEME, MOLER: 'Matrix eigensystem routines—Eispack guide' (Springer Verlag, 1974)
135 WIBERG: 'State space and linear systems' (Schaum, 1971)

Miscellaneous

136 AOKI, M.: 'Optimisation of stochastic systems' (Academic Press, 1967, Appendix II)
137 BURNSIDE, W. D., WANG, N., and PELTON, E. L.: 'Near field pattern analysis of airborne antennas', *IEEE Trans.*, 1980, **AP-28,** pp. 318–327
138 GILL, W. J.: 'A comparison of binary delay-lock tracking loop implementations', *IEEE Trans.*, 1966, **AES-2,** pp. 415–424
139 MOFFET, A. T.: 'Minimum redundancy linear arrays', *IEEE Trans.*, 1962, **AP-16,** pp. 172–175
140 MUELLER, K. W.: 'A new fast converging mean-square algorithm', *BSTJ*, 1975, **54,** pp. 143–153
141 SHELTON, G. B.: 'A near constant phase variable attenuator', *Proc. IEEE*, July 1969, pp. 1345–1346
142 WIDROW, B. *et al.*: 'Adaptive noise cancelling: Principles and applications', *Proc. IEEE*, 1975, **63,** pp. 1692–1716
143 COMPTON, R. T.: 'Improved feedback loop for adaptive arrays', *IEEE Trans.*, 1980, **AES-16,** pp. 159–168
144 HARGRAVE, P.: Research Notes. STL Labs., 1980
145 GODARA, L. C., CANTONI, A.: 'Analysis of the performance of adaptive beamforming using perturbation sequences', Tech. Report EE8015, 1980, University of Newcastle, NSW, Australia
146 ANDERSON, V. C.: 'Sidelobe interference suppression with an adaptive null processor', *J. Appl. Syst. Anal.*, 1981, **69,** pp. 185–190
147 CLAUS, A. J., KADOTU, T. T., and ROMAIN, D. M.: 'Efficient approximation of a family of noises for application to adaptive spatial processing for signal detection', *IEEE Trans.*, 1980, **IT-26,** pp. 588–595
148 KEATING, P. N.: 'The effect of array errors on frequency domain adaptive interference rejection', *J. Appl. Syst. Anal.*, 1980, **68,** pp. 1690–1695
149 KO, C. C.: 'Power inversion array in a rotating source environment', *IEEE Trans.*, 1980, **AES-16,** pp. 755–762
150 MAYHAN, J. T.: 'Thinned array configuration for use with satellite-based adaptive antennas', *IEEE Trans.*, 1980, **AP-28,** pp. 846–856
151 MILLER, K. S.: 'Complex Gaussian processes', *SIAM Rev.*, 1969, **11,** pp. 544–567
152 SCHONFELD, T. J., and SCHWARTZ, M.: 'Rapidly converging second order tracking algorithms for equalisation', *IEEE Trans.*, 1971, **IT-17,** pp. 572–579

Appendix 1

Basic vectors, matrices and statistics

A1.1 Basic vector and matrix properties

Transposition :

$(\ )^t$ represents a simple transpose without conjugation
$(\ )^*$ represents conjugation without transposition
$(\ )^H$ represents a conjugate (Hermitian) transpose

$$(\mathbf{AB})^t = \mathbf{B}^t\mathbf{A}^t \qquad (\mathbf{AB})^H = \mathbf{B}^H\mathbf{A}^H \qquad (\mathbf{AB})^* = \mathbf{A}^*\mathbf{B}^* \tag{A1.1.1}$$

$$\mathbf{A}^H\mathbf{B} = (\mathbf{A}^t\mathbf{B}^*)^* \tag{A1.1.2}$$

If $\mathbf{A} = \mathbf{A}^H$ or $\mathbf{A}^* = \mathbf{A}^t$, then square matrix $\mathbf{A}$ is Hermitian (A1.1.3)

If $\mathbf{A} = \mathbf{BB}^H$, then $\mathbf{A}$ is Hermitian for arbitary $\mathbf{B}$ (A1.1.4)

If $\mathbf{A}^H = -\mathbf{A}$ or $\mathbf{A}^* = -\mathbf{A}^t$, then $\mathbf{A}$ is skew Hermitian (A1.1.5)

If $\mathbf{Z} = \mathbf{X} + j\mathbf{Y}$ is an Hermitian matrix, where $\mathbf{X}$ and $\mathbf{Y}$ are real, then

(i) $\mathbf{X}$ is real symmetric
(ii) $\mathbf{Y}$ is skew symmetric
(iii) $j\mathbf{Y}$ is Hermitian

The diagonal elements of an Hermitian matrix are real. (A1.1.6)

If $\mathbf{Z} = \mathbf{X} + j\mathbf{Y}$ is skew Hermitian, then

(i) $\mathbf{X}$ is skew real
(ii) $\mathbf{Y}$ is symmetric real
(iii) $j\mathbf{Y}$ is skew Hermitian

The diagonal elements of a skew Hermitian matrix are imaginary. (A1.1.7)

Hermitian forms : If **A** is Hermitian and **V** is an arbitary complex vector then

$\mathbf{V}^H\mathbf{A}\mathbf{V}$ is real

$\mathbf{V}^H(\mathbf{A}+\mathbf{B})\mathbf{V} = \mathbf{V}^H\mathbf{A}\mathbf{V} + \mathbf{V}^H\mathbf{B}\mathbf{V}$ (true for arbitary **A**, **B**)

$$[(\mathbf{U}+\mathbf{V})^H\mathbf{A}(\mathbf{U}+\mathbf{V})]^{1/2} \le [\mathbf{U}^H\mathbf{A}\mathbf{U}]^{1/2} + [\mathbf{V}^H\mathbf{A}\mathbf{V}]^{1/2} \qquad (A1.1.8)$$

(**A** must be positive)

Skew Hermitian forms : If **A** is skew Hermitian then

$\mathbf{V}^H\mathbf{A}\mathbf{V}$ is imaginary (A1.1.9)

A1.2 Eigenvector representations

If **A** is Hermitian, it can be represented as

$$\mathbf{A} = \sum_{i=1}^{N} \lambda_i \mathbf{U}_i\mathbf{U}_i^H \qquad (A1.2.1)$$

where the λ_i are real and the $\mathbf{U}_i$ are complex and orthonormal: $\mathbf{U}_i^H\mathbf{U}_j = \delta_{ij}$.

If every $\lambda_i \ge 0$ then **A** is positive semidefinite (non-negative definite).

Let

$$\mathbf{Z} = [\mathbf{U}_1 \,|\, \mathbf{U}_2 \,|\cdots|\, \mathbf{U}_N]$$

be the eigenvector matrix.

Then, from eqn. A1.2.1),

$$\mathbf{A} = \mathbf{Z}\mathbf{\Lambda}\mathbf{Z}^H \qquad (A1.2.2)$$

where $\mathbf{\Lambda}$ is a diagonal matrix diag $(\lambda_1, \lambda_2, \ldots, \lambda_N)$. The rank of **A** is the number of nonzero entries in $\mathbf{\Lambda}$.

If A is non-negative definite then

$$\mathbf{V}^H\mathbf{A}\mathbf{V} \ge 0 \qquad \mathbf{V}'\mathbf{A}\mathbf{V}^* \ge 0 \qquad (A1.2.3)$$

Unitary matrices: If $\mathbf{B}^{-1} = \mathbf{B}^H$ then $\mathbf{B}^H\mathbf{B} = \mathbf{B}\mathbf{B}^H = \mathbf{I}$ and **B** is a unitary matrix (or orthogonal if real). The matrix **Z** above is unitary and, from eqn. (A1.2.2),

$$\mathbf{Z}^H\mathbf{A}\mathbf{Z} = \mathbf{Z}^H\mathbf{Z}\mathbf{\Lambda}\mathbf{Z}^H\mathbf{Z} = \mathbf{\Lambda} \qquad (A1.2.4)$$

Extrema of Hermitian forms : The maximum value of $x = \mathbf{V}^H\mathbf{A}\mathbf{V}/\mathbf{V}^H\mathbf{V}$ is the maximum eigenvalue of **A** and the minimum is the minimum eigenvalue. At these points $\mathbf{V} = k\mathbf{U}_{max}$ and $\mathbf{V} = k\mathbf{U}_{min}$ respectively where $\mathbf{U}_{max}$ and $\mathbf{U}_{min}$ are the eigenvectors associated with λ_{max} and λ_{min}.

The typical variation of $\mathbf{V}^H\mathbf{A}\mathbf{V}$ as the elements of vector **V** are varied is shown, for a real two-dimensional case, in Fig. A1.2.1. The maximum and minimum radii of the ellipse for the $\mathbf{V}^H\mathbf{A}\mathbf{V} = 1$ contour are $\lambda_{min}^{-1/2}$ and $\lambda_{max}^{-1/2}$,

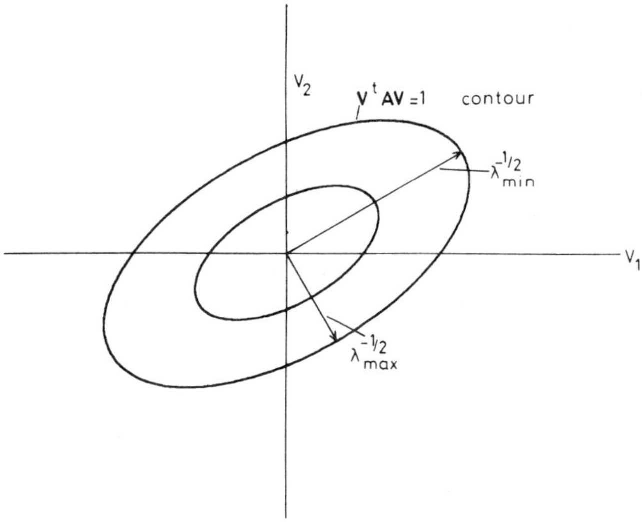

Fig. A1.2.1 *Ellipses of* $\mathbf{V}^t\mathbf{AV}$ = *constant for a symmetric matrix*

respectively, and these radii are parallel to the eigenvector associated with $\lambda_{\min}$ and $\lambda_{\max}$.

$$\mathbf{A}^{-1} = \sum_{i=1}^{N} \lambda_i^{-1}\mathbf{U}_i\mathbf{U}_i^H \qquad \text{(if there are no null } \lambda\text{s)}$$

$$\mathbf{A} + \gamma\mathbf{I} = \sum_{i=1}^{N} (\lambda_i + \gamma)\mathbf{U}_i\mathbf{U}_i^H$$

$$(\mathbf{A} + \gamma\mathbf{I})^{-1} = \sum_{i=1}^{N} (\lambda_i + \gamma)^{-1}\mathbf{U}_i\mathbf{U}_i^H \tag{A1.2.5}$$

If

$$\mathbf{B} = \sum_{i=1}^{N} \lambda_i^{1/2}\mathbf{U}_i\mathbf{U}_i^H, \quad \text{then} \quad \mathbf{BB} = \mathbf{B}^2 = \mathbf{A} \tag{A1.2.6}$$

When the positive roots are selected in this equation, then **B** is unique and is non-negative definite Hermitian.

Matrix inversion formulas:

$$(\mathbf{AB})^{-1} = \mathbf{B}^{-1}\mathbf{A}^{-1}$$

$$(\mathbf{A}^t)^{-1} = (\mathbf{A}^{-1})^t, \qquad (\mathbf{A}^H)^{-1} = (\mathbf{A}^{-1})^H, \qquad (\mathbf{A}^*)^{-1} = (\mathbf{A}^{-1})^* \tag{A1.2.7}$$

$$(\mathbf{A} + \mathbf{XX}^H)^{-1} = \mathbf{A}^{-1} - \frac{\mathbf{A}^{-1}\mathbf{XX}^H\mathbf{A}^{-1}}{1 + \mathbf{X}^H\mathbf{A}^{-1}\mathbf{X}} \tag{A1.2.8}$$

where **A** is invertible and **X** is a complex vector.

$$(\mathbf{A} + \mathbf{B})^{-1} = \mathbf{A}^{-1} - \mathbf{A}^{-1}(\mathbf{A}^{-1} + \mathbf{B}^{-1})^{-1}\mathbf{A}^{-1} \tag{A1.2.9}$$

where both **A** and **B** are invertible.

$$(\mathbf{A} + \mathbf{X}^H\mathbf{C}\mathbf{X})^{-1} = \mathbf{A}^{-1} - \mathbf{A}^{-1}\mathbf{X}^H(\mathbf{C}^{-1} + \mathbf{X}\mathbf{A}^{-1}\mathbf{X}^H)^{-1}\mathbf{X}\mathbf{A}^{-1} \tag{A1.2.10}$$

where **A** is N by N and invertible, **C** is M by M and invertible, ($M \leq N$), $\mathbf{X}^H$ is N by M and arbitary.

$$(\mathbf{I} + \mathbf{A})^{-1} = \mathbf{I} - \mathbf{A} + \mathbf{A}^2 - \mathbf{A}^3 + \mathbf{A}^4 - \mathbf{A}^5 + \cdots \tag{A1.2.11}$$

if the series converges, i.e. $|\lambda_{\text{max}}(\mathbf{A})| < 1$.

Trace and determinate : The trace of a square matrix is the sum of its diagonal elements. If **A** is Hermitian then

$$\text{Tr}(\mathbf{A}) = \sum_{i=1}^{N} \lambda_i \tag{A1.2.12}$$

In general,

$$\text{Tr}(\mathbf{A} + \mathbf{B}) = \text{Tr}(\mathbf{A}) + \text{Tr}(\mathbf{B})$$

$$\text{Tr}(\mathbf{ABC}) = \text{Tr}(\mathbf{BCA}) = \text{Tr}(\mathbf{CAB}) \qquad \text{(rotation rule)} \tag{A1.2.13}$$

$$\text{Det}(\mathbf{AB}) = \text{Det}(\mathbf{A})\,\text{Det}(\mathbf{B})$$

$$\text{Det}(\mathbf{A}^{-1}) = 1/\text{Det}(\mathbf{A}) \tag{A1.2.14}$$

If **A** is Hermitian,

$$\text{Det}(\mathbf{A}) = \prod_{i=1}^{N} \lambda_i \tag{A1.2.15}$$

For any matrix **A**

$$\text{Det}\begin{bmatrix} \alpha & \mathbf{X}^H \\ \mathbf{Y} & \mathbf{A} \end{bmatrix} / \text{Det}(\mathbf{A}) = \alpha - \mathbf{X}^H\mathbf{A}^{-1}\mathbf{Y} \tag{A1.2.16}$$

$$\text{Rank}(\mathbf{A} + \mathbf{B}) \leq \text{Rank}(\mathbf{A}) + \text{Rank}(\mathbf{B})$$

$$\text{Rank}(\mathbf{AB}) \leq \text{Rank}(\mathbf{A}), \leq \text{Rank}(\mathbf{B}) \tag{A1.2.17}$$

A1.3 Projections[119] (see Section 2.5)

If

$$\mathbf{P} = \mathbf{I} - \frac{\mathbf{V}^*\mathbf{V}^t}{\mathbf{V}^t\mathbf{V}^*}$$

where **V** is a complex N-vector, then **P** is an Hermitian matrix of rank $N - 1$.

The null eigenvalue is associated with the eigenvector $\mathbf{V}^*/\|\mathbf{V}\|$. $\mathbf{P}$ is idempotent, i.e.

$$\mathbf{P}\mathbf{P} = \mathbf{P}^2 = \mathbf{P} \tag{A1.3.1}$$

If $\mathbf{X}_1, \mathbf{X}_2, \ldots, \mathbf{X}_M$ are orthogonal N-vectors then

$$\mathbf{P}_M = \left(\mathbf{I} - \frac{\mathbf{X}_1^* \mathbf{X}_1^t}{\mathbf{X}_1^t \mathbf{X}_1^*}\right)\left(\mathbf{I} - \frac{\mathbf{X}_2^* \mathbf{X}_2^t}{\mathbf{X}_2^t \mathbf{X}_2^*}\right)(\cdots)\left(\mathbf{I} - \frac{\mathbf{X}_M^* \mathbf{X}_M^t}{\mathbf{X}_M^t \mathbf{X}_M^*}\right) = \left(\mathbf{I} - \sum_{i=1}^{M} \frac{\mathbf{X}_i^* \mathbf{X}_i^t}{\mathbf{X}_i^t \mathbf{X}_i^*}\right) \tag{A1.3.2}$$

is an Hermitian matrix of rank $N - M$. For any N-vector $\mathbf{Y}$

$$\mathbf{X}_i^t \mathbf{P}_M \mathbf{Y} = 0 \qquad i = 1, \ldots, M \tag{A1.3.3}$$

The eigenvalues of $\mathbf{P}_M$ number M of zero value and $N - M$ of unity value. The eigenvectors with unity eigenvalues are orthogonal to every $\mathbf{X}_i$:

$$\mathbf{U}_j^t \mathbf{X}_i = 0, \qquad i = 1, \ldots, M \qquad j = M + 1, \ldots, N \tag{A1.3.4}$$

If the $N - M$ vectors $\mathbf{X}_{M+1}$ to $\mathbf{X}_N$ are orthogonal to the $\mathbf{X}_i$ in (A1.3.2) then the projection complementary to $\mathbf{P}_M$ is

$$\begin{aligned}\tilde{\mathbf{P}} &= \left(\mathbf{I} - \frac{\mathbf{X}_{M+1}^* \mathbf{X}_{M+1}^t}{\mathbf{X}_{M+1}^t \mathbf{X}_{M+1}^*}\right)(\mathbf{I} \cdots)\left(\mathbf{I} - \frac{\mathbf{X}_N^* \mathbf{X}_N^t}{\mathbf{X}_N^t \mathbf{X}_N^*}\right) \\ &= \mathbf{I} - \mathbf{P}_M\end{aligned} \tag{A1.3.5}$$

A1.4 Pseudoinverses[126, 132, 133, 136]

Let $\mathbf{A}$ be an Hermitian matrix represented by eqn. (A1.2.1) with some of the λ_i zero. The pseudoinverse of $\mathbf{A}$ is

$$\mathbf{A}^+ = \sum_{i=1}^{N} \lambda_i^+ \mathbf{U}_i \mathbf{U}_i^H \tag{A1.4.1}$$

where

$$\lambda^+ = \lambda^{-1} \quad \text{if} \quad \lambda \neq 0$$

$$\lambda^+ = 0 \quad \text{if} \quad \lambda = 0$$

Let $\mathbf{A}$ be an arbitary complex matrix represented by

$$\mathbf{A} = \sum_{i=1}^{K} \gamma_i \mathbf{X}_i \mathbf{Y}_i^H$$

where the $\mathbf{X}_i$ are orthonormal N-vectors, the $\mathbf{Y}_i$ are orthonormal M-vectors, and the γ_i are complex and nonzero.

Then

$$\mathbf{A}^{+} = \sum_{i=1}^{K} \gamma_i^{-1} \mathbf{Y}_i \mathbf{X}_i^H \tag{A1.4.2}$$

Thus the pseudoinverse of a vector $\mathbf{X}$ is $\mathbf{X}^H/\mathbf{X}^H\mathbf{X}$, and if $\mathbf{Q}$ is N by M, $M < N$, and of rank M, $(\mathbf{Q}\mathbf{Q}^H)^+ = \mathbf{Q}(\mathbf{Q}^H\mathbf{Q})^{-2}\mathbf{Q}^H$.

If $\mathbf{A}$ is a full rank Hermitian matrix and $\mathbf{P}$ a projection then

$$(\mathbf{PAP})^+ = \mathbf{P}(\mathbf{PAP} + \tilde{\mathbf{P}})^{-1}\mathbf{P} \tag{A1.4.3}$$

with $\tilde{\mathbf{P}}$ being the complimentary projection.

$$\mathbf{P}(\mathbf{PAP})^+\mathbf{P} = (\mathbf{PAP})^+\mathbf{P} = \mathbf{P}(\mathbf{PAP})^+ = (\mathbf{PAP})^+ \tag{A1.4.4}$$

Eqns. (A1.4.3) and (A1.4.4) produce many new identities. For example if $\mathbf{P} = \mathbf{I} - \mathbf{C}\mathbf{C}^H$ with $\mathbf{C}^H\mathbf{C} = 1$, $\tilde{\mathbf{P}} = \mathbf{C}\mathbf{C}^H$

$$(\mathbf{PAP} + \mathbf{P}\mathbf{X}\mathbf{X}^H\mathbf{P})^+ = \mathbf{P}(\mathbf{PAP} + \mathbf{C}\mathbf{C}^H + \mathbf{P}\mathbf{X}\mathbf{X}^H\mathbf{P})^{-1}\mathbf{P}$$

Let $\mathbf{B} = \mathbf{PAP} + \mathbf{C}\mathbf{C}^H$. Then $\mathbf{P}\mathbf{B}^{-1}\mathbf{P} = (\mathbf{PAP})^+$. Substituting $\mathbf{B}$ into the eqn. above yields an equivalent to eqn. A1.2.8:

$$\mathbf{P}\left(\mathbf{B}^{-1} - \frac{\mathbf{B}^{-1}\mathbf{P}\mathbf{X}\mathbf{X}^H\mathbf{P}\mathbf{B}^{-1}}{1 + \mathbf{X}^H\mathbf{P}\mathbf{B}^{-1}\mathbf{P}\mathbf{X}}\right)\mathbf{P} = (\mathbf{PAP})^+ - \frac{(\mathbf{PAP})^+\mathbf{P}\mathbf{X}\mathbf{X}^H\mathbf{P}(\mathbf{PAP})^+}{1 + \mathbf{X}^H\mathbf{P}(\mathbf{PAP})^+\mathbf{P}\mathbf{X}} \tag{A1.4.5}$$

A1.5 Some statistical expectations for complex variates

Let a, b, c, d be independent samples from a zero mean Gaussian real distribution of variance $\frac{1}{2}$. Then the complex variates

$$u = a + jb$$

$$v = c + jd$$

have the following properties:

$$E(u^2) = E(a + jb)^2 = E(a^2) - E(b^2) + 2jE(ab) = 0$$

$$E(v^2) = 0$$

$$E(uv) = E(vu) = 0$$

$$E(uu^*) = E(|u|^2) = E(a^2) + E(b^2) - 2jE(ab) = 1$$

$$E(vv^*) = 1 \tag{A1.5.1}$$

and are typical of the complex variables found in band-pass or base-band complex signals derived from receiving arrays.

Correlated complex variates
These can be generated by the following procedure:

$$x = u \cos \tfrac{1}{2}\theta + ve^{j\phi} \sin \tfrac{1}{2}\theta$$

$$y = ue^{-j\phi} \sin \tfrac{1}{2}\theta + v \cos \tfrac{1}{2}\theta \qquad \text{(A1.5.2)}$$

Variates of this type are characteristic of the complex functions found in signal-processing application. We have

$$E(xx^*) = E(yy^*) = 1$$

$$E(xy^*) = E(x^*y)^* = e^{j\phi} \sin \theta$$

$$E(x^2) = E(y^2) = E(xy) = 0 \qquad \text{(A1.5.3)}$$

Properties of complex random vectors: Let **U** be an N-vector whose elements are samples from a complex distribution with the properties defined in A1.5.1. Then **U** has the following properties.

$$E(\mathbf{U}\mathbf{U}^t) = E(\mathbf{U}^*\mathbf{U}^H) = 0$$

$$E(\mathbf{U}\mathbf{U}^H) = \mathbf{I}$$

$$E(\mathbf{U}^H\mathbf{U}) = N$$

$$\mathrm{Tr}\,(E(\mathbf{U}\mathbf{U}^H)) = E(\mathrm{Tr}\,(\mathbf{U}\mathbf{U}^H)) = E(\mathrm{Tr}\,(\mathbf{U}^H\mathbf{U})) = \mathrm{Tr}\,(E(\mathbf{U}^H\mathbf{U})) = N$$

$$E(\mathbf{U}^t\mathbf{U}) = E(\mathbf{U}^H\mathbf{U}^*) = 0 \qquad \text{(A1.5.4)}$$

Generation of correlated distributions: If **U** is a vector of uncorrelated elements with the properties above, and if $\mathbf{X} = \mathbf{AU}$ where **A** is an arbitary complex matrix, then

$$E(\mathbf{X}\mathbf{X}^H) = E(\mathbf{A}\mathbf{U}\mathbf{U}^H\mathbf{A}^H) = \mathbf{A}\mathbf{A}^H \qquad \text{(A1.5.5)}$$

If **U** and **V** are two uncorrelated vectors the Hermitian transformation

$$\begin{bmatrix} \mathbf{X} \\ \mathbf{Y} \end{bmatrix} = \begin{bmatrix} \mathbf{A} & \mathbf{B}^H \\ \mathbf{B} & \mathbf{A} \end{bmatrix} \begin{bmatrix} \mathbf{U} \\ \mathbf{V} \end{bmatrix}$$

generates two correlated vectors **X** and **Y** such that

$$E(\mathbf{X}\mathbf{X}^H) = \mathbf{A}^2 + \mathbf{B}^H\mathbf{B} \qquad E(\mathbf{Y}\mathbf{Y}^H) = \mathbf{A}^2 + \mathbf{B}\mathbf{B}^H$$

$$E(\mathbf{X}\mathbf{Y}^H) = \mathbf{A}\mathbf{B}^H + \mathbf{B}^H\mathbf{A} \qquad \text{(A1.5.6)}$$

A1.6 Higher moments of random variates

Real: Let a, b, c, d be correlated real zero-mean Gaussian variates. Then[118]

$$E(abcd) = E(ab)E(cd) + E(ac)E(bd) + E(ad)E(bc) \qquad \text{(A1.6.1)}$$

From this result the following results can be proved. Let **A**, **B**, **C**, **D** be real N-vectors with correlated random elements and

$$\mathbf{M} = \mathbf{A}\mathbf{B}^t\mathbf{C}\mathbf{D}^t$$

then if A_n is the nth element of any vector **A**

$$M_{ik} = \sum_{j=1}^{N} A_i B_j C_j D_k \tag{A1.6.2}$$

and

$$\begin{aligned} E(M_{ik}) &= \sum_{j=1}^{N} E(A_i B_j)E(C_j D_k) + E(A_i C_j)E(B_j D_k) + E(A_i D_k)E(B_j C_j) \\ &= \sum_{j=1}^{N} (\mathbf{R}_{AB})_{ij}(\mathbf{R}_{CD})_{jk} + (\mathbf{R}_{AC})_{ij}(\mathbf{R}_{BD})_{jk} + (\mathbf{R}_{AD})_{ik}(\mathbf{R}_{BC})_{jj} \\ &= \mathbf{R}_{AB}\mathbf{R}_{CD} + \mathbf{R}_{AC}\mathbf{R}_{BD} + \mathbf{R}_{AD}\,\mathrm{Tr}\,(\mathbf{R}_{BC}) \end{aligned} \tag{A1.6.3}$$

where $\mathbf{R}_{AB} = E(\mathbf{A}\mathbf{B}^t)$ etc. From this result we obtain

$$\begin{aligned} E(\mathbf{A}^t\mathbf{B}\mathbf{C}^t\mathbf{D}) = E\,\mathrm{Tr}\,(\mathbf{B}\mathbf{C}^t\mathbf{D}\mathbf{A}^t) &= \mathrm{Tr}\,(\mathbf{R}_{BC}\mathbf{R}_{DA}) \\ &\quad + \mathrm{Tr}\,(\mathbf{R}_{BD}\mathbf{R}_{CA}) + \mathrm{Tr}\,(\mathbf{R}_{BA})\,\mathrm{Tr}\,(\mathbf{R}_{CD}) \end{aligned} \tag{A1.6.4}$$

Complex
Let u, v, x, y be complex Gaussian variates. By direct expansion of the real and imaginary parts and substitution into eqn. (A1.6.1), it can be shown that

$$E(uvxy) = E(uv)E(xy) + E(ux)E(vy) + E(uy)E(vx) \tag{A1.6.5}$$

If u, v, x, y are correlated complex variates which, pairwise, have the properties defined in eqn. (A1.5.3) then, if they are additionally Gaussian,

$$E(uv^*xy^*) = E(uv^*)E(xy^*) + E(uy^*)E(v^*x) \tag{A1.6.6}$$

Thus fourth moments of complex variates can be very different from the equivalent real variable case.

Complex vectors: Let **U**, **V**, **X**, **Y** be complex Gaussian random N-vectors with the properties defined in eqn. (A1.5.6). If

$$\mathbf{Z} = \mathbf{U}\mathbf{V}^H\mathbf{X}\mathbf{Y}^H \tag{A1.6.7}$$

then

$$Z_{ik} = \sum_{j=1}^{N} U_i V_j^* X_j Y_k^* \tag{A1.6.8}$$

and

$$E(Z_{ik}) = \sum_{j=1}^{N} \{E(U_i V_j^*)E(X_j Y_k^*) + E(U_i X_j)E(V_j^* Y_k^*) + E(U_i Y_k^*)E(V_j^* X_j)\}$$

$$= \sum_{j=1}^{N} \{(\mathbf{R}_{UV})_{ij}(\mathbf{R}_{XY})_{jk} + 0 \cdot 0 + (\mathbf{R}_{UY})_{ik}(\mathbf{R}_{VX}^*)_{jj}\} \tag{A1.6.9}$$

where $\mathbf{R}_{UV} = E(\mathbf{UV}^H)$ etc. Thus

$$E(\mathbf{UV}^H\mathbf{XY}^H) = E(\mathbf{UV}^H)E(\mathbf{XY}^H) + E(\mathbf{UY}^H)E(\mathbf{V}^H\mathbf{X}) \tag{A1.6.10}$$

Let $\mathbf{X}$ and $\mathbf{Y}$ be independent complex Gaussian N-vectors with $E(\mathbf{XX}^H) = \mathbf{R}$, $E(\mathbf{YY}^H) = \mathbf{M}$, $E(\mathbf{XY}^H) = \mathbf{0}$. We have immediately

$$E(\mathbf{XX}^H\mathbf{XX}^H\mathbf{YY}^H) = E(\mathbf{XX}^H\mathbf{XX}^H)E(\mathbf{YY}^H) \tag{A1.6.11}$$

and if $\mathbf{Z} = E(\mathbf{XX}^H\mathbf{YY}^H\mathbf{XX}^H)$ then $Z_{il} = \sum_{j=1}^{N} \sum_{k=1}^{N} X_i X_j^* M_{jk} X_k X_l^*$ where $M_{jk} = E(Y_j Y_k^*)$. Thus,

$$E(Z_{il}) = \sum_{j=1}^{N} \sum_{k=1}^{N} \{E(X_i X_j^*)E(X_k X_l^*) + E(X_i X_k)E(X_j^* X_l^*)$$

$$+ E(X_i X_l^*)E(X_j^* X_k)\}M_{jk}$$

$$= \sum_{j=1}^{N} \sum_{k=1}^{N} (R_{ij} R_{kl} + 0 + R_{il} R_{jk}^*)M_{jk}$$

$$= \mathbf{RMR} + \mathbf{R} \,.\, \mathrm{Tr}\,(\mathbf{RM}) \tag{A1.6.12}$$

Appendix 2

Differentiation of $\mathbf{W}^H\mathbf{RW}$

A2.1 Real variable case $p = \mathbf{W}^t\mathbf{RW}$, $\mathbf{R}$ symmetric

If the ith element of $\mathbf{W}$ is W_i, and the ith row, jth column element of $\mathbf{R}$ is R_{ij} then

$$p = \mathbf{W}^t\mathbf{RW} = \sum_{i=1}^{N} \sum_{j=1}^{N} W_i R_{ij} W_j \tag{A2.1.1}$$

and

$$\partial p/\partial W_k = \sum_{i=1}^{N} \sum_{j=1}^{N} \left[\frac{\partial W_i}{\partial W_k} R_{ij} W_j + W_i R_{ij} \frac{\partial W_j}{\partial W_k} \right] \tag{A2.1.2}$$

Now

$$\frac{\partial W_i}{\partial W_k} = \delta_{ik} \qquad \text{and} \qquad \frac{\partial W_j}{\partial W_k} = \delta_{jk} \tag{A2.1.3}$$

Thus (A2.1.2) reduces to

$$\frac{\partial p}{\partial W_k} = \sum_{j=1}^{N} R_{kj} W_j + \sum_{i=1}^{N} W_i R_{ik} \tag{A2.1.4}$$

As $\mathbf{R}$ is symmetrical these two summations have the same value, thus

$$\frac{\partial p}{\partial W_k} = 2 \sum_{j=1}^{N} R_{kj} W_j \tag{A2.1.5}$$

and if we define $\partial p/\partial W_k$ as the elements G_k of a first-order gradient vector $\mathbf{G}$ then

$$G_k = \frac{\partial p}{\partial W_k} \qquad \text{and} \qquad \mathbf{G} = 2\mathbf{RW} \tag{A2.1.6}$$

A2.2 Complex variable case[32] $p = \mathbf{W}^H\mathbf{R}\mathbf{W}$ with $\mathbf{R}$ Hermitian

Define

$$\mathbf{W} = \mathbf{W}_r + j\mathbf{W}_j, \qquad \mathbf{R} = \mathbf{R}_r + j\mathbf{R}_j \qquad j = \sqrt{-1} \tag{A2.2.1}$$

$$\begin{aligned} p &= \mathbf{W}_r^t(\mathbf{R}_r + j\mathbf{R}_j)\mathbf{W}_r - \mathbf{W}_j^t(\mathbf{R}_r + j\mathbf{R}_j)\mathbf{W}_j \\ &\quad + j\mathbf{W}_r^t(\mathbf{R}_r + j\mathbf{R}_j)\mathbf{W}_j - j\mathbf{W}_j^t(\mathbf{R}_r + j\mathbf{R}_j)\mathbf{W}_r \end{aligned} \tag{A2.1.2}$$

Noting that a quadratic form over a skew symmetric matrix is zero the $\mathbf{R}_j$s in the first two terms of (A2.2.1) can be removed. Also

$$\mathbf{W}_r^t\mathbf{R}_r\mathbf{W}_j = \mathbf{W}_j^t\mathbf{R}_r\mathbf{W}_r \tag{A2.2.3}$$

which allows the $\mathbf{R}_r$s in the second two terms to be removed. We have remaining

$$p = \mathbf{W}_r^t\mathbf{R}_r\mathbf{W}_r + \mathbf{W}_j^t\mathbf{R}_r\mathbf{W}_j - \mathbf{W}_r^t\mathbf{R}_j\mathbf{W}_j + \mathbf{W}_j^t\mathbf{R}_j\mathbf{W}_r \tag{A2.2.4}$$

We will later make use of the result

$$(\mathbf{W}_r^t\mathbf{R}_j\mathbf{W}_j)^t = \mathbf{W}_j^t\mathbf{R}_j^t\mathbf{W}_r = -\mathbf{W}_j^t\mathbf{R}_j\mathbf{W}_r \tag{A2.2.5}$$

which invokes the skew symmetric property of $\mathbf{R}_j$. The real quadratic forms in (A2.2.4) can now be differentiated using the results in Section A2.1. Differentiating with respect to the real parts of $\mathbf{W}$ first we substitute (A2.2.5) into (A2.2.4) to get

$$p = \mathbf{W}_r^t\mathbf{R}_r\mathbf{W}_r + \mathbf{W}_j^t\mathbf{R}_r\mathbf{W}_j - 2\mathbf{W}_r^t\mathbf{R}_j\mathbf{W}_j \tag{A2.2.6}$$

Writing the bilinear form as a double summation we have

$$\frac{\partial}{\partial W_r^k}\mathbf{W}_r^t\mathbf{R}_j\mathbf{W}_j = \frac{\partial}{\partial W_r^k}\sum_{n=1}^{N}\sum_{m=1}^{N} W_r^n R_j^{nm} W_j^m = \sum_{m=1}^{N} R_j^{km} W_j^m = (\mathbf{R}_j\mathbf{W}_j)^k \tag{A2.2.7}$$

and, by analogy, with (A2.1.6)

$$\frac{\partial}{\partial W_r^k}(\mathbf{W}_r^t\mathbf{R}_r\mathbf{W}_r) = 2(\mathbf{R}_r\mathbf{W}_r)^k \tag{A2.2.8}$$

Thus, finally,

$$\frac{\partial p}{\partial W_r^k} = 2(\mathbf{R}_r\mathbf{W}_r)^k - 2(\mathbf{R}_j\mathbf{W}_j)^k = 2\ \mathrm{Re}\ (\mathbf{RW})^k \tag{A2.2.9}$$

For the imaginary derivatives we have from (A2.2.4) and (A2.2.5),

$$p = \mathbf{W}_r^t\mathbf{R}_r\mathbf{W}_r + \mathbf{W}_j^t\mathbf{R}_r\mathbf{W}_j + 2\mathbf{W}_j^t\mathbf{R}_j\mathbf{W}_r \tag{A2.2.10}$$

and, as above,

$$\frac{\partial p}{\partial W_j^k} = 2(\mathbf{R}_r\mathbf{W}_j)^k + 2(\mathbf{R}_j\mathbf{W}_r)^k = 2\ \mathrm{Im}\ (\mathbf{RW})^k \tag{A2.2.11}$$

These two components of the gradient, (A2.2.9) and (A2.2.11) are sometimes combined to give a complex gradient

$$\mathbf{G} = \frac{\partial p}{\partial W_r^k} + j\frac{\partial p}{\partial W_j^k} = 2\mathbf{R}\mathbf{W} \tag{A2.2.12}$$

and this is the cross correlation vector measured if the output y of a weighted antenna array is correlated with the individual element signals $\mathbf{X}$:

$$E(\mathbf{X}^*y) = E(\mathbf{X}^*\mathbf{X}'\mathbf{W}) = \mathbf{R}\mathbf{W} \tag{A2.2.13}$$

Definition (A2.2.12) is only sensible for real valued functions of complex variables such as quadratic forms or the real parts or imaginary parts of complex valued functions if taken separately.

Appendix 3

Minimum norm property of projections

We wish to show that vector $\mathbf{A}_p$ in eqn. (2.5.16):

$$\mathbf{A}_p = \left(\mathbf{I} - \sum_{i=1}^{M} \frac{\mathbf{X}_i^* \mathbf{X}_i^t}{\mathbf{X}_i^t \mathbf{X}_i^*}\right)\mathbf{A} \qquad \text{(A3.1.1)}$$

is the vector with the smallest norm having a given projection on $\mathbf{A}$, $\mathbf{A}^H\mathbf{A}_p$ = constant, and orthogonal to $\mathbf{X}_1, \ldots, \mathbf{X}_M$.

Consider the addition of some arbitary vector $\mathbf{V}$ to $\mathbf{A}_p$

$$\mathbf{A}_p' = \mathbf{A}_p + \mathbf{V} \qquad \text{(A3.1.2)}$$

$\mathbf{V}$ must be orthogonal to $\mathbf{A}$ else

$$\mathbf{A}^H(\mathbf{A}_p + \mathbf{V}) = \mathbf{A}^H\mathbf{A}_p + \mathbf{A}^H\mathbf{V} \neq \mathbf{A}^H\mathbf{A}_p \qquad \text{(A3.1.3)}$$

$\mathbf{V}$ must also be orthogonal to every $\mathbf{X}_i$ or

$$\mathbf{X}_i^t(\mathbf{A}_p + \mathbf{V}) = \mathbf{X}_i^t\mathbf{A}_p + \mathbf{X}_i^t\mathbf{V} = \mathbf{X}_i^t\mathbf{V} \neq 0 \qquad i = 1, \ldots, M \qquad \text{(A3.1.4)}$$

But $\mathbf{A}_p$ lies in the subspace spanned by $\mathbf{A}$ and the $\mathbf{X}_i$ and from eqns. (A3.1.3) and (A3.1.4) we conclude that $\mathbf{V}$ must be orthogonal to $\mathbf{A}_p$. Therefore

$$\|\mathbf{A}_p + \mathbf{V}\|^2 = \|\mathbf{A}_p\|^2 + \|\mathbf{V}\|^2 \geq \|\mathbf{A}_p\|^2 \qquad \text{(A3.1.5)}$$

and conclude that $\mathbf{A}_p$ cannot be changed without increasing its norm.

Appendix 4

Eigenvalue solution for norm-bounded power minimisation

The output power minimisation problem with a linear constraint on the weighting vector together with a norm bound introduced in Section 6.5 can be solved efficiently in eigenspace. Let $\mathbf{Z}$ be the eigenvector matrix of $\mathbf{R}$ (Appendix A1.2) such that $\mathbf{R} = \mathbf{Z}\boldsymbol{\Lambda}\mathbf{Z}^H$. Define

$$\mathbf{w} = \mathbf{Z}^H\mathbf{W}$$

$$\mathbf{c}^* = \mathbf{Z}^H\mathbf{C}^* \tag{A4.1.1}$$

where $\mathbf{W}$ is the array space weighting vector and $\mathbf{C}$ is the constraint vector such that $\mathbf{C}^t\mathbf{W} = N$ where N is the number of elements. We have

$$\mathbf{C}^t\mathbf{W} = \mathbf{C}^t\mathbf{Z}\mathbf{Z}^H\mathbf{W} = \mathbf{c}^t\mathbf{w} = \sum_i c_i w_i \tag{A4.1.2}$$

$$\mathbf{W}^H\mathbf{W} = \mathbf{W}^H\mathbf{Z}\mathbf{Z}^H\mathbf{W} = \mathbf{w}^H\mathbf{w} = \sum_i |w_i|^2 \tag{A4.1.3}$$

It can be shown by the method of Lagrange that the minimisation of $p = \mathbf{W}^H\mathbf{R}\mathbf{W}$ subject to the linear constraint $\mathbf{C}^T\mathbf{W} = N$ and equality constraint $\mathbf{W}^H\mathbf{W} = \mathbf{B}$ has the solution

$$\tilde{\mathbf{W}} = \frac{N(\mathbf{R} + \lambda_0\mathbf{I})^{-1}\mathbf{C}^*}{\mathbf{C}^t(\mathbf{R} + \lambda_0\mathbf{I})^{-1}\mathbf{C}^*} \tag{A4.1.4}$$

where the value of λ_0 is determined to satisfy the norm condition. Eqn. (A4.1.4) can be transformed into eigenspace to give

$$\mathbf{Z}^H\tilde{\mathbf{W}} = \tilde{\mathbf{w}} = \frac{N(\boldsymbol{\Lambda} + \lambda_0\mathbf{I})^{-1}\mathbf{c}^*}{\mathbf{c}^t(\boldsymbol{\Lambda} + \lambda_0\mathbf{I})^{-1}\mathbf{c}^*} = \frac{N\{c_i^*/(\lambda_i + \lambda_0)\}}{\sum_i |c_i|^2/(\lambda_i + \lambda_0)} \tag{A4.1.5}$$

and eqn. (A4.1.3) becomes, using this result,

$$\tilde{\mathbf{W}}^H\tilde{\mathbf{W}} = \tilde{\mathbf{w}}^H\tilde{\mathbf{w}} = \frac{N^2 \sum_i |c_i|^2/(\lambda_i + \lambda_0)^2}{\left\{\sum_i |c_i|^2/(\lambda_i + \lambda_0)\right\}^2} = B \tag{A4.1.6}$$

This last equation can be solved by a standard root finding algorithm as it is essentially a scalar problem, i.e. we determine the root of

$$f(\lambda_0) - B = \tilde{\mathbf{w}}^H\tilde{\mathbf{w}} - B = 0 \tag{A4.1.7}$$

The method of *regula falsi* has been successfully used and, working in logs to linearise the function $f(\lambda_0)$ somewhat, produces an acceptibly accurate value of λ_0 in less than 10 iterations. This value of λ_0 is then substituted into eqn. (A4.1.5) to find $\mathbf{w}$ and subsequently $\tilde{\mathbf{W}}$. If the linear solution ($\lambda_0 = 0$) has a weighting vector norm which does not exceed the bound B then no action is taken and this weight vector is taken as the optimal one. Thus a logical check for this condition is performed before commencing the solution (A4.1.7) for λ_0.

Although the technique is a prolific user of software space, in a computer program, due to the eigenvector routines[134] the overall speed is comparable to matrix inversion. In practice the determination of λ_0 is a relatively trivial task in comparison to the solution for the eigenvalues and eigenvectors of $\mathbf{R}$ and when a large number of steering directions are required with one covariance matrix sample the load is shared among a large number of look directions and the method becomes extremely efficient.

Index